DEVELOPMENTS IN
EARTH SURFACE PROCESSES, 10

MOUNTAINS WITNESSES OF GLOBAL CHANGES RESEARCH IN THE HIMALAYA AND KARAKORAM: SHARE-ASIA PROJECT

DEVELOPMENTS IN EARTH SURFACE PROCESSES

Volumes 1 and 3 are out of print

2. WEATHERING, SOILS & PALEOSOLS
I.P. MARTINI and W. CHESWORTH (Editors)

4. ENVIRONMENTAL GEOMORPHOLOGY
M. PANIZZA

5. GEOMORPHOLOGICAL HAZARDS OF EUROPE
C. EMBLETON and C. EMBLETON-HAMANN (Editors)

6. ROCK COATINGS
R.I. DORN

7. CATCHMENT DYNAMICS AND RIVER PROCESSES
C. GARCIA and R.J. BATALLA (Editors)

8. CLIMATIC GEOMORPHOLOGY
M. GUTIÉRREZ

9. PEATLANDS: EVOLUTION AND RECORDS OF ENVIRONMENTAL AND CLIMATE CHANGES
L.P. MARTINI, A. MARTINEZ CORTIZAS and W. CHESWORTH (Editors)

Developments in Earth Surface Processes, 10

MOUNTAINS WITNESSES OF GLOBAL CHANGES RESEARCH IN THE HIMALAYA AND KARAKORAM: SHARE-ASIA PROJECT

Edited by

Renato Baudo
Italian National Research Council, Institute of Ecosystem Study (CNR-ISE),
Verbania Pallanza, Italy

Gianni Tartari
Italian National Research Council, Water Research Institute (CNR-IRSA),
Brugherio, Italy and Ev-K^2-CNR Committee, Bergamo, Italy

Elisa Vuillermoz
Ev-K^2-CNR Committee, Bergamo, Italy

ELSEVIER

Amsterdam – Boston – Heidelberg – London – New York – Oxford
Paris – San Diego – San Francisco – Singapore – Sydney – Tokyo

Elsevier
Radarweg 29, PO Box 211, 1000 AE Amsterdam, The Netherlands
Linacre House, Jordan Hill, Oxford OX2 8DP, UK

First edition 2007

Library of Congress Cataloguing in Publication Data
A catalog record for this book is available from the Library of Congress

British Library Cataloguing in Publication Data
A catalogue record for this book is available from the British Library

ISBN: 978-0-444-52990-9
ISSN: 0928-2025

Printed and bound by CPI Group (UK) Ltd, Croydon, CR0 4YY

Transferred to Digital Print 2012

Foreword

This new volume on *Mountains, Witnesses of Global Changes* in our book series, *Developments in Earth Surface Processes*, is a departure from our dominant focus on geomorphology in that we address here a predominance of atmospheric and environmental factors in some alpine environments, particularly those of the Himalaya. Inasmuch as many of these atmospheric and environmental processes directly or indirectly interact with, control, or are controlled by landforms of the high mountains, the linkages of the papers of this book with the underlying geomorphological themes of our book series are obvious. These linkages are especially exemplified by the great snow and ice resources of the Himalaya, the sources of the downstream melt-waters that are so vital to the millions of people who depend upon such waters for irrigation throughout the region. Any major disturbance to the westerly and monsoon sources of such an elemental product of atmospheric processes must be scientifically assessed with very great precision and originality in order to be better able to forecast future changes.

Anthropogenic and natural aspects of global change, including climatic warming from buildup of greenhouse gases and global dimming from aerosol emissions, are climate forcings of commonly opposite sign that complicate interpretations. Issues of effects on the hydrological cycle are paramount, especially in South and East Asia where billions of people live. Most simply, climate warming alone could lead to melting away of small glaciers, with concomitant decrease in vital melt waters downstream. But increased heating over oceans could lead similarly to greater evaporation and increased monsoonal precipitation over land, thus potentially leading to glacier growth. Similarly, certain aerosols can absorb solar radiation and increase warming, while others reflect incoming solar radiation, or increase cloud cover to cool the Earth's surface. Increased clouds can increase precipitation, but aerosol-induced clouds have smaller droplets that reduce precipitation. Thus, strong questions of the direction of future change exist that must be addressed by robust research and new methods of data collection.

The Himalaya, the so-called 'water towers' of Asia, is marked by a great paucity of primary data collection points that can be used for predicting future trends. The Chinese government, however, has recently established the world's highest climate station, at an altitude of 5200 m, on the Tibetan side of Mount Everest (Qomolangma in Chinese; Sagarmatha in Nepali). Forty more automatic and satellite-linked weather stations across Tibet will aid greatly in data collection across this roof of the world. Coupled with the new data-collecting sites described in the volume herein, much better forecasting tools will become available. But because of the state of adversity between India and Pakistan, hydrological forecasting south of the Himalayan chain is regarded as a classified strategic asset by both countries. This short-sighted treatment of what would normally be treated as the common resource heritage of humankind is a reflection of only some of the difficulties faced by scientists in the region. As the competition for water heats up in this new century the reshaping of national economies and new geopolitical alliances will likely result. Thus, the importance of research in mountains as witness of global change could not be greater. It is the business of this book to offer some insights to facilitate such inquiries.

This book is divided into the five main sections that were the divisions of the conference in Rome sponsored by the Government of Italy in November 2005; atmospheric brown clouds (ABC), the Italian Ev–K^2–CNR Committee in Project ABC, *SHARE* – Asia (Stations at High Altitude for Research on the Environment of Asia) scientific fields of atmospheric physics and chemistry, and global change, environmental indicators of global changes, and commitments to environmental monitoring at altitude in Asia. The 35 papers presented here are by some of the scientists who are expert in various aspects of the high-altitude environments of South Asia and elsewhere. Several of the papers are presented only as abstracts because their authors chose not to contribute longer versions of their work; we included these for a sense of completeness from the original conference.

The overall impression one is left with after reading over these works is that impressive understandings of environmental changes in the mountains of South Asia and elsewhere in the world have been acquired, but that far more needs to be done. Issues of atmospheric pollution, changes to alpine lakes, shrinking glaciers, diminishing water supplies, and other related problems are clearly presented in these papers. The future may be grim for some people in the mountains of the world unless attention is brought to bear upon some of these problematic issues and solutions sought.

Like the proverbial canary in the coal mine, the harsh environments of high alpine terrains are quite delicate enough to show changes as a kind of natural early-warning system, have we but the wit to observe and understand such changes as they happen. By dint of careful long-term monitoring, we scientists of the mountains of the world hope to alert the world's people of imminent problems associated with such areas. As one reads through these papers, one may see how the problems are being studied at present, and perhaps one may be helped to greater awareness of possible solutions to problems that are developing. The wider this information is disseminated, the more pointed the future research can become, as people become aware of possibilities. We offer this latest volume in our book series on *Developments in Earth Surface Processes* as state-of-the-art surficial geoscience in South Asia and a few other mountain areas in hopes that others will be drawn to continue such studies in the magnificent, but changing, Himalaya and other mountains of the world.

John F. Shroder, Jr.
Editor-in-Chief
Developments in Earth Surface Processes

Preface

From an environmental point of view, mountains are particularly sensitive and important for monitoring the state of health of our planet. Only through distribution of meteorological climatological and atmospheric composition monitoring points in mountain regions, coupled with modeling simulations will we be able to thoroughly analyze complex pollutant transport mechanisms and better understand imminent global changes. The Himalaya–Karakoram Range, because of its elevation and geographic location, represents one of the ideal places for studying long-range pollutant transport systems on a regional scale and for monitoring changes induced by mechanisms that act on a global scale through monsoon circulation.

The Ev–K^2–CNR Committee promotes interdisciplinary remote-area research in environmental and the earth sciences. Recently, it launched the project *SHARE* – Asia (Stations at High Altitude for Research on the Environment in Asia) for development of an integrated system of measurements that will contribute to increasing general scientific knowledge on climatic and pollution related processes while helping build local capacity for monitoring the relevant phenomena. *SHARE* – Asia currently includes the Pyramid Meteo Network (PMN), a climate monitoring network founded in 1994 by the Ev–K^2–CNR Committee, comprising six stations installed in Nepal's Sagarmatha National Park (SNP), and two stations in Pakistan on the Baltoro Glacier.

The meeting that generated the papers in this volume was held in Rome on 16–17 November 2005. It was organized by the Ev–K^2–CNR Committee and promoted by the Italian National Research Council (CNR) in collaboration with the Italian National Mountain Institute (IMONT). The purpose of the meeting and this book is to highlight the uniqueness of the scientific work of Ev–K^2–CNR in important international projects like Coordinated Enhanced Observing Period (CEOP), Atmospheric Brown Clouds (ABC), International Global Atmospheric Chemistry (IGAC), Global Atmospheric Watch (GAW), and Global Land Ice Measurements from Space (GLIMS).

The Ev–K^2–CNR Committee thus aims to create a unique opportunity for dialogue between major environmental scientists and experts, highlighting the close relationship between diverse themes with a common thread: in-depth comprehension of the environmental phenomena that are determining the health of our planet.

Renato Baudo, Gianni Tartari and Elisa Vuillermoz
Editors

Contents

SHARE-Asia Scientific Fields: Atmospheric Physics & Chemistry, & Global Change

SHARE-Asia Scientific Fields: Environmental Indicators of Global Changes

Acknowledgments

The Promoting Committee of SHARE – Asia that initiated and guided the Rome meeting and this subsequent book consisted of Honorary President F. Pistella, who is president of CNR, the Italian National Research Council. The president of the Promoting Committee was G. Arnoldi, who was also the vice president of the Italian Parliament Group, "Friends of the Mountains." In addition, the Promoting Committee also included A. Da Polenza, president of the Ev-K^2-CNR Committee, G. Deodato, director of the Italian Development Corporation, and E. Mensi, president of the Italian National Mountain Institute (IMONT).

The Scientific Committee that guided the Rome meeting and this resulting book consisted of R. Passino, president of the Earth and Environment Department of CNR, B. Banmali Pradhan of the International Centre for Integrated Mountain Development (ICIMOD) from Kathmandu, Nepal, L. Barrie from the Global Atmosphere Watch (GAW) of the World Meteorological Organization from Geneva, Switzerland, R. de Bernardi (CNR-ISE), S. Fuzzi (CNR-ISAC/IGAC), T. Koike from the University of Tokyo who is the head of the Coordinated Enhanced Observing Period (CEOP), under the framework of the World Climate Research programme (WCRP), W. Lau, (NASA – Goddard Space Flight Center), F. Prodi (CNR-ISAC), V. Ramanathan (Scripps Institute of Oceanography), and G. Tartari (CNR-IRSA/Ev-K^2-CNR).

The Organizing Committee that made everything happen in an orderly fashion was headed by Coordinator R. Baudo (CNR-ISE), who was assisted by A. Lami (CNR-ISE), B. Schommer (Ev-K^2-CNR Committee), and F. Sernia (CNR – Headquarters).

We thank all who attended the meetings in Rome and who contributed their technical and organizational efforts to make the meeting and this book come to fruition. Without the direct help of the legions of people who were involved in all these processes, such productive meetings and useful scientific volumes would not be possible.

The Editors

List of Corresponding Authors

Ivo Allegrini CNR – Institute for Atmospheric Pollution, Rome, Italy, *allegrini@iia.cnr.it*

Saraju K. Baidya Department of Hydrology and Meteorology, Kathmandu, Nepal, *sarjo@dhm.gov.np*

Leonard A. Barrie Environmental Division, World Meteorological Organization, Geneva, Switzerland, *LBarrie@wmo.int*

Renato Baudo CNR – Institute of Ecosystem Study, Verbania Pallaza, Italy, *r.baudo@ise.cnr.it*

Sam Benedict CEOP International Coordinator, Coronado, USA, *sam.benedict@gewex.org*

Laura Bertolani Epson Meteo Centre, Research and Development Division, Cinisello Balsamo, Milan, Italy, *laura.bertolani@epson-meteo.org*

Michael P. Bishop Department of Geography and Geology, University of Nebraska at Omaha, Omaha, USA, *mpbishop@mail.unomaha.edu*

Paolo Bonasoni CNR – Institute for Atmospheric Sciences and Climate, Bologna, Italy, *p.bonasoni@isac.cnr.it*

Alton C. Byers Research and Education, The Mountain Institute, Elkins, USA, *abyers@mountain.org*

Giuseppe Calori ARIANET, Milan, Italy, *g.calori@aria-net.it*

Giulio Cozzi CNR – Institute for the Dynamics of Environmental Processes, University of Venice, Venice, Italy, *cozzig@unive.it*

Agostino Da Polenza Ev-K^2-CNR Committee, Bergamo, Italy, *evk2cnr.org@evk2cnr.org*

Harald Egerer UNEP Vienna – Interim Secretariat of the Carpathian Convention, Vienna, Austria, *harald.Egerer@unvienna.org*

Sandro Fuzzi Institute for Atmospheric Sciences and Climate, National Research Council (CNR), Bologna, Italy, *s.fuzzi@isac.cnr.it*

Mario Gallorini National Institute of Metrological Research (I.N.R.I.M), Unit of Radiochemistry and Spectroscopy c/o University of Pavia, Pavia, Italy, *gallorin@unipv.it*

Gian Paolo Gobbi CNR – Institute for Atmospheric Sciences and Climate, Rome, Italy, *g.gobbi@isac.cnr.it*

Gregory B. Greenwood Mountain Research Initiative, Bern, Switzerland, *greenwood@scnat.ch*

Angelo Guerrini, General Director, Italian National Research Council (CNR), Rome, Italy

Syed Iqbal Hasnain HIGH ICE, New Delhi, India, *iqbalhasnain@hotmail.com*

Nai-Yung Christina Hsu NASA Goddard Space Flight Center, Greenbelt, USA, *hsu@climate.gsfc.nasa.gov*

Giovanni Kappenberger MeteoSwiss, Locarno Monti, Switzerland, *giovanni.kappenberger@meteoswiss.ch*

Toshio Koike University of Tokyo, Tokyo, Japan, *tkoike@hydra.t.u-tokyo.ac.jp*

Andrea Lami CNR – Institute for Ecosystem Study, Verbania – Pallanza, Italy, *a.lami@ise.cnr.it*

William K. M. Lau Laboratory for Atmospheres, NASA/Goddard Space Flight Center, Greenbelt, USA, *lau@climate.gsfc.nasa.gov*

Yaoming Ma Institute of Tibetan Plateau Research, Chinese Academy of Sciences, Beijing, China, *ymma@itpcas.ac.cn*

Fabio Malaspina Department for Aeronautical Meterology Experimentations, Italian Air Force (Re.S.M.A.), Bracciano, Rome, Italy, *f.malaspina@meteoam.it*

Stefano Polesello Water Research Institute, Italian National research Council (IRSA–CNR), Brugherio, (MI), Italy, *polesello@irsa.cnr.it*

Bidya Banmali Pradhan International Centre for Integrated Mountain Development (ICIMOD), Kathmandu, Nepal, *bbanmali@icimod.org*

Veerabhadran Ramanathan Scripps Institution of Oceanography, University of California at San Diego, La Jolla, USA, *vram@fiji.ucsd.edu*

Ghulam Rasul Meteorological Service of Pakistan, Islamabad, Pakistan, *grmet@yahoo.com*

Rakhshan Roohi Water Resources Research Institute, National Agricultural Research Center, Islamabad 45500, Pakistan, *drroohi_gis@yahoo.com*

Raffaele Salerno Epson Meteo Centre, Research and Development Division, Cinisello Balsamo, Milan, Italy, *raffaele.salerno@epson-meteo.org*

John F., Shroder Jr. Department of Geography and Geology, University of Nebraska at Omaha, Omaha, USA, *jshroder@mail.unomaha.edu*

Claudio Smiraglia "Ardito Desio" Earth Sciences Department, University of Milan, Milan, Italy, *claudio.smiraglia@unimi.it*

Ruben Sommaruga Laboratory of Aquatic Photobiology and Plankton Ecology, Institute of Ecology, University of Innsbruck, Innsbruck, Austria, *ruben.sommaruga@uibk.ac.at*

Gianni Tartari CNR – Water Research Institute, Brugherio and Ev-K^2-CNR Committee, Bergamo, Italy, *tartari@irsa.cnr.it*

Si-Chee Tsay NASA Goddard Space Flight Center, Greenbelt, USA, *si-chee.tsay-1@nasa.gov*

Elisa Vuillermoz Ev-K^2-CNR Committee, Bergamo, Italy, *elisa.vuillermoz@evk2cnr.org*

List of Acronyms

4DDA	Four-dimensional data analyses
AAR	Accumulation area ratio
ABC	Atmospheric Brown Clouds Project
ABL	Atmospheric boundary layer
AERONET	Aerosol robotic network
AGAGE	Advanced Global Atmospheric Gases Experiment
AGU	American Geophysical Union
AL:PE	Acidification of mountain lakes: palaeolimnology and ecology
ANN	Artificial neural networks.
AOC	Advisory and Oversight Committee
AOD	Aerosol optical depth
AOT	Aerosol optical thickness
APN	Asia-Pacific Network for Global Change
ASM	Asian summer monsoon
ASTER	Advanced Spaceborne Thermal Emission and Reflection Radiometer
AWSs	Automatic weather stations
BANG	Bangladesh
BCR	Community Bureau of Reference
BENG	West Bengal
BHUT	Bhutan
BIHA	Bihar and Jharkhand
BMRC	Bureau of Meteorology Research Centre
BOB	Bay of Bengal
BR	Biosphere reserve
BS	Bias score
BTC	Bilateral Technical Committee
CACGP	Commission on Atmospheric Chemistry and Global Pollution
CALIPSO	Cloud-Aerosol Lidar and Infrared Pathfinder Satellite Observation
CAMP	CEOP (Coordinated Enhanced Observing Period) Asian–Australia Monsoon Project
CCCM	Canadian Climate Centre Model
CDA	CEOP (Coordinated Enhanced Observing Period) Central Data Archive
CEM	Epson Meteo Centre

CEOP	Coordinated Enhanced Observing Period
CEOP-ICTS	CEOP (Coordinated Enhanced Observing Period) –Inter-CSE (Continental-Scale Experiment) Transferability Study
CEOS	Committee on Earth Observation Satellites
CEOS/IGOS-P	Committee on Earth Observation Satellites /Integrated Global Observing Strategy Partnership
CESVI	Cooperazione e Sviluppo (Cooperation and Development)
CGIAR	Consultative Group on International Agricultural Research
CGRER	Centre for Global and Regional Environmental Research
CIMS	CEOP (Coordinated Enhanced Observing Period) Inter-Monsoons Model Study
CIRES-CDC	Cooperative Institute for Research in Environmental Science – Climate Diagnostics Center
CKNP	Central Karakorum National Park
CLIVAR	Climate variability and predictability
CMORPH	CPC (Climate Prediction Center) MORPHing technique
CNR	Consiglio Nazionale delle Ricerche (Italian National Research Council)
CNR-IDPA	Consiglio Nazionale delle Ricerche—Istituto per la Dinamica dei Processi Ambientali (Italian National Research Council – Institute for the Dynamics of Environmental Processes)
CNR-IIA	Consiglio Nazionale delle Ricerche – Istituto sull' Inquinamento Atmosferico (Italian National Research Council – Institute for Atmospheric Pollution)
CNR-IRSA	Consiglio Nazionale delle Ricerche – Istituto di Ricerca sulle Acque (Italian National Research Council – Water Research Institute)
CNR-ISAC	Consiglio Nazionale delle Ricerche – Istituto di Scienze dell'Atmosfera e del Clima (Italian National Research Council – Institute of Atmospheric Sciences and Climate)
CNR-ISE	Consiglio Nazionale delle Ricerche – Istituto per lo Studio degli Ecosistemi (Italian National Research Council – Institute of Ecosystem Study)
CNRS	Centre National de la Recherche Scientifique (French National Center for Scientific Research)
CNRS-LaMP	Centre National de la Recherche Scientifique –Laboratoire de Météorologie Physique (French National Center for Scientific Research – Laboratory of Meteorology Physic)
CPC	Climate Prediction Center
CPTEC-INPE	Centro de Previsão de Tempo e Estudos Climáticos – Istituto Nacional de Pesquisas Espacias (Weather Forecast and Climate Study Center – Brazilian National Institute of Spatial Research)
CRM	Certificate reference material

CSE	Continental scale experiment
DCG	Debris-covered glaciers
DCP	Data collection platform
DEMs	Digital elevation models
DGPS measurements	Differential GPS measurements
DHM	Department of Hydrology and Meteorology
DHSVM	Distributed Hydrology-Soil-Vegetation Model
DLR	Deutsches Zentrum für Luft-und Raumfahrt (German Aerospace Center)
DM	Dimming effect
DMAS	Directorate of Mountaineering and Allied Sports
DMF	Divergence of moisture flux
DMPS/SMPS	Differential/Scanning Mobility Particle Sizer
DN	Digital numbers
DOC	Dissolved organic carbon
DOD	US Department of Defense
DODS	Distributed Oceanographic Data System
DORIS	French satellite positioning system
DSS	Decision support systems
DTM	Digital terrain models
DTR	Diurnal temperature range
DWD	Deutsche Wetterdienst (German Weather Service)
EAOP	Enhanced automated observing period
ECMWF	European Center for Medium range Weather Forecasting
ECPC	Experimental Climate Prediction Center
EF	Enrichment factors
EHIM	Indian eastern Himalaya
EHP	Elevated (atmospheric) heat pump effect
ELA	Equilibrium line altitude
EMERGE	Integrated Project to Evaluate the Impacts of Global Change on European Freshwater Ecosystems
ENSO	El Niño Southern Oscillation
EOP-1/2/3/4	First/second/third/fourth enhanced observing period
EOS	Earth Observation Satellite
ERA40	Forty-year European Re-Analysis data of the global atmosphere
ESA	European Space Agency
ETHZ	Eidgenössische Technische Hochschule Zürich (Swiss Federal Institute of Technology, Zurich)
ETM +	Enhanced Thematic Mapper Plus
ETS	Equitable threat score
EURAC	European Academy
EUSAAR	EUropean Supersites for Atmospheric Aerosol Research
Ev–K^2–CNR Committee	Everest–K2–Italian National Research Council Committee
FAO	Food and Agriculture Organization (United Nations)

FAR	False alarm ratio
FWHM	Full-width half-maximum
FYROM	Former Yugoslav Republic of Macedonia
GAME/Tibet	GEWEX (Global Energy and Water cycle Experiment) Asian Monsoon Experiment on the Tibetan Plateau
GAW	Global Atmosphere Watch
GAW SAGs	Global Atmosphere Watch Scientific Advisory Groups
GAWSIS	Global Atmosphere Watch Station Information System
GCM	General circulation model
GC-MS	Gas chromatography – mass spectrometry
GCPs	Ground control points
GDAS	Global gridded analysis
GDOP	Geometric dilution of precision
GEN	Glaciological Expedition to Nepal
GEO	Group on Earth Observations
GEOS	Goddard Earth Observing System
GEOSS	Global Earth Observation System of Systems
GEWEX	Global Energy and Water Cycle Experiment
GFAAS	Graphite furnace atomic absorption spectrometry
GF-AAS	Graphite furnace – absorption spectroscopy
GFD3	Geophysical Fluid Dynamics Laboratory R-30 Model
GHGs	Greenhouse gases
GHP	Global Energy and Water Cycle Experiment Hydrometeorological Panel
GIScience	Geographic information science
GISS	Goddard Institute for Space Sciences
GLIMS	Global Land Ice Measurements from Space
GLOCHAMORE	Global Change and Mountain Regions
GLOF	Glacier lake outburst flood
GLP	Global Lands Project for mountains
GPCP	Global Precipitation Climatology Project
GPS	Global positioning system
GSL	Growing seasonlength
GTOS	Global Terrestrial Observing System
HARY	Haryana (including Delhi)
HEIFE	Heife river field experiment
HIGHICE, India	Society for Himalayan Glaciology, Hydrology, Ice, Climate and Environment, India
HKH	Hindu Kush–Himalaya
HKKH	Hindu Kush–Karakorum–Himalaya
I.N.RI.M.	Istituto Nazionale di Ricerca Metrologica (Italian National Institute of Metrology Research)
I.PUNJ	Indian Punjab
ICIMOD	International Centre for Integrated Mountain Development
ICP-SFMS	Inductively coupled plasma sector field mass spectrometry

ICSI	International Commission of Snow and Ice
ICSU	International Council for Science
ICSU-WDCC	International Council for Science – World Data Center for Climate
ICTP	International Centre for Theoretical Physics
IGAC	International Global Atmospheric Chemistry
IGACO	Integrated Global Atmospheric Chemistry Observations
IGBP	International Geosphere-Biosphere Programme
IGBP-PAGES	International Geosphere-Biosphere Programme – Past Global Changes
IGWCO	Integrated Global Water Cycle Observations
IHDP	International Human Dimensions Programme on global environmental change
IILE	Italian Isotopic Lead Experiment
IMONT	Istituto Nazionale della Montagna (Italian National Mountain Institute)
INAA	Instrumental neutron activation analysis
INDOEX	Indian Ocean Experiment
IOP	Intensive observation period
IPCC	Intergovernmental Panel of Climate Change
IPY	International Polar Year
IR	Infrared
IRD	Institut de Recherche pour le Développement (Development Research Institute)
ITPCAS	Institute of Tibetan Plateau research, Chinese Academy of Sciences
IUCN	International Union for the Conservation of Nature and Natural Resources (The World Conservation Union)
IUFRO	International Union Forestry Research Organisations
JA	July, August
JAXA	Japan Aerospace Exploration Agency
JMA	Japan Meteorological Agency
JWGV	Joint Working Group on Verification
KEO	Carpathian Environment Outlook
KKH	Karakoram Highway
LAI	Leaf area index
LAS	Live access server
LCN	Lake cadastre number
LEAFS	Laser excited atomic fluorescence spectrometry
LIS	Limnological information system
LPI	Lake Piramide Inferiore (Lower Pyramid Lake)
LPS	Lake Piramide Superiore (Upper Pyramid Lake)
LSM	Land surface model
LTER	Long term ecological research
MAAP	Multi-angle absorption photometer
MAB	Man and the Biosphere Program

MAPD	Mean absolute percent difference
MAPR	Madhya Pradesh and Chhattisgarh
MBRs	Mountain Biosphere Reserves
MCS	Meso-scale convective system
MEA	Multilateral Environmental Agreement
MEI	Multivariate ENSO Index
MENRIS	Mountain Environment and Natural Resources Information System (ICIMOD)
MICS-Asia	Model Intercomparison Study – Asia
MISR	Multiangle imaging spectroradiometer
MIUR	Ministero dell'Istruzione, dell'Università e della Ricerca. (Italian Ministry of Education, University and Research)
MJ	May, June
MM5	Meso-scale Meteorological Model
MODIS	Moderate resolution imaging spectroradiometer
MOHPREX	Monsoon Himalayan Precipitation Experiment
MOLAR	Mountain Lake Research project
MOLTS	Model output location time series
MORP	Monitoring and research platform
MPI	Max Planck Institute
MRI	The Mountain Research Initiative
MSAVI	Modified Soil Adjusted Vegetation Index
NASA-GSFC	National Aeronautic and Space Administration – Goddard Space Flight Center (USA)
NASA's ESAS	National Aeronautic and Space Administration's Earth Science Enterprise Strategy (USA)
NASA-GLDAS	National Aeronautic and Space Administration – Global Land Data Assimilation Systems (USA)
NASA-GMAO	National Aeronautic and Space Administration –Global Modeling and Assimilation Office (USA)
NASDA	National Space Development Agency (USA)
NCAR	National Centre for Atmospheric Research (USA)
NCEP	National Centers for Environmental Prediction (USA)
NCMRWF	National Centre for Medium Range Weather Forecasting (USA)
NDVI	Normalized Difference Vegetation Index
NEPA	Nepal
NetCDF	Network Common Data Form
NGO	Non-governmental organization
NIST	National Institute of Standards and Technology (USA)
NMWP	Pakistan North West Frontier Province (usually written NWFP)
NOAA	National Oceanic and Atmospheric Administration (USA)
NWP model	Numerical Weather Prediction model
OC	Organochlorine compound
OPC	Optical particle counter

OPeNDAP	Open-source Project for a Network Data Access Protocol
OSU	Oregon State University
P.PUNJ	Pakistani Punjab
PAH	Polycyclic aromatic hydrocarbon
PCB	Polychlorinated biphenyls
PEP II	Pole-Equator-Pole II
PGE	Platinum-group element
PMD	Pakistan Meteorological Department
PMN	Pyramid MeteoNetwork
PMW	Passive microwave
POC	Particulate organic carbon
POD	Probability of detection
POP	Persistent organic pollutant
PR	Precipitation radar
QPF	Quantitative precipitation forecast
RAINS Project	Regional Air Pollution INformation and Simulation Project
RAJO-MEGHA	Radiation, Aerosol Joint Observations – Monsoon Experiment in Gangetic-Himalayan Area
RASS	Radio acoustic sounding system
RATEAP	Remote areas trace elements atmospheric pollution
RCM	Regional atmospheric climate models
ReSMA	Reparto Sperimentazioni di Meteorologia Aeronautica (Italian Air Force Meteorological Service – Department for Aeronautical Meteorology Experimentations)
RI	Rainfall intensity
RMSE	Root mean square error
RONAST	Royal Nepal Academy of Science and Technology
RS	Remote sensing
RSD	Relative standard deviation
SACEP	South Asia Co-operative Environment Programme
SAS	Simplified Arakawa-Shubert
SD	Size distribution
SeaWiFS	Sea-viewing wide field-of-view sensor
SEBS	Surface energy balance system
SHARE – Asia	Stations at High Altitude for Research on the Environment in Asia
SIO	Scripps Institution of Oceanography
SMTMS	Soil temperature and soil moisture network
SRMs	Standard reference materials
SSA	Single scattering albedo
SST	Sea surface temperature
START	Global Change System for Analysis, Research and Training
STH	Sub-tropical high

TARA	Tibet Assistance to Remote Areas
TE	Trace elements
TIMS	Thermal ionization mass spectrometry
TMI	The Mountain Institute
TOMS	Total ozone mapping spectrometer
TPS	Total particle size
TRMM	Tropical Rainfall Measuring Mission
TSP	Total suspended particles
UCAR-JOSS	University Corporation for Atmospheric Research/Joint Office for Science Support
UK89	United Kingdom Meteorological Office Model
UKMO	United Kingdom Meteorological Office
UNCED	United Nations Conference on Environment and Development
UNEP	United Nations Environmental Programme
UNEP RRC.AP	United Nations Environment Programme Regional Resource Centre for Asia and Pacific
UNEP Vienna–ISCC (Austria)	United Nations Environment Programme Vienna—Interim Secretariat. of the Carpathian Convention (Austria)
UNEP-ROE	United Nations Environment Programme – Regional Office for Europe
UNESCO-MAB	United Nations Educational, Scientific and Cultural Organization – Man and Biosphere
UNFCCC	United Nations Framework Convention for Climate Change
UNOmaha	University of Nebraska at Omaha
UT	University of Tokyo
UTC	Coordinated universal time
UTM	Universal transfer macerator
UTPR	Uttar Pradesh and Uttaranchal
UV	Ultraviolet
WCRP	World Climate Research Program
WCRP-CliC	World Climate Research Program – Climate and Cryoshpere
WCRP-COPES	World Climate Research Program – Coordinated Observation and Prediction of the Earth System
WEC-IIASA	World Energy Council – International Institute of Applied Systems Analysis
WESP	Water and Energy Simulations and Prediction
WG-EPAC	Working Group on Environmental Pollution and Atmospheric Chemistry
WGISS	Working Group on Information Systems and Services
WGS84	World Geodetic System 1984
WHIM	Indian western Himalaya
WMO	World Meteorological Organization

WMO-CAS	World Meteorological Organization – Commission for Atmospheric Science
WRRI-NARC	Water Resources Research Institute, National Agricultural Research Center (Pakistan)
WSSD	World Summit for Sustainable Development
WTF-CEOP	Working Group on Information Systems and Services Test Facility for CEOP
WWF	World Wildlife Fund
WWRP-WGNE	World Weather Research Program – Working Group on Numerical Experimentation

Mountains Witnesses of Global Changes
R. Baudo, G. Tartari, and E. Vuillermoz (Editors)

1

Introduction

Angelo Guerrini, Agostino Da Polenza and Harald Egerer

1. A. Guerrini

At the outset we would like to emphasize the long-term attention of the Italian National Research Council (CNR), under the direction of President Fabio Pistella, to the fields of environmental research. Many CNR institutions participate in national and international scientific research projects in remote areas with significant results, developing expertise in extreme environmental research. The Ev-K^2-CNR – SHARE-Asia (Stations at High Altitude for Research on the Environment in Asia) Project is closely related to those CNR activities that contribute to global change studies. Several institutions, in fact, have found a field-specific opportunity for working within the SHARE-Asia paradigm, such as the Institute of Atmospheric Sciences and Climate, and the Water Research Institute.

The creation of a high altitude monitoring network in the Himalaya–Karakoram region to increase general knowledge on environmental and Earth sciences in these regions will be able to directly benefit from the participations of CNR experts, such as those who are dedicated to ongoing research activities in the Arctic and Antarctic. We must also bear in mind the high altitude research bases CNR has set up in Italy, such as the Mt. Cimone Station of the Apennines and the "Testa Grigia" laboratory on Plateau Rosà in the Italian Alps, where our researchers have honed their skills and contributed appreciably to understanding environmental processes. Our tradition in this research field is long and the knowledge we have developed is comprehensive.

By having recently constituted an "Earth and Environment Department," CNR has consecrated its aims to create a more consolidated and organized structure that is able to provide answers to some of our planet's most pressing issues. We hope that the dedication of our researchers, combined with an appropriate political and governmental commitment, may truly aid us in concretely making the requisite contributions.

The relationships of CNR with the scientific community in general are excellent, and efforts are underway to engage other kinds of academic expertise. CNR is in fact

ISSN: 0928-2025 DOI: 10.1016/S0928-2025(06)10001-2

reinforcing relationships with the Italian Mountain Institute (IMONT) and with the organizers of the initiative of this book, the Ev-K^2-CNR Committee, in order to operate more effectively as a national system. The aspects of SHARE-Asia that aim to stimulate technology transfer and capacity-building processes, now very important within environmental policies, are in particular a distinctive component of this type of approach.The many factors expressed in this volume allow us to be full of hope for the future.

The conference, "Mountains, witnesses of global changes. Research in the Himalaya and Karakoram: SHARE-Asia Project," created an exceptional opportunity for dialog between major environmental scientists and experts. This highlights the close relationship between diverse themes with a common underlying thread, the in-depth comprehension of the environmental phenomena that are determining the health of our planet.

Thanks to contributions from delegates of some of the most important international global change research programs, particularly WMO–CEOP (Coordinated Enhanced Observing Period), UNEP–ABC (Project Atmospheric Brown Clouds), WMO–GAW (Global Atmosphere Watch) and GLIMS (Global Land Ice Measurements from Space), we were able to create a unique forum for discussing climate and natural ecosystems. The participation of Italian institutional representatives from the government, parliament and the Italian CNR, along with members of key decision-making and scientific institutions from Asia, demonstrated the potential that science has to positively influence policy and exemplified the condition that such processes must be further reinforced in order to effectively mitigate or reverse hazardous environmental trends.

2. A. Da Polenza

For nearly 20 years, Ev-K^2-CNR has been promoting interdisciplinary remote area research in Environmental and Earth sciences. This tradition has been maintained since the organization's inception by Prof. Ardito Desio, whose name is closely linked to the history of Himalayan exploration and scientific research, as well as to the first successful summit of the second highest mountain in the world, K2 Mountain, in 1954. Following in the footsteps of this great leader, Ev-K^2-CNR specialized in research in high altitude remote areas, and particularly in the Himalaya–Karakoram range. Adhering to the strategies and commitments of CNR regarding research, knowledge development, promotion of research networks, support of public administrations, and technological solutions, the aim is to meet the needs of mountain populations through capacity building and international scientific cooperation.

Due to the importance of mountain ecosystems in global change, the elevation and geographic location of the Himalaya–Karakoram region make it an ideal place for studying long-range pollutant transport systems on a regional scale and for monitoring changes induced by mechanisms that act on a global scale through monsoon circulation. Ev-K^2-CNR research began in Nepal's Khumbu Valley, near Mt. Everest, with the installation of the Pyramid International Laboratory-Observatory at 5050 m a.s.l. in 1990. Environmental and Earth science researchers soon became intent on understanding certain mechanisms unique to the mountain area. This interest gave rise to a climate-monitoring network comprising six stations in Nepal's Khumbu Valley and two automatic weather stations (AWSs) in Pakistan, in the Baltoro Valley of the Karakoram.

The SHARE-Asia Project is aimed at the development of an integrated system of measurements throughout the Himalaya Karakoram range, which will contribute to increasing general scientific knowledge on climatic and pollution-related processes while helping build local capacity for monitoring-related phenomena and impacts. Within this framework, the Ev-K^2-CNR Committee is now managing one of the world's higher atmospheric observatories, the ABC-Pyramid, operational at 5079 m a.s.l. in Nepal. Many glaciers in the region, such as the Khumbu, Changri–Nup, and Karakoram are periodically monitored by Ev-K^2-CNR contributing to the SHARE-Asia knowledge base. Limnological and paleolimnological research in the Sagarmatha National Park area of Nepal is performed annually, while geophysical investigations in Pakistan and Nepal complete the basic structure of the SHARE-Asia research paradigm. In coming years, this paradigm will be replicated and tailored in other regions, to create the SHARE environmental monitoring network. The Rome Conference represented the ideal launching pad for all that SHARE can and will represent in this context.

3. H. Egerer

The United Nations Environment Program (UNEP) and its Executive Director, Prof. Klaus Toepfer, have taken special positive notice of the Italian support of UNEP activities. In this regard and considering the growing attention being paid to mountain-related issues at the global level, one could say that together we could move mountains or, to put it more appropriately, we can help to protect and manage the fragile mountain ecosystems that provide so many critical environmental services to humanity. The highest appreciation of UNEP goes to the Italian Ministry of Environment and Territory for having done its best to support Carpathian countries since the developing of the idea of the Carpathian Convention until its final signature, as well as for the ongoing support.

Recently UNEP also signed a Memorandum of Cooperation with EURAC, a long-term Italian partner in Bolzano, to allow further close work together. EURAC, as an international center of excellence for sustainable mountain development, is providing continuous scientific backstopping to the Carpathian Convention, including scientific contributions to the Carpathian Environment Outlook (KEO) currently conducted by UNEP GRID Geneva together with scientific institutions in the seven Carpathian countries.

UNEP has strong views on capacity-building processes based on mountain potentials. As the world's leading intergovernmental environmental organization, UNEP is the authoritative source of knowledge on the current state of affairs and trends shaping the global environment. The mission of UNEP is to provide leadership and encourage partnership in caring for the environment by inspiring, informing, and enabling nations and peoples to improve their quality of life without compromising that of future generations. As the principal environmental body in the UN system, UNEP has a lead role in promoting Chapter 15 of Agenda 21 on the Conservation of Biological Diversity, as well as those adopted in other relevant chapters. UNEP has initiated negotiations for many international environmental conventions that address major environmental issues of the day.

On a regional level, concern for the environment has led to the establishment of new regional groupings based on shared natural resources, including the Baltic Sea, Danube Basin, Rhine, and, more recently, the Black and the Caspian Seas. Affected countries have made use of regional MEAs to create protection regimes.

As acknowledged by the Earth Summit in Rio and by the World Summit on Sustainable Development (WSSD), mountains are vital to all life on earth and to the well being of people everywhere. Mountains are indeed the source of many services and benefits upon which the quality of our life and the health of our planet depend. Mountain ecosystems are an important source of water, energy, and biological diversity. Furthermore, they are a source of such key resources as minerals, forest and agricultural products, and of recreation.

Mountains are evidently mighty, lofty places and over the centuries we have allowed ourselves to see them, in our mythology and folklore, as both impregnable and capable of taking care of themselves. Mountains are seen as providers of abundant natural resources, free of charge, and without regard to the value of the goods and services generated by their natural world. Consequently socio-economic conditions have deteriorated in the mountain regions of all continents, as exemplified by poverty, limited employment opportunities, depopulation, environmental and natural resources degradation, retreat of glaciers, loss of ethno-cultural traditions, interethnic tensions, conflicts, and lack of access to information. The conservation of ecosystems, prevention and mitigation of disasters, and sustainable development of mountain regions are a common emerging concern of our countries and the international community. However, nowadays the goods and services provided by mountain ecosystems have little or no market value despite their importance in the economic lives of communities and nations. These trends need to be reversed by further ecosystem valuation leading decision-makers to reinvest in nature's capital of mountain ecosystems. Research indicates that investing in nature can provide an excellent rate of return and help to meet internationally agreed-upon development goals.

The Mountain Partnership, launched at WSSD in 2002, is a broad alliance of countries, intergovernmental organizations and major groups working toward improving the lives of mountain people and protecting mountain environments around the world. Its secretariat is hosted by the FAO in Rome, with the participation of UNEP. The Merano Conference in Italy on October 2003 was a landmark event for the Mountain Partnership; it was the first global meeting of members and a unique forum in which to identify common needs, priorities, and concerns and to explore key issues related to the structure, membership, and governance of the partnership. Countries in Central and Eastern Europe as well as in Central Asia have taken this opportunity to propose initiatives for the protection and sustainable management of major trans-boundary mountain ranges in Europe and in Central Asia. The Cuzco Conference in Peru confirmed mentioned initiatives in 2004.

The Alpine Convention, adopted in 1991, which brought all Alpine countries to work together on mountain development and protection, has provided much inspiration in this regard, particularly in Europe, Asia, Latin America, and Africa. Already in the International Year of the Mountains 2002, UNEP/ROE launched the European Mountain Initiative, assisting governments of the region in facilitating increased cooperation for the protection and sustainable management of trans-boundary mountain ranges. At the outset, in 2001, UNEP/ROE was requested by the Government of Ukraine to service a regional cooperation process for the Carpathians, a major trans-boundary mountain range shared by the seven different countries in Central and Eastern Europe. In 2002, the Alpine–Carpathian partnership was initiated by the Ministry of the Environment and Territory of Italy. At the Fifth Ministerial Conference "Environment for Europe" held in Kyiev, Ukraine, in May 2003, the Carpathian countries adopted the Framework Convention on the Protection and Sustainable Development of the Carpathians, consequently signed by all seven countries. The Interim Secretariat for the Convention was inaugurated in Vienna, Austria, on 15 July 2004.

At the initiative of Armenia, a first "Meeting of the Authorized Representatives on the Development of a Legal Instrument for the Protection of the Caucasus Mountain Ecosystem" was held from 26 to 27 June 2001 in Yerevan, Armenia. The meeting adopted a resolution, recognizing the need for a legal instrument, e.g. the Caucasus Convention, and requesting UNEP's further assistance in the development of such instrument.

Significant steps toward closer cooperation in the Balkan region are being made within the context of the Mountain Partnership on the proposal of FYROM. The process began at the Second Global Meeting of the Mountain Partnership, or Cuzco Conference in 2004 with discussion of the opportunities to develop a collaborative mechanism for countries in the Balkan region that would be similar to those already existing for the Alpine and Carpathian regions. This so-called "Balkan process," speaking mountain languages, peak to peak with the "Caucasus process," was further explored at a meeting of representatives from the Balkan and Caucasus countries in Bolzano, Italy, in December 2005, with a view toward launching a formal process leading to closer cooperation in mountain development in the Balkans and Caucasus.

One can only hope that sharing experiences in capacity building on mountain potentials may serve as a wake-up call to policy makers of South Asian countries to build trans-boundary platforms for sustainable development of South Asian mountain regions that will give even more impetus to national, regional and global mountain-related initiatives. These efforts might one day be crowned by a Convention for the Protection and Sustainable Development of the Himalaya Hindu Kush Karakoram.

Regional mountain cooperation supports, and is supported by, science, as exemplified by UNEP involvement in Atmospheric Brown Cloud (ABC) Project. The WSSD in 2002 was the first time that the ABC really received international attention. The ABC Project aimed at the development of science and capacity to study the issues of aerosols by assessing the impacts of ABC on health, ecosystem and agriculture, climate change and water budget under one common framework, and by raising awareness on the issue and promoting actions for mitigation. It goes without saying that the ABC Project is an example of the application of advanced technologies to mountain research that aims to contribute to sustainable management processes. UNEP plans to establish a network of ground-based monitoring stations throughout Asia to study the composition and seasonal pattern of the haze. Ev-K^2-CNR's input will be to capture traces of aerosols and particulates accumulating at sensitive high altitude AWS sites in Himalayas, allowing climatic, atmospheric, chemical, and ecological measurements to be carried out above the cloud's level.

Special emphasis should be placed on the role of Italian support regarding the ABC Project. According to the latest briefing from UNEP-Asia, Italian support will go toward additional monitoring equipment for the Himalaya, which will enable existing stations to monitor vital data on aerosols and particulates. These stations will then join a network of stations across the region. This general assistance in strengthening the scientific base of ABC, and in capacity building for the scientists from the Hindu Kush Himalayan region, are excellent examples of how quality research, promoted in cooperation with regional intergovernmental organizations, can directly influence capacity-building efforts and bring scientific results to the notice of governments.

Atmospheric Brown Clouds (ABC)

Mountains Witnesses of Global Changes
R. Baudo, G. Tartari, and E. Vuillermoz (Editors)

2

Global and regional climate change: the next few decades

Veerabhadran Ramanathan

Abstract

Approximately 27 years ago, scientists predicted that global warming due to man-made greenhouse gases (GHG) would be detectable by the year 2000 (e.g., Madden and Ramanathan, 1980) and that this warming would be accompanied by a large amplification of warming at high altitudes and a rise in sea level (See IPCC, 2007 for a historical perspective). There is now a general scientific consensus about the veracity of these predictions and a growing societal awareness that global warming is the most vexing environmental issue facing the planet. The newly released IPCC summary for policy makers (IPCC, 2007) concludes that the radiative forcing due to the increase in the greenhouse gases (carbon dioxide, methane, nitrous oxide, tropospheric ozone, and several halo carbons) from the pre-industrial to the year 2004 is about $3\,W/m^2$ area of the earth's surface. The energy added to the planet by this radiative forcing is equivalent to burning 25 trillion light bulbs (thousand billion bulbs distributed around the planet) of 60-W power each, every second of the day and night, throughout the year. What is less recognized, however, is a comparably major global problem dealing with air pollution. New data have revealed that, due to fast long-range transport, trans-oceanic plumes of atmospheric brown clouds (ABCs) containing tiny particles intercept sunlight, cause surface dimming, cool the surface, warm the air and disrupt regional rainfall patterns and lead to large-scale drying (Ramanathan et al., 2005, 2007; also see http://www-abc-asia.ucsd.edu/). For the Indian sub-continent and the adjacent N Indian Ocean including Arabian Sea and Bay of Bengal, it has been shown (Ramanathan et al., 2001) that the brown clouds from S and SE Asia have decreased the solar radiation at the surface by about 7% (for the 6 month long dry season) and increased atmospheric solar heating by as much as 50%. When the observationally constrained 3-dimensional radiative forcing by the brown clouds was introduced into a coupled ocean-atmosphere model, it led to a deceleration of the summer monsoon circulation and helped account for the observed reduction of 5–7% in the summer rainfall since the 1950s (Ramanathan et al., 2005). The primary mechanism, as discussed in Ramanathan et al. (2005), is that the dimming by brown clouds suppressed the sea surface temperatures in the N Indian Ocean from responding to the greenhouse warming; whereas, south of the

ISSN: 0928-2025 DOI: 10.1016/S0928-2025(06)10002-4

equator (where the brown cloud effects are minimal) the sea surface warmed by as much as 0.7 K since the 1950s. The net effect is a reduction in the north–south sea surface temperature gradient which in turn led to a deceleration of the monsoon circulation. This finding has since been confirmed by two other independent studies (Lau et al. in this issue; and another by Meehl et al., submitted). The simulated reduction in rainfall was incorporated in an Agro-economic model by Aufhammer et al. (2006). This first study of its kind, demonstrated that the effect of greenhouse warming on minimum temperatures, when combined with reduction in rainfall by brown clouds, have decreased rice production in Indian by as much as 14% since the 1950s. The important aspect of this finding is that the rate of growth of rice harvest in India has slowed down considerably (from over 3% in the 1960s to less than 1% now). The Aufhammer et al. study provides an explanation for this anomalous behavior. GHGs on the other hand, warm the surface and the atmosphere and make the planet wetter with correlative effects, such as melting of glaciers and altering the strengths of storms, among others.

New studies (Chung et al., 2005; Yu et al., 2006), using observationally constrained aerosol observations, have clearly shown that anthropogenic aerosols (i.e., brown clouds) have led to a global averaged dimming of about 3–5 W/m^2, as of 2000–2003. There is a comparably large atmospheric solar heating, and the sum of the two, the radiative forcing at the top of the atmosphere, is about -0.6 to $-2.7\,W/m^2$.

It now seems that the surface cooling effect of ABCs may have masked as much as 50% of the global warming due to GHGs. This presents a dilemma for the global community because efforts to curb air pollution, which has to happen to mitigate their negative effects on human health and the eco-system, may unmask the ABC cooling effect and could lead to a large amplification of the surface warming in the coming decades. On the other hand, if GHGs are curbed due to concerns about global warming, which also has to happen in view of their large global negative impacts, ABCs from air pollution may weaken the monsoon rainfall in parts of South and East Asia, thus presenting conflicting options between those regions that are negatively impacted by global warming and those by air pollution. Another potentially major negative impact of ABCs is that the solar heating of the atmosphere by soot can accelerate the atmospheric warming by greenhouse gases and thus contributes significantly to the observed retreat of the Himalayan glaciers. The uncertainties in our understanding of these effects are large, but the deeper we delve into the science we are discovering new ways in which human activities are changing the environment during the Anthropocene (the time period of humanity), and the ethical and scientific dilemmas are becoming more formidable.

Further Reading

Auffhammer, M., Ramanathan, V., and Vincent, J.R., 2006. Integrated model shows that atmospheric brown clouds and greenhouse gases have reduced rice harvests in India. PNAS, 10.1073/pnas.0609584104.

Chung, C.E., Ramanathan, V., Kim, D., and Podgorny, I.A., 2005. Global anthropogenic aerosol direct forcing derived from satellite and ground-based observations. J. Geophys. Res. 110, D24207, doi:10.1029/2005JD006356.

IPCC, 2007. Climate Change 2007: The Physical Science Basis. Summary for Policy Makers, The Intergovernmental Panel on Climate Change, p. 21.

Lau, K.M. and Kim, K.M. Does aerosol weaken or strengthen the Asian monsoon water cycle? (Chapter 3). In: Shroder, J.F. Jr. (Ed.), Mountains: Witnesses of Global Changes – Research in the Himalaya and Karakoram. Elsevier publishing, Amsterdam, Vol. 10, pp. 13–22.

Madden, R.A. and Ramanathan, V., 1980. Detecting climate change due to increasing CO_2 in the atmosphere. Science 209, 763–768.

Ramanathan, V., Chung, C., Kim, D., et al., 2005. Atmospheric brown clouds: impacts on south asian climate and hydrological cycle. PNAS, Vol. 102, No. 15, 5326–5333.

Ramanathan, V., Crutzen, J., and Lelieveld, J., 2001. The Indian Ocean experiment: an integrated assessment of the climate forcing and effects of the Great Indo-Asian Haze. J. Geophys. Res. Atm. 106 (D 22), 28,371–28,399.

Ramanathan, V., Li, M.V., Ramana, M.V., et al., 2007. Atmospheric brown clouds: hemispherical and regional variations in long range transport, absorption and radiative forcing, In Review, J. Geophys. Research, submitted.

Yu, H., Kaufman, Y.J., and Chin, M., 2006. A review of measurement-based assessments of the aerosol direct radiative effect and forcing. Atmos. Chem. Phys. 6, 613–666.

Mountains Witnesses of Global Changes
R. Baudo, G. Tartari, and E. Vuillermoz (Editors)

3

Does aerosol weaken or strengthen the Asian monsoon?

William K.M. Lau and Kyu-Myong Kim

Abstract

In this paper, the relative roles of solar dimming (SDM) effects vs. atmospheric heating effects by absorbing aerosols on Asian monsoon rainfall and circulation are investigated using an atmospheric general circulation model coupled to a mixed layer ocean, forced by prescribed global aerosol forcing. Comparison between experiments with and without aerosol forcing shows that equilibrium condition is attained in the Asian monsoon atmosphere-land-ocean system after approximately 30 years of integration, by which time the model sea surface temperatures (SST) in the Indian Ocean and western Pacific will have dropped by more than 1°C. In spite of the reduction in SST, rainfall is found to be increased in northern India and the Tibetan Plateau in late spring and early summer (May–June). This increase is attributed to the "elevated-heat-pump (EHP) effect", i.e., heating of the middle and upper troposphere induced by absorbing aerosols (dust and black carbon) piling over the southern slopes of the Himalayas, coupled with feedback with the deep convection and the large-scale circulation. In July–August, when the aerosol loading is substantially diminished, rainfall over all the Asian monsoon regions, except over and in the vicinity of the Tibetan Plateau, is reduced. The overall reduction is due to a spin-down of the large-scale monsoon circulation, stemming from the cooler earth surface and diminished land–sea thermal contrast induced by aerosol SDM. The results are supported by preliminary observations of increased loading of absorbing aerosols over the Indo-Gangetic Basin, and enhanced monsoon rainfall in May–June over northern India during the last two decades.

1. Introduction

Aerosol-related health hazards and floods and droughts associated with monsoon variability are two of the most pressing environmental threats confronting many Asian countries. Traditionally, aerosol and monsoon were studied as separate disciplines – the former as a local air pollution problem and the latter as a purely natural phenomenon. However, there has been shift in this paradigm as a result of recent studies, which show that the global and regional water cycles can be

ISSN: 0928-2025 DOI: 10.1016/S0928-2025(06)10003-6

profoundly affected by the presence of aerosols in the atmosphere (Rosenfeld, 2000; Ramanathan et al., 2001; and many others).

Aerosols scatter and/or absorb solar radiation, reducing solar radiation reaching the earth surface, through the so-called solar dimming (SDM) effect. SDM can lead to surface cooling that results in overall increased stability of the lower atmosphere, reduced surface evaporation, suppressed convection, and hence a weakened global water cycle. Another way aerosols can affect the water cycle is through cloud microphysics, whereby aerosol increases the number of cloud condensation nuclei, prolongs lifetime of clouds, and inhibits the growth of cloud drops to raindrops. This leads to more clouds, increased reflection of solar radiation, further cooling at the Earth's surface, and weakening of the water cycle. It is important to point out that the aforementioned aerosol-induced changes on aspects of the water cycle are only local effects applicable to an urban area, or a small cluster of cloud and rain types. For the global water cycle, the large-scale circulation plays an important role in aerosol transport and in determining the environment under which aerosols interact with clouds and precipitate locally. The continuous and collective action of the aerosol local effects may induce a host of feedback processes that generate atmospheric teleconnection that conveys the aerosol impacts far away (thousands of kilometers) from the source regions. The feedback processes can further spur new diabatic heat sources and sinks, and alter the general circulation which in turn modulates the local effects (Kim et al., 2006).

In monsoon regions the aerosol-water cycle dynamic interaction is highly complex not only because of the presence of, and possible interactions among, diverse mixtures of aerosol species with vastly different radiative properties, but also because the monsoon is strongly influenced by ocean and land surface processes, land use, land change, as well as greenhouse warming effects. Thus, sorting out the impacts of aerosol forcing and interaction with the monsoon water cycle is a very challenging problem.

Up to now, much of the aerosol–monsoon water cycle research has focused on determination of the radiative properties of anthropogenic aerosols and estimation of the magnitude of radiative forcing of the local effects of aerosols (Kaufman et al., 1991; Takemura and Nakajima, 2002; Sekiguchi et al., 2003; and many others). Recently, there have been more studies devoted to the dynamical response of the entire monsoon water cycle to aerosol forcing using general circulation models (GCM). Menon et al. (2002) showed from idealized aerosol forcing experiments that absorption by black carbon over industrial pollution regions of South and East Asia may induce rainfall anomalies that may cause the long-term north-dry and south-wet pattern over China. Ramanathan et al. (2005) suggested from coupled GCM experiments that the SDM effect due to industrial pollution, especially the increased loading of black carbon over India, may lead to a long-term (order of several decades) weakening of the South Asian monsoon, by spinning down the global water cycle through reduced surface evaporation, and decrease surface thermal contrast. Based on experiments with the NASA GEOS model, Lau et al. (2006, hereafter referred to as LKK) proposed that the elevated-heat-pump (EHP) effect – solar absorption by elevated dust aerosols piling up against the southern slopes of the Tibetan Plateau may cause an intensification of the Indian monsoon. Aspects of the

EHP effects have been shown to be consistent with observations (Lau and Kim, 2006). The aforementioned studies point to various possible scenarios of aerosol impacts on the Asian monsoon, and set off an earnest scientific debate on whether aerosols are weakening or strengthening the monsoon. In this paper, we plan to shed some new light on this debate by demonstrating the relative role of the Tibetan Plateau heating vs. sea surface temperature (SST) cooling in regulating the aerosol-monsoon water cycle.

2. The elevated-heat-pump effect

In this section, we recapture the salient features of the EHP effect as proposed by LKK. Absorption of solar radiation by dust and black carbon aerosols, which stack up against the southern slope of the Tibetan Plateau, produces anomalous warming in the upper troposphere, and induces a positive feedback associated with increased deep convection and enhanced moisture inflow from the south, leading to increased monsoon rainfall during the early monsoon season.

As shown in Fig. 3.1a the anomalous warming in the upper troposphere over the Tibetan Plateau begins in March and becomes well established in April and May (Fig. 3.1b and c). The warming forces an anomalous local meridional overturning

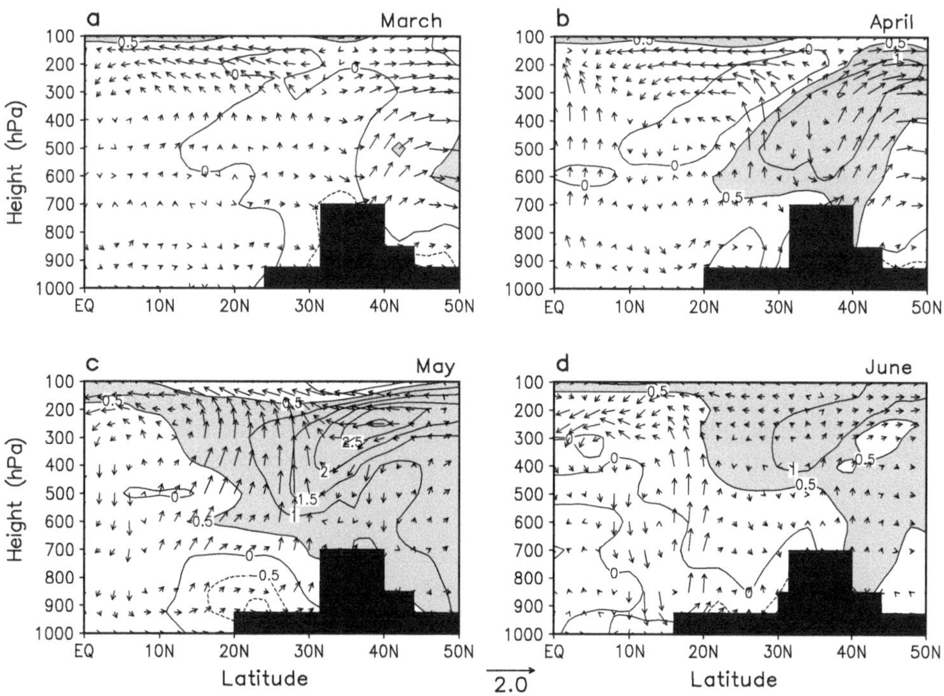

Figure 3.1. Vertical cross-section of temperature and wind anomalies over the Indian subcontinent in the NASA GCM simulations, showing the anomalous meridional overturning, near surface cooling, and upper troposphere heating as a result of aerosol-induced circulation changes (Lau et al., 2006).

with rising motion over northern India and subsidence over southern India, which becomes very pronounced in May and June (Fig. 3.1c and d). Notice that in May, the land surface over the Indian subcontinent is anomalously cold due to the SDM effect. However, the EHP draws in warm, moist air (relative to the inland air) above the colder air mass over the land surface. Over the cold surface, rainfall is suppressed as predicted by the semi-direct effect (Hansen et al., 1997). Thus, the air mass retains most of its moisture, as it approaches the foothills of the Himalayas, producing a delayed rain effect, which increases convection over northern India and intensifies the anomalous monsoon overturning circulation. This overturning persists into June and July (not shown) resulting in an overall increase in rainfall over northern India and suppressed rainfall in southern India and the northern Indian Ocean.

The increased rainfall over northern India and reduced rainfall in southern India are only a part of a large-scale response of the entire Asian monsoon to aerosol-induced forcing, featuring extensive rainfall reduction in southern China and the East China Sea, an east–west oriented band of enhanced rainfall along 30–35°N in central China, southern Korea and Japan, and reduced rainfall over northeastern China (Fig. 3.2a). The suppression of rainfall over southern China is due to surface cooling spurred by industrial pollution mainly from sulfate and black carbon. In contrast to the Indian monsoon region, the absorption heating due to black carbon over southern China remains largely within the lower troposphere due to the lack of orographic forced ascent, and consequentially is not efficient as an atmospheric heat source to initiate new convection. As a result, the stabilizing influence due to semi-direct effect of aerosols prevails, as is evident in the widespread suppression of rainfall over southern China.

Figure 3.2b shows that the aerosol-induced rainfall anomaly is associated with the development of a large-scale surface pressure and wind anomaly pattern, represented by an eastward extension and strengthening of the western subtropical high, which appears to be connected to a large-scale anticyclonic circulation anomaly over southern India and the Indian Ocean. The increased low-level westerlies over northern India and the Bay of Bengal indicate a strengthening of the Indian monsoon. The climatological southwesterly flow over the northern South China Sea is replaced by low-level easterlies. Anomalous southwesterlies occur over central China, South Korea and Japan, signally a northward shift and weakening of the *Mei-Yu* rainfall regime over East Asia. Analyses of further experiments with various combinations of aerosol forcing suggest that the anomaly patterns are due to the combined effects of dust and black carbon aerosols, with dust playing a primary role in instigating these patterns.

3. "Elevated heat pump" vs. "solar dimming"

As indicated in the previous section, the EHP effect intensifies and shifts the South Asian monsoon rain belt northward toward the foothills of the Himalayas, and leaves southern India and the northern Indian Ocean anomalously dry in the early monsoon season. However, the model experiments by LKK were conducted under prescribed SST conditions, so that the SDM effect was operative only over land.

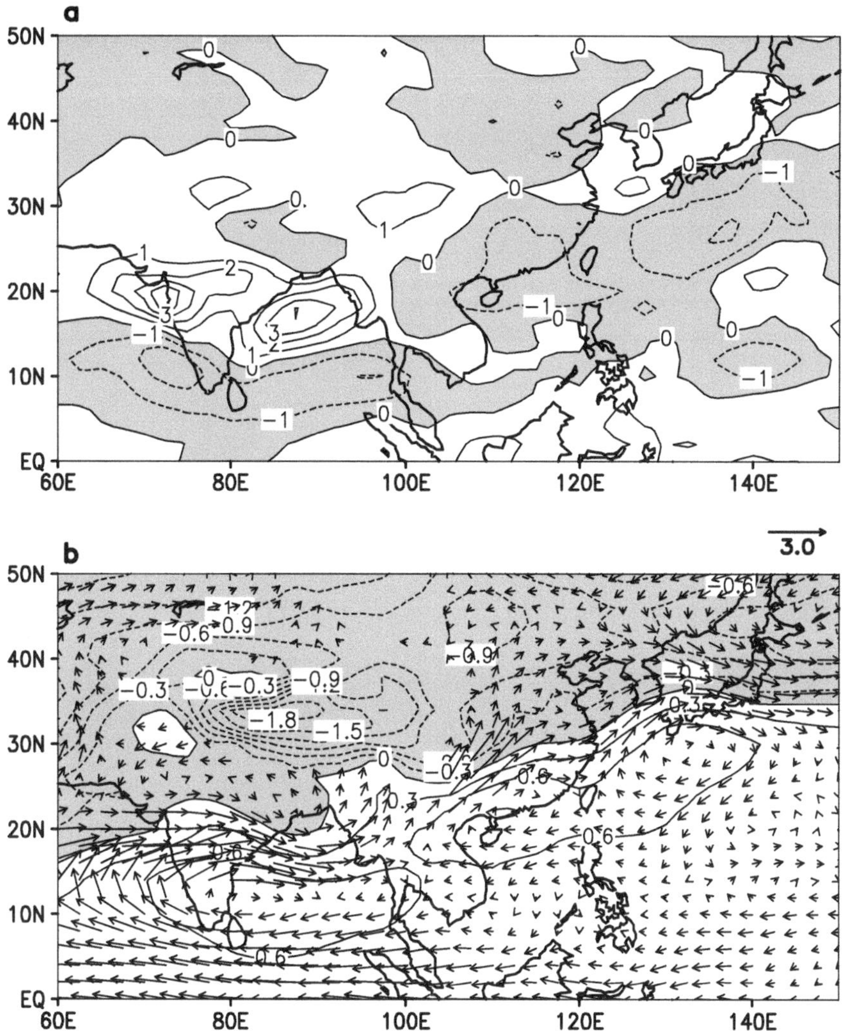

Figure 3.2. Spatial pattern of aerosol-induced seasonal mean (JJA) anomalies of (a) rainfall and (b) sea level pressure and 850 hPa wind, based on the experiments with the NASA GEOS model (Lau et al., 2006).

To examine the relative importance of EHP vs. SDM over ocean and land, we have carried out the same experiments as LKK, except that the GEOS GCM is now coupled to a mixed layer ocean model. The coupled model is integrated for 90 years, with and without global aerosol forcing. The latter is defined as the control. The anomaly is defined as the difference between the experiment with aerosol and with the control. Figure 3.3 shows the time–latitude cross-section of the aerosol-induced precipitation anomaly over the Indian subcontinent for the early monsoon season, May–June (MJ) and for the late monsoon season, July–August (JA), based on the difference between the aerosol and no aerosol experiments.

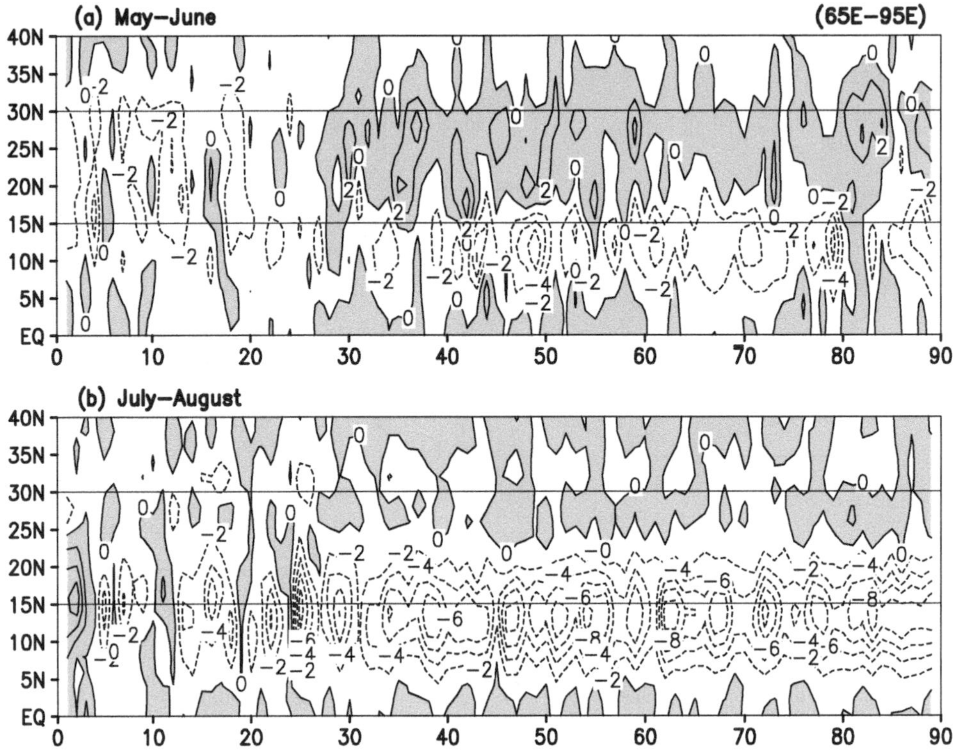

Figure 3.3. Time–latitude cross-section of aerosol-induced rainfall anomaly (mm/day) in a 90-year simulation using a coupled mixed layer ocean – GCM for (a) May–June and (b) July–August over the longitudinal sector of the Indian subcontinent. The region of northern India (15–30°N) is highlighted with horizontal lines.

For MJ, it is clear the SDM effect dominates for the first 20 years, with rainfall suppression over the entire Indian subcontinent. However, as the ocean atmosphere adjusts to the imposed aerosol forcing, the EHP effect seems to have gained an upper hand beginning at year 30, and the final equilibrium state shows the characteristic EHP signal – enhanced monsoon rainfall over northern India, and reduced rainfall to the south. For JA, the EHP effect, albeit much weaker, is still detectable over northern India, but SDM effect seems to have dominated with overall reduction in monsoon rainfall due to spinning down of the monsoon Hadley circulation (not shown).

The distribution of rainfall and 850 hPa wind anomalies in the new equilibrium is shown in Fig. 3.4 as the difference of the 30-year mean (years 30–60) between the aerosol experiment and the control. In May–June (Fig. 3.4a), increased rainfall is found over the Tibetan Plateau, northern India and the Arabian Sea, and the western Pacific between 10 and 20°N, while reduced rainfall is found over the Bay of Bengal, IndoChina, South China Sea, and East Asia. The wind pattern indicates anomalous easterlies over the Bay of Bengal, northeasterlies over Japan and East Asia, and westerlies flow over northeastern India and Bangladesh. The wind

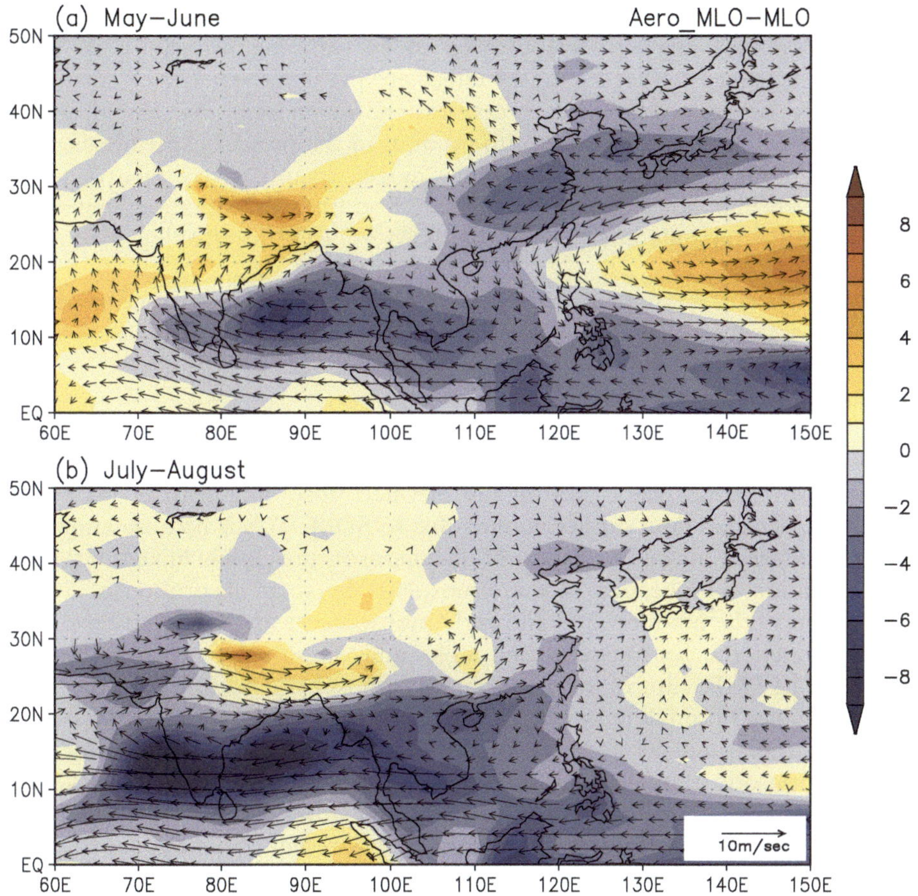

Figure 3.4. Spatial pattern of aerosol-induced rainfall and 850 hPa wind anomalies for (a) May–June and (b) July–August from the experiments with the NASA GEOS model coupled with mixed layer ocean model.

pattern signals a weakening of the large-scale monsoon circulation, except over the southern Tibetan Plateau and northeastern India. The reduced rainfall is associated with the increased cooling in the land surface relative to the adjacent ocean due to SDM, thus reducing the surface moisture flux and thermal contrast between land and ocean. On the other hand, the increased rainfall over the Tibetan Plateau and northern India is maintained by the increased thermal gradient in the upper troposphere over the Tibetan Plateau and the regions to the south (figure not shown) maintained by the EHP effect. In July–August, the aerosol forcing is diminished; suppressed rainfall and anomalous low-level easterlies are found over much of the Indian Ocean and the Bay of Bengal, indicating a weakened Asian monsoon. Residual signal of the EHP effect can still be seen as increased rainfall over the Tibetan Plateau, and low-level westerly flow to its south near 25°N. It should be noted that by the end of the 90-year integration, the SST has cooled by more than 1°C over a large span of the open tropical ocean. In the real world, the tropical oceans are actually observed to be warming by ~0.5°C in the last 50 years. This suggests that the SDM effect on SST may actually be small, either balanced by oceanic ventilation processes and/or masked by global warming effects. These effects are not taken into account in the present experiments. Hence, the results in Figs. 3.3 and 3.4 cannot be judged quantitatively relative to other factors influencing monsoon rainfall and SST in the real situation, but only to demonstrate the possible interplay of the SDM vs. EHP in affecting the monsoon regions.

4. Concluding remarks

Our results suggest that on subseasonal to interannual timescales, the EHP effect associated with atmospheric heating by elevated absorbing aerosol may have a stronger impact than land surface cooling due to SDM. This is because atmospheric loading, residence time and transport of dust aerosols are closely linked to the large-scale circulation. The heating of the atmosphere by solar absorption of dust and black carbon in the pre-monsoon and early monsoon season appear to have a regulatory effect on the heat sources and sinks and on subsequent evolution of the monsoon. The EHP features increased upper tropospheric heating over the Tibetan Plateau, enhanced rainfall in northern India and the Bay of Bengal, reduced rainfall in southern India and the northern Indian Ocean in May–June, and subsequently increased monsoon rainfall over all India. As such, the EHP is a plausible and testable hypothesis that needs to be verified with observations and further experiments. Aspects of these features are confirmed in a recent observational study (Lau and Kim, 2006). Better observations of aerosol composition, radiative properties and distribution over the Indo-Gangetic Plain and the Himalayas are required to fully validate the EHP hypothesis.

On the decadal and longer timescales, both SDM and EHP may have important impacts, with EHP being more evident in the late spring and early summer, and SDM in mid and late summer. This result is also consistent with recent GCM experiments by Chung and Ramanathan (2006) which demonstrated that atmospheric heating effect by absorbing aerosols can lead to increased June–July–August (JJA) monsoon rainfall, but when the effect of weakened SST gradient over the northern Indian Ocean and the Arabian Sea, presumably due to SDM effect, is included, the Indian monsoon is weakened. As stated before, the aerosol signal in monsoon rainfall is likely to be confounded by other forcing agents such as El Nino, land surface processes, interdecadal variability, and possibly global warming. Long data records of rainfall and aerosol concentration and types over the South Asian monsoon region are necessary to isolate the aerosol signals from other forcing agents. This will be challenging because while reasonably reliable long-term rainfall record exists for India, and other monsoon regions, the same cannot be said for aerosol data. Before the satellite era, there were simply no global aerosol data. Most aerosol data are

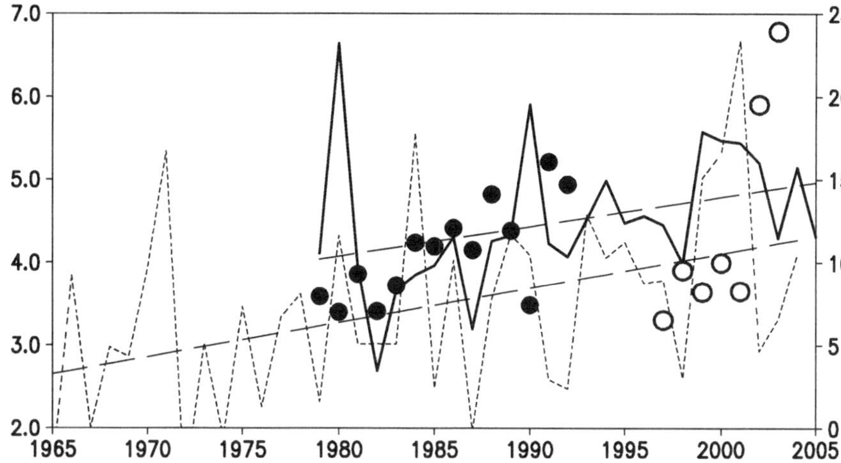

Figure 3.5. Time series of GPCP rainfall (solid line), and IITM rainfall for the Central Northeast subdivision (dotted line) over the Indo-Gangetic Plain (15–25 N, 70–90 E) in May–June. Black and white dots indicate different satellite estimates of TOMS aerosol index over the same region. The dashed straight line indicates linear trend of GPCP, and IITM rainfall respectively. Unit of rainfall is in mm per day. TOMS aerosol index is non-dimensional.

retrieved from air pollution inventories from local municipalities and are of non-research quality at best.

As an example, Fig. 3.5 shows time series of rainfall anomalies from the global precipitation climatology project (GPCP) and aerosol index from the total ozone mapping spectrometer (TOMS) over the Indo-Gangetic Plain 15–25°N for May–June from 1979 through 2004. Also included is rainfall anomalies from the Indian Institute of Tropical Meteorology (IITM) for the period 1965–2004. The latter is the only multi-decadal global satellite data for absorbing aerosols that have ever existed. The increased rainfall trend from 1982 through 1992 in GPCP rainfall and the longer term trend from 1965 through 2004 are quite obvious and appear to correlate well with the increase in aerosol loading during that period. Unfortunately the TOMS AI data have a gap from 1993 to 1996. For the latter period (1997–present), the aerosol data may have a different reference calibration compared to the previous record, and the alignment with the rainfall data is not as good. Hence, the result in Fig. 3.5 while encouraging as a possible indication of the EHP effect, cannot be construed as a validation of the effect. Clearly, the determination of possible long-term effect of aerosol effects on monsoon climate and water cycle requires extensive data-mining efforts into historical, local archives of surface temperatures, solar radiation, visibility, or other long-term proxy records of temperature from tree rings and aerosols from ice cores in land glaciers over the Himalayas and the Tibetan Plateau. Aerosol observations from satellite observations such as MODIS, Cloudsat-Calipso will provide data for validating and improving the realism of representation of aerosol radiation and aerosol-cloud microphysical processes, which are critical in reducing the uncertainties of model projections of aerosol effects on water cycle of the Asian monsoon.

Acknowledgement

This research is supported by the NASA Interdisciplinary Investigation and the Modeling and Analysis Program. Dr. F. Yang provided support in conducting the coupled model integration.

References

Chung, C.E. and Ramanathan, V., 2006. Weakening of north India SST gradients and the monsoon rainfall in India and the Sahel. Journal of Climate 19, 2036–2045.

Hansen, J., Sato, M., and Ruedy, R., 1997. Radiative forcing and climate response. Journal of Geophysical Research 102, 6831–6864.

Kaufman, Y.J., Fraser, R.S., and Mahoney, R.L., 1991. Fossil fuel and biomass burning effect on climate: heating or cooling?. Journal of Climate 4, 578–588.

Kim, M.K., Lau, K.M., Chin, M., et al., 2006. Atmospheric teleconnection over Eurasia induced by aerosol radiative forcing during boreal spring. Journal of Climate 19, 4700–4718.

Lau, K.-M. and Kim, K.-M., 2006. Observational relationships between aerosol and Asian monsoon rainfall, and circulation. Geophysical Research Letters, 33, L21810, doi: 10.1029/2006GL027546.

Lau, K.-M., Kim, M.-K., and Kim, K.-M., 2006. Asian summer monsoon anomalies induced by aerosol direct forcing: the role of the Tibetan Plateau. Climate Dynamics 26 (7–8), 855–864.

Menon, S., Hansen, J., Nazarenko, L., and Luo, Y., 2002. Climate effects of black carbon aerosols in China and India. Science 297, 2250–2253.

Ramanathan, V., Chung, C., Kim, D., et al., 2005. Atmospheric brown clouds: impacts on South Asian climate and hydrologic cycle. Proceedings of the National Academy of Science USA 102, 5326–5333.

Ramanathan, V., Crutzen, P.J., Kiehl, J.T., and Rosenfeld, D., 2001. Aerosols, climate and the hydrologic cycle. Science 294, 2119–2124.

Rosenfeld, D., 2000. Suppression of rain and snow by urban and industrial air pollution. Science 287, 1793–1796.

Sekiguchi, M., Nakajima, T., Suzuki, K., et al., 2003. A study of the direct and indirect effects of aerosols using global satellite data sets of aerosol and cloud parameters. Journal of Geophysical Research 108 (D2), 4699.

Takemura, T. and Nakajima, T., 2002. Single-scattering albedo and radiative forcing of various aerosol species with a global three-dimensional model. Journal of Climate 15, 333–352.

Mountains Witnesses of Global Changes
R. Baudo, G. Tartari, and E. Vuillermoz (Editors)

4

Global retrieval of aerosol properties from sources to sinks by MODIS

Nai-Yung Christina Hsu

Abstract

Mineral dust and smoke aerosols play an important role in both climate forcing and oceanic productivity throughout the entire year. Due to the relatively short lifetime (a few hours to about a week), the distributions of these airborne particles vary extensively in both space and time. Consequently, satellite observations are needed over both source and sink regions for continuous temporal and spatial sampling of dust and smoke properties. However, despite their importance, the high spatial resolution satellite measurements of these aerosols near their sources have been lacking. In this paper, we demonstrate the capability of a new satellite algorithm to retrieve aerosol optical thickness and single scattering albedo over bright reflecting surfaces such as urban areas and deserts. Such retrievals have been difficult to perform using previously available algorithms that use wavelengths from the mid-visible to the near IR because they have trouble separating the aerosol signal from the contribution due to the bright surface reflectance. The new algorithm, called Deep Blue, utilizes blue-wavelength measurements from instruments such as MODIS and SeaWiFS to infer the properties of aerosols, since the surface reflectance over land in the blue part of the spectrum is much lower than for longer wavelength channels. We have validated the satellite-retrieved aerosol optical thickness with data from AERONET sun-photometers over land, including desert and semi-desert regions. The comparisons show reasonable agreements between these two. Our results show that the dust plumes lifted from the deserts near India/Pakistan border, over Afghanistan, and the Arabian Peninsula are often observed by MODIS to be transported along the Indo-Gangetic Basin and mixed with the fine-mode pollution particles generated by anthropogenic activities in this region, particularly during the pre-monsoon season (April–May). These new satellite products will allow scientists to determine quantitatively the aerosol properties near sources using high spatial resolution measurements from SeaWiFS- and MODIS-like instruments.

ISSN: 0928-2025 DOI: 10.1016/S0928-2025(06)10004-8

Mountains Witnesses of Global Changes
R. Baudo, G. Tartari, and E. Vuillermoz (Editors)

5

Radiation, aerosol joint observations – monsoon experiment in Gangetic-Himalayan area (RAJO-MEGHA): synergy of satellite-surface observations

Si-Chee Tsay and Brent N. Holben

Abstract

Monsoon rainfalls sustain the livelihood of more than half of the world's population. Understanding the mechanism that drives the water cycle and fresh water distribution is highlighted as one of the major short-term goals in NASA's Earth Science Enterprise Strategy, and the interaction between natural/anthropogenic aerosols, clouds, and precipitation is a critical component of that mechanism. In Asia, sheer population density presents a major environmental stress. In addition, economic expansion in this region is accompanied by increases in biomass/biofuel burning, industrial pollution, and land cover and land use changes. With a growth rate of ~8% per year for Indian economy, more than 600 million people from Lahore, Pakistan to Calcutta, India over the Indo-Gangetic Basin have particularly witnessed increased frequencies of floods and droughts, as well as a dramatic increase in atmospheric loading of aerosols (i.e., anthropogenic and natural aerosols) in recent decades. Continuous sun-photometry observations (2001–2004) at Kanpur, India also reveal high values of monthly mean aerosol optical thickness of 0.4–0.8 year-round. The Asian monsoon is a dominant component of the global water and energy cycle, and provides the critical fresh water supply to the Indo-Gangetic Basin. Meltwater from the Himalayas sustains the regional agriculture throughout the dry season. However, recent observations indicate that glaciers are rapidly shrinking, jeopardizing the long-term water supply over the region. The A-Train satellite constellation, Aqua, CALIPSO, CloudSat and Aura are, and will be, deliberately placed in orbit to take synergistic measurements to help provide a better understanding of climate forcing due to trace gases, aerosols, and clouds. As a complement to these satellite capabilities, an initiative to deploy NASA SMART-COMMIT facilities and an array of AERONET sun-photometers, in concert with the A-Train/Terra, will be presented. The GSFC SMART-COMMIT facilities will provide aerosol and radiation measurements from a suite of radiometers, micropulse lidar, trace-gas concentration analyzers, particle sizers, mass analyzers, nephelometers, aethalometer, and standard

ISSN: 0928-2025 DOI: 10.1016/S0928-2025(06)10005-X

meteorological probes. These valuable observations will be utilized to address the following scientific questions:

(1) What are the spatial and temporal distributions of aerosol properties (e.g., chemical, microphysical, optical and radiative) in the RAJO-MEGHA region during the pre-monsoon and monsoon season?
(2) What are anthropogenic aerosols in the regions, and can they be remotely sensed?
(3) How accurately can we determine aerosol radiative forcing over the regions?
(4) How do the cloud properties evolve as a result of interaction with anthropogenic and natural (or aggregate) aerosols?
(5) What are the impacts of aerosol–cloud interactions on the regional hydrological cycle during the pre-monsoon season and break period?

The expected close collaboration of RAJO-MEGHA with various research projects in the region (e.g., ABC, CLIVAR, GEWEX, and CEOP) will definitely provide a better understanding of the role that absorbing aerosols (dust and black carbon) play in affecting interannual and intraseasonal variability of the Indian monsoon, in particular, and of global water cycle, in general.

Mountains Witnesses of Global Changes
R. Baudo, G. Tartari, and E. Vuillermoz (Editors)

6

Contribution of the WMO global atmosphere watch to high mountain atmospheric chemistry observations

Leonard A. Barrie

Abstract

The Global Atmosphere Watch (GAW) of the World Meteorological Organization (WMO) is set up to systematically monitor and analyse atmospheric composition world wide in order to provide assessments and predictions for policy development. The GAW-WMO global monitoring programme is linked to many other world data centres, and systems to trade data, to provide measurement standards, and to assist in quality assurance through world standard calibrations of data.

1. Introduction

The Global Atmosphere Watch (GAW) Programme of the World Meteorological Organization (WMO) was established in 1989. It is focused upon the role of atmospheric chemistry in global change. Consisting of a partnership of managers, scientists and technical expertise from 80 countries, GAW is coordinated by the WMO Secretariat in Geneva, Switzerland, and the Working Group on Environmental Pollution and Atmospheric Chemistry (WG-EPAC) of the WMO Commission for Atmospheric Science (CAS).

The focus, goals and structure of WMO-GAW are outlined in detail in the Strategic Implementation Plan (Anonymous, 2001) and its addendum (Anonymous, 2004) available at http://www.wmo.ch/web/arep/gaw/gaw_home.html. Recognizing the need to bring scientific data and information to bear in the formulation of national and international policy, the mission is threefold:

(a) Systematic monitoring of atmospheric chemical composition and related physical parameters on a global to regional scale.
(b) Analysis and assessment in support of environmental conventions and future policy development.
(c) Development of a predictive capability for future atmospheric states.

ISSN: 0928-2025 DOI: 10.1016/S0928-2025(06)10006-1

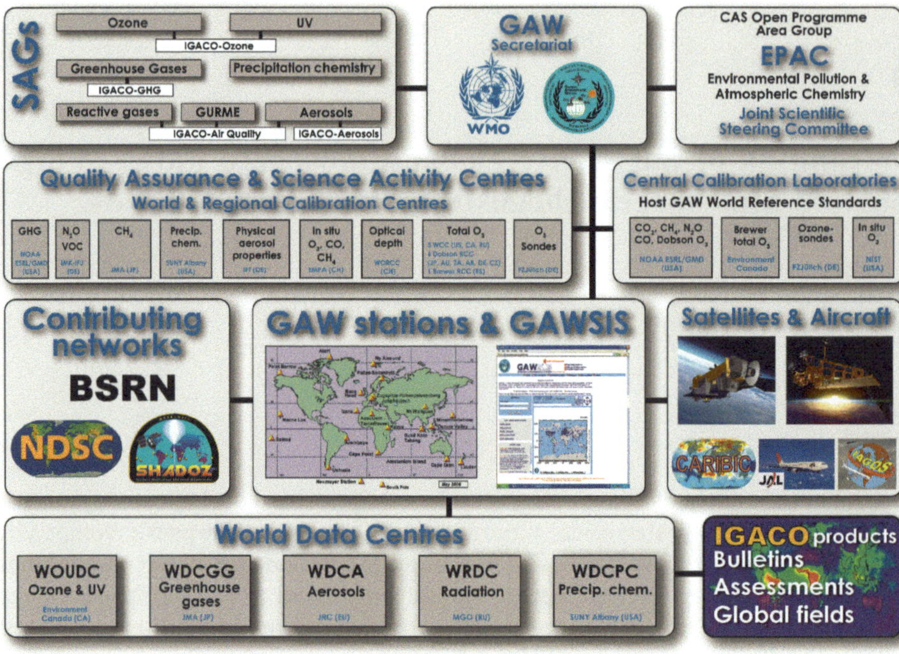

Figure 6.1. Components of the WMO-GAW global monitoring programme.

2. GAW monitoring

The components of the WMO-GAW monitoring programme are summarized in Fig. 6.1. Global GAW networks focus on six measurement groups: greenhouse gases, UV radiation, ozone, aerosols, major reactive gases (CO, VOCs, NO_y and SO_2), and precipitation chemistry. The GAW Station Information System (GAWSIS) was developed and is maintained by the Swiss GAW programme. It is the host of all GAW metadata on observatory managers, location and measurement activities. According to GAWSIS there are 23 Global, 640 Regional and 73 Contributing stations operating or have submitted data to a GAW World Data Centre. GAW Scientific Advisory Groups (SAGs) for each of the six measurement groups establish measurement standards and requirements, while calibration and quality assurance facilities ensure valid observations. Five GAW World Data Centres collect, document and archive data and quality assurance information and make them freely available to the scientific community for analysis and assessments. Note the linkages of GAW to contributing partner networks and to satellite observations that contribute to Integrated Global Atmospheric Chemistry Observations (IGACO). In the past decade, the emphasis of the WMO-GAW community on standardization, calibration, quality assurance, data archiving/analysis and building the air chemistry monitoring networks has resulted in major advances.

The workshop in Rome of November, 2005, and this book focus upon the theme 'Mountains, Witnesses of Global Change'. The key high altitude long-term atmospheric chemistry observatories in GAW are shown in Fig. 6.2. They are superimposed upon a best estimate of the global distribution of annual average tropospheric aerosol optical depth (AOD) (courtesy of S. Kinne, MPI, Hamburg, Germany). It was compiled by combining data from six satellites (operating for limited periods between 1979 and 2004). Observations for a region were selected using ground-based AOD observations as guidance. It should be emphasized that these observatories are operated through the strong commitment of national organizations in several countries including NOAA USA (South Pole, Mauna Loa, Summit), MeteoSwiss (Jungfraujoch, Mt. Kenya, Assekrem), the Kenyan Meteorological Department (Mt. Kenya), the Algerian Office National de la Météorologie (Assekrem), the Austrian Meteorological Service (Sonnblick), the German Weather Service (DWD) and Federal Environment Office (UBA) (Zugspitze-Hohenpeissenberg), the Institute of Atmospheric Sciences and Climate Italy (Mt. Cimone) and the China Meteorological Agency (Mt. Waliguan).

There is a clear lack of long-term high altitude air chemistry observatories in the Himalaya and some of the western mountains of the Americas. In building a long-term observatory in the Everest and K2 region, the government agencies and research institutes of Italy are clearly filling a major observational gap. Observational results from these observatories will be highlighted in a future paper to be presented at the Joint CACGP/IGAC/WMO Symposium on Atmospheric Chemistry at the

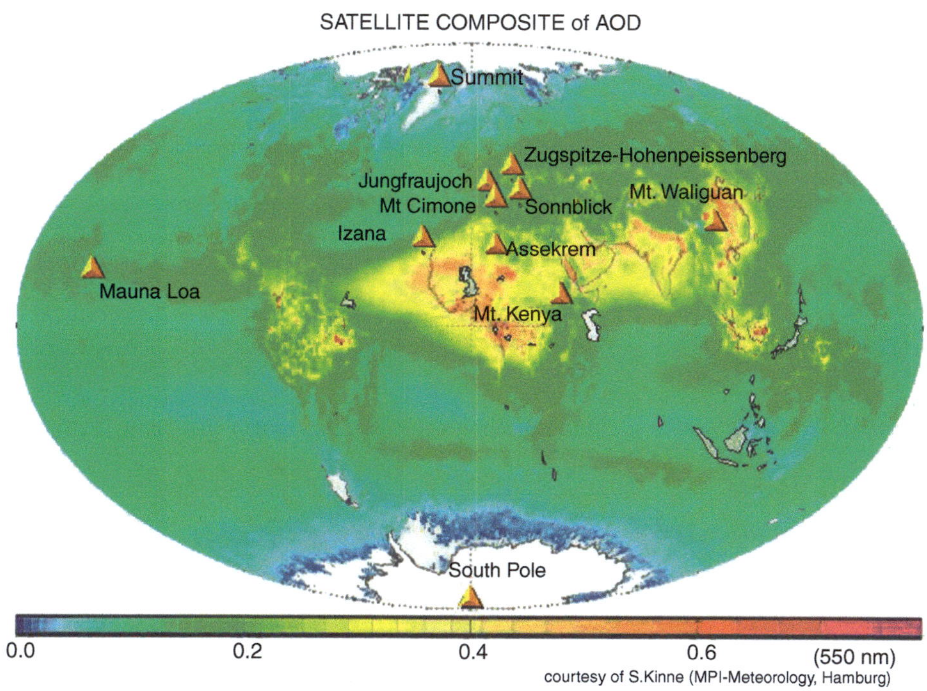

Figure 6.2. High altitude observatories of the WMO Global Atmosphere Watch atmospheric chemistry.

Interfaces 2006, 17–23 September 2006 in Cape Town, South Africa, (http://www.atmosphericinterfaces2006.co.za/) and a subsequent publication.

References

Anonymous, 2001. GAW: Strategic Plan, Strategy for the Implementation of the Global Atmosphere Watch Programme (2001–2007), A Contribution to the Implementation of the Long-Term Plan. GAW Report 142.

Anonymous, 2004. GAW: Strategic Plan Addendum for the period 2005–2007 to the Strategy for the implementation of the Global Atmosphere Watch Programme (2001–2007). GAW Report 142.

Ev-K^2-CNR in Project ABC

Mountains Witnesses of Global Changes
R. Baudo, G. Tartari, and E. Vuillermoz (Editors)

7

From Himalaya to Karakoram: the spreading of the project Ev-K^2-CNR

Renato Baudo, Beth Schommer, Chiara Belotti and Elisa Vuillermoz

Abstract

The Pyramid International Laboratory-Observatory is the symbol of the Ev-K^2-CNR Project. The project actually began in 1987, when Prof. Ardito Desio, 90 years old at the time, enthusiastically launched a new geological and geodetic research campaign in the Himalayan area. However, it was only with the building of the Pyramid International Laboratory-Observatory, inaugurated by Prof. Desio himself in 1990, that the project acquired a unique "logistic base" for its scientific research. The laboratory, located at 5050 m a.s.l. in the Khumbu Valley, on the Nepali side of Mount Everest, is in fact the first high-altitude scientific research center of its kind. It is self-sufficient in its energy supply and contains all common scientific instrumentation, making it a suitable place for studying climatic and environmental changes, medicine and human physiology in extreme conditions, geology, geodesy and seismic phenomena. Over time, a wealth of knowledge, initiatives and international relationships have been accumulated and continue to be added to by Ev-K^2-CNR through research in the fields of medicine and physiology; environmental sciences, earth sciences, anthropological sciences and clean technologies. The Ev-K^2-CNR Committee has been able to play a strategic role in the framework of collaboration amongst institutions, governments and organizations for the exchange and transfer of experiences, technologies and scientific and cultural knowledge. The increasingly interdisciplinary approach to research by the team has also led to the development of integrated programs for promoting the socio-economic development of local populations and environmental safeguarding in the region, such as the international Partnership initiative created through the Italian government around Ev-K^2-CNR's expertise, or the regional Ev-K^2-CNR Project "Stations at High Altitude for Research on the Environment in Asia" (SHARE-Asia), aimed at the establishment of a network of research and monitoring stations for the long-term study of evolutionary environmental processes in the Himalayan–Karakoram region,

ISSN: 0928-2025 DOI: 10.1016/S0928-2025(06)10007-3

with a strong technology transfer and capacity-building component to the benefit of local populations and research institutions.

1. History

The Italian project Ev-K^2-CNR was launched in Central Asia in 1987, founded on the ingenuity and scientific inquisitiveness of Prof. Ardito Desio, who had already explored the Shaksgam Valley on the Chinese side of the Himalayas in 1929 during the Duke of Spoleto expedition. In 1954, the famous expedition led by Prof. Desio was the first to succeed in reaching the top of Mount K2, the second highest mountain in the world. Thirty-three years later, when an American scientist surprisingly announced that Mount K2 was actually 11 m taller than Mount Everest according to the latest astronomical calculations, the leader of the 1954 Italian expedition could not help but wonder if they had actually conquered the highest peak on Earth instead of the second highest, as previously assumed. Prof. Desio lost no time in organizing a new scientific expedition, aimed at providing an accurate measurement of Mount Everest. With funding from the Italian National Research Council and in collaboration with mountaineer and manager Agostino Da Polenza, the expedition (unfortunately for us Italians) confirmed that Mount Everest is effectively the highest mountain in the world, and Mount K2 remains only second. The team of Prof. Poretti, using the same state-of-the-art techniques applied to establish the true height of Mount Everest (8846.10 m a.s.l.), later went on to re-measure Mount K2 (8611 m a.s.l.), hence confirming the validity of the previous measurements of both mountains. The measurement techniques used, combining traditional surveying applications with Global Positioning System (GPS) technology, became established as the internationally recognized standards for mountain measurements. Moreover, during these expeditions, an intensive program of geodetic and geophysical data collection contributed new information on this important area where the Indian and Asiatic plates collide.

In 1989, two Italian companies offered Prof. Desio a pyramid-shaped structure made of glass, aluminum and steel, to be used as a sort of alpine hut and research laboratory for carrying out studies in fields such as meteorology, hydrology, medicine, ethnography, zoology and botany; thus saving researchers the difficulty of performing high-altitude research from precarious tents. This Pyramid was originally supposed to have been installed on the Tibetan side of Mount Everest, in the Tingri Valley, but unfortunately, after Prof. Desio had already reached an agreement with the Chinese Academy of Sciences, the student demonstrations in Tien An Men Square prevented fulfillment of the plan.

In 1990, thanks to an agreement with the Nepal Academy of Science and Technology (NAST, ex RONAST), the Pyramid was transferred to Lobuche, a mountain pasture in the Sagarmatha (Nepali name of Mount Everest) National Park. The first permanent high-altitude scientific laboratory was thus established at 5050 m a.s.l. and it has since evolved to become completely self-sufficient in its energy supply, has controlled temperature, telecommunications systems and all the usual equipment of a research laboratory.

2. The Pyramid International Laboratory-Observatory

The pyramid shape of the structure (with a square base of 13.22 m/side and a height of 8.40 m) provides great stability and natural resistance against atmospheric agents such as rain, snow and wind. Its external covering in mirrored glass provides a suitable integration within the surrounding environment, while also limiting inner heating from the sun. Inside, the Pyramid is divided into three floors, housing the research laboratories and the radio-satellite and computer communication equipment station. On the external southern side, there is an independent annexed lodge, built in wood and stone according to the traditional architecture of Nepali lodges, which hosts the people in the studies performed at the Pyramid. Up to 20 people (scientists, technicians and logistic support personnel) can receive high (European mountain hut) standard food and lodging there at any given time throughout the year.

The complex is self-sufficient, thanks to a hybrid power supply comprising a micro hydroelectric plant (fed by a water pipeline tapping a small lake above the Pyramid), and a system of solar panels (4 independent fields on the southern side and 2 on the eastern side of the lab). Wind generators were also tested as a power source, but with unsatisfactory results. In emergencies, a reduced-emission catalytic electric generator is also available.

To minimize environmental impact, any materials to be transported to the Pyramid are carefully examined and selected. Solid wastes produced there are sorted for either local disposal (in cooperation with the Sagarmatha Pollution Control Committee) or transported down valley to be properly disposed of. If necessary, solid waste may even be shipped back to Italy, as was the case for the old lead batteries of the photovoltaic system, when the accumulators were replaced in 2002.

Initially, the Ev-K^2-CNR Project was mainly devoted to studies in the field of earth sciences, but with the International Laboratory-Observatory in place, the project expanded to cover other fields as well. Today studies include the fields of medicine and physiology, environmental sciences, anthropological sciences, clean technologies and environmental management systems.

The project is entirely managed by the Ev-K^2-CNR Committee, a non-profit organization, and activities are developed through its Scientific Committee made up of a Chairman and scientific coordinator for each of the aforementioned disciplines. This Scientific Committee has the duty of defining a 3-year program of studies, after having evaluated research proposals from the dozens of scientists involved in the project. As this project is also considered a joint venture between the Ev-K^2-CNR Committee and NAST (ex RONAST), all projects to be carried out in Nepal must be further authorized by a Bilateral Technical Committee (BTC), made up of representatives of both organizations. The BTC is furthermore in charge of selecting Nepali experts to collaborate with Italian scientists in all studies.

It must be emphasized that collaboration in the Ev-K^2-CNR Project has always been open to foreign scientists as well. In fact, experts from several European (Austria, France, Germany, England and Switzerland) and non-European (Australia, Canada, Japan and United States of America) countries have taken advantage of the opportunity to stay at the Pyramid International Laboratory-Observatory for performing their studies.

3. The Ev-K^2-CNR Project

In brief, the Ev-K^2-CNR Project on the one hand aims to study the effects of the environment on different functions of the human body and on the instruments that make living and working in extreme conditions at high altitudes possible; on the other hand it takes advantage of the rare opportunity for studying the effects of man on the environment on a global scale that the Himalayan environment offers, given its particular geographic characteristics (high altitude and remoteness). Profiting from the experience accumulated through these studies, by advancing scientific knowledge on human, environmental, climatic, geological and geophysical features of remote areas at high altitude, the Ev-K^2-CNR Project provides specialized support that contributes to the sustainable development and environmental protection of these areas, and to improving the quality of life of mountain populations. The studies carried out in the 15+ years of the project can be thus summarized as follows (AA.VV., 2003).

3.1. Earth sciences

Research in the various disciplines of the earth sciences has traditionally represented the core activity of the Ev-K^2-CNR Project. Since 1988, under the leadership of Prof. Desio, important studies have been performed in the geodetic, geophysical and geological fields. In-depth studies were carried out in the regions of Mount K2 and Mount Everest in order to acquire new gravimetric and geological data on one of the main collision areas between the Indian and the Asian plates (AA.VV., 2005).

Among the various activities in the Himalayas over the last few years, measurement and survey campaigns have been carried out in Northern Karakorum (Sinkiang, China), Northern Pakistan, Makalu, Dolpo, Shaksgam Valley, Nanga Parbat, Shimsal Pass and the Khumbu and Arun Valleys. Of particular importance was the re-measurement of some of the world's highest peaks using state-of-the-art geodetic and satellite – GPS – instrumentation. Measurements carried out in this project so far include Mount Everest (1992) and Mount K2 (1996), Mount Cervino (1999), Mount Rosa (2000) and the highest peak in Latin America, Mount Aconcagua (2001).

Another important field of research within the area earth sciences concerns gravimetric and seismic activity of the Himalayan chain. The installation of a seismic station in one of the most active regions of the world provides researchers with access to exceptional information which can be used for the analyses of complex geological phenomena, providing a reasonable degree of predictability for future events.

Furthermore, the Pyramid can be considered a point of reference for the study of leveling lines between India and Tibet with the construction of the first GPS network in the Himalayas and installation of a reference station within the French DORIS satellite positioning system. Data acquired during Ev-K^2-CNR Project expeditions has also been used to create an initial geographic information system (GIS) of the Khumbu valley, integrated by satellite data. Finally, magnetism and gravimetric studies have allowed researchers to determine the thickness of the Earth's crust and to "observe" the subduction of India beneath Tibet. These data were further used to calculate the local and global geoids in the Himalayan area.

3.2. *Medical and physiological sciences*

The Pyramid International Laboratory-Observatory provides a privileged location for research in this field, offering the opportunity for studying physiological adaptations at elevations between 5000–8000 m a.s.l., representing a valid point of reference for what happens in everyday life at more common and easily accessible altitudes. The Pyramid, located at 5050 m, is reachable only after a trek of at least 5 days. It is possible to stop for a few days at lower altitudes (generally at 3500 m or 4200 m) and study the same subjects in various states of hypoxia and at different levels of exposure (acute versus chronic) and acclimatization. This provides opportunities for examining the effects of limited availability of oxygen on various human physical functions, such as modified reaction times; automatic and controlled mental processes; memory efficiency; bioenergetics of high-altitude exercise; structural and functional modifications of the cardiovascular system; variations in body composition and related endocrine and metabolic parameters; food absorption efficiency; effects of UV rays on the epidermis.

Not only is the Pyramid located at a higher altitude than all other similar alpine observatories even in Europe, it also provides a unique chance to study the local populations (Tibetan and Sherpa) who have lived for centuries at high and extremely high altitudes, distinguishing the Pyramid from another, similar high-altitude laboratory in Bolivia. In this way, numerous physiological studies have been carried out comparing reactions of the following subjects: subjects born and raised at high altitudes (Sherpa), their descendants raised at lower altitudes, Europeans with or without previous mountain experience, etc. Research results offer valuable indications on the particular role of the genetic component and on the adaptation of physical functions to extreme conditions. It can also find immediate applications in various pathologies, which can afflict living beings at much lower altitudes. Results can be used to the benefit of patients suffering from hypoxia, for various reasons (fetal and neonatal hypoxia), or for those who, for pathological reasons, have a reduced afflux of blood to the brain. Subjecting healthy subjects to acute and/or chronic hypoxia is an excellent experimental model for the study of reactions to hypoxia in patients with cardio-respiratory pathologies, a problem an increasing number of patients must deal with (around 0.3% of the world population). In conclusion, knowledge acquired through medical and physiological research activities at the Pyramid International Laboratory-Observatory can be seen as contributing to our understanding of not only what happens to patients with particular dysfunctions but also the effects on our organism when we, for professional and/or recreational reasons, visit mountains at altitudes above 2500 m a.s.l.

3.3. *Environmental sciences*

High-altitude areas in the Himalayas are an ideal location for environmental studies. Ev-K^2-CNR's main fields of activity in this sector can be divided into six areas:

climate change; lake cataloguing, analysis and classification; transport of pollutants; glaciers and glaciology; high-altitude vegetation and fauna; assessment of environmental impacts for improved land use.

3.3.1. *Climate change*

During early mountaineering and scientific expeditions in the Himalayan regions over the last century, research on climatic characteristics was carried out with outcomes being first widely diffused in the 1960s. Only in 1994, however, when the first automatic weather station with bi-hourly continuous monitoring of data was installed within the framework of the Ev-K^2-CNR Project, did regular climate measurements at high altitudes (>5000 m a.s.l.) become possible. This monitoring process has since been known for the exceptional regularity of its measurements and marked the starting point of the Ev-K^2-CNR Project's recent wider involvement in international climate and meteorological research (Baudo et al., 1998).

A landmark moment in this evolution is represented by recognition of the Pyramid Meteo Network (PMN) in the World Meteorological Organization (WMO) Coordinated Enhanced Observing Period (CEOP) – Asia–Australia Monsoon Project (CAMP)-Tibet, as of 2002.CEOP is part of the WMO program aimed at a more thorough analysis of global climatic phenomena, World Climate Research Program (WCRP). Thanks to the unique characteristics of the location of the Pyramid, the PMN has been inserted as the Himalayas Reference Point for the CEOP/CAMP-Tibet sub-project. Following its inclusion in the CEOP Project, Ev-K^2-CNR researchers set the following scientific goals (see also section on SHARE-Asia below):

- carry out sophisticated meteo-climatic research at a high international level;
- maintain a high profile role within the CEOP International Scientific Panel, which coordinates activities on the interaction between the monsoon and the Himalayan range;
- insert the PMN within an international circuit of highly qualified weather stations, recognized as such by major scientific institutions (WMO, NOAA, NASDA, etc).

3.3.2. *Lake cataloguing, analysis and classification*

Morphological classification of the lakes of the Everest National Park began in 1990, the purpose being identification of all permanent lakes and collection of information on their size and geographic location (Lami and Giussani, 1998). A series of expeditions have since been organized, for the collection of hydro-chemical and biological data, leading to the creation of the Sagarmatha National Park "Limnological Information System" (LIS). The LIS, which forms the nucleus for a GIS of the Khumbu Valley developed by the Ev-K^2-CNR Committee, has been completed in early 2007, covering the entire Sagarmatha National Park area. The most advanced Nepalese national cartography and the latest information technologies are being used to ensure the high quality of the final product. These efforts in lake analysis and classification are part of the Ev-K^2-CNR SHARE-Asia Project (below). They are also aimed at contributing to:

- cartographic verification with field measurements in remote areas;
- comprehension of the relationship between chemical composition and altitude, the contribution of chemical species from the atmosphere, rock composition and the presence of anthropogenic alterations in water basins that can cause euthophication phenomena;
- identification of diffused pollution phenomena on Earth (pollution transport from near and distant emissions can be traced in atmospheric depositions which end up in lake waters and sediments, giving us information on minute environmental alterations caused by such pollutants);
- study of water biocenosis (plankton, benthos, etc.). Himalayan lakes contain a large number of various endemic species, contributing greatly to the enrichment of the Earth's biodiversity. Water life forms also vary according to altitude and latitude and can be different according to their location along the Himalayan chain.
- paleolimnology research. Lake sediments provide unique information for better understanding life, biology, climate, the morphological characteristics of a given area over the past centuries and millennia. Lake sediments in fact efficiently record the history of our planet and its evolutions.

3.3.3. Transport of pollutants

Geographic isolation in the areas of the Himalayas, Karakorum and Hindu Kush mountains facilitates measurement of base levels of pollutants. It is also possible to carry out measurements on atmospheric depositions (snow and rain) in the Himalayan regions, similarly to that done in Polar areas, with the added advantage of easier access throughout the year. Also similar to Polar research, studies in the Himalayas have indicated the presence of certain pollutants that demonstrate a clear relationship between the concentration of these elements and cold regions of the planet. Further investigations are needed on this topic, which forms an integral part of the Ev-K^2-CNR SHARE-Asia Project, described below. So far, comparative studies of rain and snow chemistry in the Khumbu Valley have led to the conclusion that, while long distance transport of pollutants is evident, the phenomenon has not yet reached levels able to produce evident environmental alterations. Identification of the "Atmospheric Brown Clouds" (ABC) phenomenon in Asia, through the INDOEX program, has however given rise to new interest in understanding the mechanisms and impacts of pollution transport. Effects of pollutants on rainfall and solar irradiation have already begun to show negative trends (increase in intensity of precipitation episodes, provoking natural disasters; decrease in overall precipitation in other areas, leading to desertification; decrease in productivity of rice cultivations in India due to lessened solar radiation). Air sampling for analysis of atmospheric composition, combined with deposition studies (lake waters, rain samples, snow) at high altitude, will surely contribute to understanding key aspects of this phenomenon and Ev-K^2-CNR has been invited to take part in the UNEP Project ABC to this end.

3.3.4. Glaciers and glaciology

An important contribution to understanding global change is provided by nearly a decade of research on the glaciers of the Everest region, which supplies priceless information to help quantify the greenhouse effect. One primary example is the activity carried out on the Changri Nup glacier where, using GPS satellite techniques, the constant retreat of the uncovered glacier terminus and a significant increase in the size of small seasonal lakes has been documented. Such glacial lakes, in fact, can cause disastrous floods, known as glacier lake outburst floods or GLOFs, when their natural (moraine or ice) barriers burst, leading to a sudden outpouring of water mixed with ice and debris into the valleys below. One of the most catastrophic events of this type occurred in 1941 in Peru, when over 4500 people were victims of a debris-filled flood that invaded the city of Huaràz.

A recent study has indicated that 48 potentially dangerous glacial lakes are currently found throughout the Himalayas. In Nepal, memories live on of a terrible flood of mud and debris caused by a GLOF on August 4, 1985. This GLOF of the Dig Tscho lake (Langmoche) destroyed the nearly completed Namche Small Hydropower Plant, along with everything else down the valley. From 1977 to 1998 five major GLOFs occurred in Nepal, while an ICIMOD study estimated six GLOFs between 1935 and 1981 in Tibet, which have been registered or are visible from satellite images. Unfortunately, the increase in global temperatures over the past 50 years is causing a noticeable increase in the formation and expansion of glacier lakes – many in Nepal that are now potentially dangerous did not exist or were only in their beginning stages 50 years ago. Continuous monitoring and careful research is thus increasingly crucial in order to avoid new tragedies and to better comprehend similar phenomena occurring closer to home in the Alps.

3.3.5. Study of high-altitude vegetation and fauna

Situated in a protected area, the Pyramid Laboratory-Observatory provides a unique opportunity for studying regional flora and fauna. Apart from important contributions regarding botanic and zoological systems, specific projects have also concentrated on reproductive ethology, ungulate conservation, phytosociology, biochemistry, vegetal physiology, etc. Research has, for example, provided an opportunity for the elaboration of a management program of wild fauna as a possible economic resource in the Himalayan and Karakoram mountain regions. Another project investigates the morpho-physiological characterization of high-altitude genetic vegetal resources and their valorization within the framework of environmental conservation and their agricultural use. Recent discoveries by Ev-K^2-CNR researchers have also led to the elaboration of a long-term monitoring program on the endangered snow leopard and its prey species.

3.3.6. Assessment of environmental impacts for improved land use

The interest of environmental science in remote areas should not be seen as a mere scientific speculation. Human presence at very high altitudes, the exploitation of unstable land by such settlements and use of landscape resources by trekkers and mountaineers, all pose urgent questions connected to the conservation and stewardship of natural resources. In order to implement sustainable development for local populations, assessments of environmental impact need to be carried out to help identify sustainable strategies and understand the ecological and economic aspects of human activities. Without adequate comprehension of these processes, the often little-known cultures of these regions risk being lost as a consequence of economic development and importation of new social lifestyles and the natural resources, which are part of global human heritage, are in jeopardy.

3.4. Anthropological science, communication and development

Researchers in this field study the cultures and traditions of the populations native to the Himalayan range, particularly focusing on certain related communities in the Autonomous Region of Tibet–People's Republic of China, Qinghai, Nepal, Vietnam and Laos. Research activities will soon be expanded to include the mountain populations of Pakistan's northern areas. All studies place emphasis on cultures or cultural aspects of local communities that are at risk of dying out. Comparisons can also be made with ethnic groups living in other parts of the world in similar environmental conditions. Up to now, studies have essentially aimed at reconstructing the cultural history of the mountain areas in question, on the basis of written documentation and oral traditions. Bearing in mind more practical needs, research results also attempt to address problems arising from modernization and development processes affecting these regions and civilizations.

In recent years, an exhaustive anthropological study of certain Tibetan and Tibeto–Burmese populations in the Mount Everest region has been carried out. Research on Tibetan cultures in particular has focused on the areas of Porong and Henan, the historical seats of nomadic principalities which were partially autonomous from both Tibetan and Chinese governments. These innovative studies were possible, despite a relative lack of anthropological research in the Tibet Autonomous Region, thanks to collaboration with Tibetan organizations and associations such as Tibetan Academy of Social Sciences in Lhasa, Tibet University and Tibet Assistance to Remote Areas (TARA). These research programs have also striven to identify areas of intervention where acquired knowledge can be applied to solving specific education and livelihood problems faced in the Himalayan areas today (strategies to mitigate degradation of mountain pastures; educational projects and cultural promotion of rural areas). Tibetan area studies are furthermore integrated with research carried out in Nepal on traditional shamanic rituals and possession rituals. Such studies, highlighting particular cultural traditions like dancing and music, favor the use of photographic and video documentation in a Visual Anthropology tradition. The objectives of the Ev-K^2-CNR Research Area "Anthropological Science, Communication and Development" may be summarized as follows:

- Contribute to cultural conservation through documentation, supporting the initiatives of local communities in this sense.

- Gather knowledge that allows a more correct evaluation of development strategies in rural areas with special reference to management of mountain pastures.
- Cooperate with local institutions and staff, favoring dialogue and an exchange of research sources and methods, promoting in particular the capacity building of local researchers. Joint publications involving both European and local authors and the participation of local researchers in international conferences both represent important steps towards the attainment of an appropriate international profile for the local experts.
- Disseminate research products which provide greater visibility. This project aims to extensively employ the video medium to reach a wider public, both via scientific circuits and ordinary television channels.

3.5. Clean technologies and environmental management systems

Reduced atmospheric pressure, daily high range of temperature, extreme meteorological conditions, difficult access, etc. characterize the environment where the international Pyramid Laboratory-Observatory is located. The Laboratory thus becomes an ideal place for the study and experimentation of new technological solutions to support use of renewable power resources with minimum environmental impact and maximum efficiency. The main purpose of this research and its applications is to contribute to the improvement of the social, economic and environmental conditions of local populations. The main goals of the research area are to introduce new clean technologies and elaborate methodologies for environmental management of mountain areas, in synergy with other research activities of the Ev-K^2-CNR Project.

Expected results include elaboration of a management system for remote mountain areas that can be applied and replicated in other similar regions. Current areas of research include: analysis of renewable energy power stations (photovoltaic, water and wind) and customization of electrical/electronic systems, such as computers and scientific equipment. Structures to provide comfort and conserve energy are also designed, while materials are produced for testing in extreme conditions. Solid-waste recycling and management form a fundamental part of the management system, as does the purification and supply of drinking water.

Project objectives include the following:

- Clean technologies
  - Analysis of state-of-the-art clean technologies (production of renewable energy, waste disposal and systems for the purification of waste water).
  - Evaluation of possibilities to introduce/experiment priority technologies based on results of research in various fields in collaboration with local stakeholders.
- Environmental management systems
  - Definition of guidelines and outline of international experiences in design and implementation of environmental management systems in landscape-related tourist destinations.

- Identification of a pilot area and classification of major environmental characteristics in a sample area, in collaboration with experts of other disciplines and local actors.
- Elaboration of a plan in collaboration with local actors and introduction of an environmental management system: definition of roles, actions and commitments.
- Implementation of an environmental management system which can be replicated in similar mountain regions, comprising an interdisciplinary methodology (including anthropological, medical and environmental components).

4. The Ev-K^2-CNR multidisciplinary approach

The most recent example of an Ev-K^2-CNR team of researchers from distinct disciplines working in synergy within a specific field project is from 2004, when Ev-K^2-CNR scientists launched an ambitious research program within the "K2 2004 – 50 years later" mountaineering project. This combination of scientific research in extreme environments with the commemorative mountaineering expedition was aimed at celebration of the Golden Jubilee of the 1954 successful first summit of K2 by the Italian team led by Prof. Ardito Desio. While researchers in the field of medicine and physiology studied the physiological adaptation of the native Tibetan populations to high altitude (an ideal model for understanding how to help patients at sea level adapt to chronic hypoxia), others recorded the cardio-respiratory parameters of the climbers involved in the expeditions and the effects of chronic hypoxia on the endocrine-metabolic system of healthy subjects. Environmental experts installed a new automatic meteorological station at 4000 m a.s.l. at Urdukas in Pakistan, as part of the interdisciplinary SHARE-Asia program (see below). Climbers also collected snow samples at various altitudes, aimed at tracing evidence of environmental contaminants there. A team of glaciologists measured the movements of Baltoro glacier as an indicator of climatic changes. Prior to climbing K2, the mountaineers summited Mt. Everest to carry out an innovative measurement using a custom geo-radar to accurately calculate the thickness of the summit snow layer. For the first time, the actual height of the rock summit of Everest was thus determined. The entire expedition used an environmental management system developed by Ev-K^2-CNR to minimize environment impacts of the significant human presence of the team at both Mount Everest and Mount K2.

Multidisciplinary field work can easily lead to the development of research across and within distinct disciplines (interdisciplinary research), the most beneficial way to help safeguard complex ecosystems like those in high mountains and improve the quality of life of the local populations. One of the main characteristics of the Ev-K^2-CNR Committee is in fact its ability to combine scientific knowledge and a strong interactive involvement with the local environment and populations. Over the years it has became apparent that, beyond its basic scientific role dictated by the research it carries out, Ev-K^2-CNR is also in a position to help address problems related to the transfer of research results and the exchange of technical know-how on a local scale.

In order to make a significant impact in this sense, Ev-K^2-CNR has emphasized a process that would minimize the fragmented specificity of research, favoring a more interdisciplinary approach. This process has developed spontaneously over time and today represents the true innovation in the Ev-K^2-CNR Committee's approach to its research activities. Integrated projects of study, whether carried out at the Pyramid International Laboratory-Observatory in the Khumbu Valley or in other areas of the Himalayan–Karakoram mountain complex, today represent the core of the Ev-K^2-CNR Committee's efforts and are paving the way towards the future.

5. The future of the Ev-K^2-CNR Project

All countries in the Hindu Kush–Karakoram–Himalaya (HKKH) region face physical, social and economic vulnerabilities: food insecurity, high levels of poverty, growing demand for and improper use of natural resources, destruction of high mountain forests, reduction of availability of water, loss of biodiversity, air pollution, natural hazards and disasters, slope instability and landslides and a lack of communication and coordination. Such vulnerabilities are likely to threaten the ecosystems and compromise the development of these mountain areas. For a long time, stakeholders did not want to consider the effects of human and natural dynamics in mountain regions and the very few actions that were carried out in the field of conservation have underestimated the origins of the problems and HKKH mountain development potentials. Current problems such as fragile political situations and inadequate legal instruments and procedures for ensuring sustainable natural resources management, combined with a lack of coordination and exchange of key geophysical, biological and hydro-meteorological data, make a systemic approach to mountain-area management and planning a real challenge. Few strategic options are known for managing fragile mountain ecosystems, but it is generally agreed that coordinated monitoring actions, foreseeing the collection and analysis of reliable and consistent data on long term evolutionary processes, are of great urgency.

5.1. DSS–HKKH partnership initiative

Ev-K^2-CNR plays a leading role in an international partnership initiative launched by the Italian Government as a practical outcome of the World Summit on Sustainable Development (Johannesburg, 2002). This Partnership, "*Institutional Consolidation for the Coordinated and Integrated Monitoring of National Resources towards Sustainable Development and Environmental Conservation in the Hindu Kush–Karakoram–Himalaya,*" is being implemented by IUCN through executing Partners Ev-K^2-CNR, the International Center for Integrated Mountain Development (ICIMOD), the Italian NGO, CESVI. It aims to involve various stakeholders in the sustainable development of mountain populations in HKKH region by introducing decision support systems (DSS) for sustainable management of the natural resources of several high-altitude protected areas.

Without adequate measures for monitoring and control, in fact, the poverty of populations living in these areas would be destined to worsen. An accurate analysis of the environmental effects and of their impact on local development therefore is a high priority. The Partnership initiative thus represents the willingness of stakeholders in beneficiary countries to jointly tackle the environmental, political and socio-economic vulnerability of their mountain regions. Ev-K^2-CNR and partners will contribute to the consolidation of institutional capacity for systemic planning and management at the local, national and regional levels, focusing on the creation of a coordinated system for collection and analysis (monitoring) of natural resource data; knowledge about local populations (socioeconomic conditions); development of participatory, decisional and operational capacities of the beneficiary populations; transfer of technologies for the systemic management of data to be used by decision makers at a local, national and regional levels to improve systemic planning and management of mountain region.

The Partnership will provide tools and instruments to facilitate the consistency of various actions for management of mountain ecosystems, within the framework of a DSS, a computer-aided tool for supporting systemic land-use planning, management and monitoring that can help understand and assess the relationships among various management issues, enabling policy makers to make informed decisions and have a better understanding of the consequences of the actions they take. The transboundary planning and mountain management issues that will be supported by the creation of a DSS base will at first be prepared for three pilot sites in the HKKH: Nepal (Sagarmatha National Park), Pakistan (Central Karakoram National Park) and Tibet Autonomous Region of China (Quomolongma Nature Preserve). These pilot DSSs will focus in detail on the operational activities towards the implementation and monitoring of protected area management in high-altitude regions.

5.2. Sagarmatha National Park GIS project

A first attempt to create a GIS for the Khumbu Valley was carried out in the mid-nineties, based on geological, glaciological, limnological, hydrographic and topographic data of the area surrounding the Pyramid. Scarce spatial resolution of existing topographic maps, however, strongly compromised the quality of the final product. The evolution of satellite survey technologies, new data gathered in the field and the availability of new and more complete topographic maps, have now allowed for creation of a more complete and reliable GIS. Through this project, combining expertise and data gathered over a lengthy period of research experience, the Ev-K^2-CNR Committee can thus maximize the functional application of its research outputs, and further strengthen the link with the region in which the projects are carried out.

The creation of a GIS for Sagarmatha National Park, one of the building blocks of the DSS–HKKH Partnership initiative (above), will combine scientific results and images from across various sectors of Ev-K^2-CNR research activity. Its aim is to provide the international scientific community with an instrument that can be used as an "interactive database" for management of multisource data with non-homogeneous formats and contents. With a high degree of flexibility, the GIS project supports researchers in elaboration of data and allows them to obtain a representation of results in their required format. Such developments are particularly relevant since the degree of information and content they provide is much higher than that of the raw data available before elaboration and integration. The characteristics of this GIS project will also favor a multidisciplinary output, allowing various researchers to introduce their data within a common base set of geographic coordinates, on a single geo-referenced map of the area in question. Finally, the database can be easily updated with new and more sophisticated information, through the application of mathematical models, such as numerical climate models, which allow information on climate-related effects to be "superimposed" on existing geo-referenced information.

5.3. *Karakorum Trust*

Ev-K^2-CNR has been developing more innovative strategies for using scientific knowledge to promote sustainable management in mountain regions together with the local authorities. In Pakistan, building upon the Italian legacy in the region of K2, Ev-K^2-CNR has been entrusted with the lead of an Italian government program called "Karakorum Trust". The program's aim is to contribute with scientific knowledge and development cooperation initiatives to a full and proper implementation of the Central Karakoram National Park (CKNP) that includes the summit of K2. This idea was already proposed by Desio in 1991, to help face the environmental degradation and low quality of life, which can be easily seen in the Baltoro region of Pakistan's northern areas.

The program will contain aspects linked to both coordination and monitoring of a systemic package of activities proposed by and with local stakeholders, and capacity-building actions aimed at helping the local populations better manage their fragile mountain environment. The 3-year program will put into practice the Italian political will expressed during 2004 – to give back to Pakistan's northern areas and help preserve their (and the world's) unique natural and cultural heritage. To do so, activities will be developed and carried out in the following fields:

- health and sanitation,
- sustainable tourism,
- natural resource management,
- agriculture and trade,
- education and professional development.

5.4. *Stations at high altitude for research on the environment: SHARE-Asia project*

The Himalayan–Karakoram mountain chain is well-known as an important barrier between the Indian ocean basin and the highland of Tibet. The powerful monsoon and the strong high pressure field known as the "Tibetan high" determine climatic effects that highlight the mechanisms of transport and fall-out of pollutants at high altitude. Furthermore, the continual growth of the highest mountains of the word creates an intense geo-dynamic effect, causing in turn a growing interest in monitoring the region's seismic activity both for understanding the scientific aspects of the movement of the Earth's surface and for helping foresee and control the occurrence of natural disasters. Precise geophysical measurements are also needed for both geodetic research and monitoring of frequent movements of unstable mountain lands, like glacier flow and landslides. The interest amongst Ev-K^2-CNR researchers in various fields for creation of an environmental-climatic monitoring network is evident. Such a network can also be used to actively contribute to capacity building of the local scientific community, with potential outputs for interesting scientific developments in international cooperation between the West and the developing countries of the region.

The idea for the SHARE-Asia Project has naturally evolved out of the expertise in high-altitude and remote area investigations accumulated within the Ev-K^2-CNR Project. Contributing to the concept is not only the multidisciplinary experience of the Pyramid Laboratory-Observatory, but also the understanding of the importance of a geographical expansion of monitoring activities and the potential of a system developed from a single point into a network. The local PMN climatic monitoring network forms the basis of the SHARE-Asia expansion, including the first high-altitude meteorological station that has recorded continuous data at 5050 m since 1994 and the five automatic weather stations subsequently installed along the vertical axis of the Khumbu Valley. Ev-K^2-CNR researchers have already been using the data collected there to develop a climatic model for the area, in collaboration with the WMO CEOP/CAMP program.

The same philosophy is now driving the interest of the international scientific community to establish a more advanced monitoring network horizontally across the highest mountains of the world, from the Alps to the Himalayas. Ev-K^2-CNR has taken up this challenge and will inaugurate the SHARE-Asia network in 2006. Development of the monitoring network will go through three phases. In the first stage, SHARE-Asia will install and operate a network of six automatic stations in the mountainous regions from Georgia, Pakistan, Nepal and Bhutan, at altitudes between 2500 and 5000 m a.s.l. Each monitoring site will be equipped with an advanced automatic system for data management and transmission of data and instruments for performing surveys on:

- *Atmosphere and Climate:* assessment of natural climatic conditions and modifications, and analysis of pollutant transport mechanisms. Installation of a fully equipped climatic monitoring system of key meteorological parameters and energy exchange. Installation of an air sampling system with specific sensors for the analysis of gases in relation to the greenhouse effect (CO_2) and the assessment of polluting dusts and metals.
- *Earth sciences:* monitoring of activities of the Earth's crust and geophysical measurements. Installation of highly sensitive digital geophysical monitoring equipment (seismic, GPS). Installation of GPS Master reference stations for more precise geodetic measurements.

In the second stage, researchers and technicians will be dedicated to sampling activities and data collection at the stations. Specific data will also be collected on high-altitude lakes, waters, snow and glaciers in order to assess of their intrinsic characteristics and sample for pollutant traces, as well as to perform studies aimed at understanding the effects of climate change (i.e., on glaciers).

The third stage is an evolution of the system in all its potential: acquisition of standard parameters for retrieval of specific data (to meet the requirements of specific international environmental programs); systematic research campaigns in limnology, glaciology, and geology; enhancement of technology transfer to the benefit of local scientific bodies, including all activities of monitoring, analysis and field investigations. Outputs will be made available internationally, to improve general understanding of high altitude and remote environments, as well as to contribute to studies on pollution, climate change and earth sciences. The unprecedented excellence of all these data continues to provide the Committee with the opportunity of productively developing additional scientific collaborations within international programs like the UNEP Project "Atmospheric Brown Clouds (ABC)" for study of the impacts of atmospheric pollution on a continental scale in Asia. The SHARE-Asia site at the Pyramid Laboratory-Observatory has in fact already been included as a Complimentary Site within Project ABC. Installation of similar observing and sampling equipment at the second SHARE-Asia site in Pakistan is foreseen and has already raised the interest of members of the Project ABC Science Team, given the potentials for creating a dedicated network of ABC stations at medium and high altitudes. Other potential international collaborations that SHARE-Asia aims to develop are within programs such as: the Long Term Ecological Research (LTER) network, the Global Atmospheric Watch (GAW), the International Global Atmospheric Chemistry Project (IGAC) and Global Land Ice Measurement from Space (GLIMS).

6. Conclusions

In conclusion, Ev-K^2-CNR represents a unique heritage of multidisciplinary high-altitude scientific and technological research, accumulated over 15 years of intense work and contributing increasingly to capacity building of local researchers and scientific institutions. Founded on the tradition of Italian exploration and research in the Karakoram and Himalaya, led by Prof. Ardito Desio, Ev-K^2-CNR has developed their innovative approach in and across the fields of medicine and physiology; environmental sciences; earth sciences; anthropological sciences and clean technologies, first around the International Pyramid Laboratory-Observatory at 5050 m a.s.l. in Nepal. They are now expanding their framework of activities to include other high-altitude research sites stretching across Asia's mountain chains, to better comprehend the global mechanisms affecting these fragile mountain ecosystems on a local level, while at the same time contributing to improving the quality of life and preservation of culture of the often challenged local communities. Ev-K^2-CNR has clearly not become less active with age. The spreading of their project activities and the specialization of their skills and expertise will ensure that they continue to look forward to new, exciting studies across the entire HKKH range.

References

AA. VV., 2005. Il K2 cinquant'anni dopo. La Ricerca Scientifica negli ambienti estremi. Proceedings. Rome, Italy, 17 December, 2004. Il Veltro Editrice, Rome, Italy, 312 pp.

AA. VV., 2003. Proceedings of the International Seminar on Mountains, Kathmandu, Nepal, 6–8 March, 2002. Royal Nepal Academy of Science and Technology, 647 pp.

Baudo, R., Tartari, G., and Munawar, M., 1998. Top of the World Environmental Research: Mount Everest – Himalayan Ecosystem, Ecovision World Monograph Series, Backhuys Publication, Leiden, The Netherlands, 293 pp.

Lami, A. and Giussani, G., (Guest Eds)., 1998. Limnology of high altitude lakes in the Mt. Everest regions (Himalaya, Nepal). Mem. Ist. Ital. Idrobiol., International Journal of Limnology, 57, 235.

Mountains Witnesses of Global Changes
R. Baudo, G. Tartari, and E. Vuillermoz (Editors)

8

SHARE-Asia contributions to ABC research

Gianni Tartari

Abstract

The Ev-K^2-CNR Project has been promoting and developing research at high altitude (>2,500 m a.sl.) in the Himalaya and Karakoram since 1987. Recently, activities have been focused on development of a monitoring network (stations at High Altitude for Research on the Environment in Asia: SHARE-Asia) to increase the environmental and geophysical scientific knowledge in these mountain regions. Research and monitoring activity at high altitude require a particular experience and a well-organized network. Ev-K^2-CNR has accumulated a significant experience in managing a high altitude network of automatic weather stations along the Khumbu Valley (Nepal) and in northern Pakistan (Baltistan region) in the framework of the Coordinated Enhanced Observing Period Project. With the installation of an Atmospheric Brown Clouds Project (ABC) monitoring station near the Pyramid Laboratory-Observatory in 2006, near the base of Mount Everest, SHARE-Asia can also contribute to the study of atmospheric circulation of pollutants. The "ABC-Pyramid" is the first of a network of stations that are planned to be installed at altitudes between 2,500 and 5,000 m a.s.l. along the Himalayan–Karakoram chain. These stations will be operated under active cooperation with the local scientific community, creating ample cooperation between western countries and developing countries in the region.

1. Introduction

The Himalayan mountain chain is a barrier between the Indian Ocean basin and the highland of Tibet for air mass circulation. The role of the Himalaya and the Tibetan plateau within the Asian monsoon systems has been well reported by Bollasina and Benedict (2004), with regard to research priorities in modelling observations, and by Lau (2005), regarding the implications of aerosols in hydrologic cycle interactions in monsoon systems. Rainfall analysis shows that the Asian monsoon is also influenced by the Southern Oscillation (Shrestha, 2000) and indicates the complexity of the interaction of air masses and the lowlands/highlands in south–central Asia. These arguments have a key role in recent understanding of the mechanisms that are influencing the regional variations of climate.

ISSN: 0928-2025 DOI: 10.1016/S0928-2025(06)10008-5

The Himalayan–Karakoram chain also affects the mechanism of long-range transportation of anthropogenic compounds such as gas and particulates in aerosols, at high altitude and directly influencing the wet and dry deposition of inorganic ions and trace compounds, both by natural (O'Dowd et al., 2004) and xenobiotic origin (Wania and Mackay, 1993). Aerosols influence radiative forcing that is an important and critical component of global climate. The lack of detailed knowledge of the optical properties of aerosols makes them one of the largest uncertainties in climate forcing assessments (Dubowik et al., 2002).

The Himalayan–Karakoram geologically is a recent chain undergoing continuing growth. The highest mountains of the world create an intense geo-dynamic effect, causing in turn a growing interest in monitoring seismic activity and understanding the scientific aspects of the movement of the Earth's surface. These phenomena need to be better understood, especially in the event of, or to prevent natural disasters, using technology such as global positioning systems to perform extremely precise land measurements. This satisfies both geodetic needs and contributes to monitoring the movement of unstable land, such as in the regular flow of glaciers but also in the case of landslides, which occur commonly in young, steep mountains and directly influence the hydrological networks. The geological and environmental features of the Himalayan and Karakoram chains need to be closely associated with the neighbouring areas; the previously mentioned Tibetan plateau to the north and the Indian peninsula to the south. This is especially true considering that India and China actually represent the main socioeconomic phenomena influencing the environment of the region – the rapid growth of the developing countries.

The links between atmosphere, climate, environment, geomorphology and hydrogeology in research therefore represent a fundamental challenge in the coming decades for central-eastern Asia, if understanding of the potential effects on populations living in these countries is to become possible. As reported recently by the Nordic Institute of Asian Countries (http://www.nias.ku.dk/research/strategy.asp?l0 = 4&who =), poverty, as well as economic growth and urbanization, present Asia with unprecedented environmental challenges. A cloud of smog covers Asia where 16 of the world's 20 most polluted cities are found. This situation affects many of Asia's natural resources that are rapidly deteriorating and millions of Asians live under environmentally stressed conditions that have a serious impact on living standards and quality of life.

The needs for research are then evident, to support both the local stakeholders and end-users, using a partnership approach. Also necessary is more knowledge of interactions on a global scale, to understand mitigation strategies and/or to elaborate possible future scenarios.

1.1. The SHARE-Asia project

The idea for the Stations at High Altitude for Research on the Environment in Asia (SHARE-Asia) Project grew out of expertise in high altitude and remote area investigations of the Ev-K^2-CNR Project (Baudo et al., 1998; Lami and Giussani, 1998). The scientific committee which coordinates this project established the Pyramid

Laboratory-Observatory in the early 1990s, the key installation for multi-disciplinary studies (environmental and earth sciences, medicine, etc.) at 5,050 m a.s.l. at the base of Mount Everest. During its 15 years of activity, Italian and international researchers have since understood the importance of expanding monitoring activities and increasing the potential of a system from a single point into a network. A multi-purpose local climatic monitoring network, the Pyramid Meteo Network, has been established in the early years of the new century. The same philosophy is now driving the interest of the international scientific community to expand the monitoring network across other high mountains of the world. An environmental-climatic monitoring network can be of interest to researchers in different fields. A network of stations may in fact operate in active cooperation with local scientists, creating a wide scientific milieu that may produce interesting developments in international scientific cooperation between western countries and developing countries of the region.

The general aims of the SHARE-Asia Project are focused on developing a monitoring network to increase the environmental and geophysical scientific knowledge at high altitudes in mountain regions of the Himalaya and Karakoram. SHARE-Asia is, however, a part of a more general project (SHARE) that looks at high mountains of the world for the installation of monitoring stations, such as automatic weather stations, aerosol monitoring instruments, global positioning system (GPS) seismic monitors, and so forth, in order to aid in understanding local impacts of global climate change and other Earth phenomena. The general objectives of SHARE, and the local subprojects like SHARE-Asia, are to develop an integrated system of measurements and to activate technology transfer and capacity-building processes through direct involvement of local organizations and experts.

The scientific objectives of SHARE-Asia monitoring network are summarized below:

- *Meteorological-climate research:* analysis, study and understanding of Himalayan range interactions with large scale Asian monsoon circulation; development of climatic models linked with the atmospheric pollutants.
- *Atmospheric chemistry:* understanding the biogeochemical circulation of pollutants; study the background concentrations of ozone, aerosols and greenhouse gases.
- *Glaciological research:* study the climate pressure and morphological changes of glaciers in the Himalaya and Karakoram.
- *Limnological and paleolimnological measurements:* study the distribution dynamics of pollutant species in relation to transport, fallout and re-suspension mechanisms so as to trace contamination sources; understanding the ecological features of high altitude lacustrine environments.
- *Earth surface coordinate research:* implementation of GPS Master Stations for high precision measurements.

It is important underline the strategic approach of each issue is to link its monitoring activities with others, in an attempt to give an integrated answer in the fields of environmental and geophysical studies.

2. Data collection and data sharing

Research and monitoring activities at high altitude require a particular experience and a well-organized network. The Ev-K^2-CNR Committee in more than a decade has developed a wide experience in managing a high altitude research activities, in particular using a network of automatic weather stations (Tartari et al., 1998a,b; Bollasina et al., 1999; Bertolani et al., 2000a,b; Bollasina et al., 2002) along the Khumbu Valley (Nepal) and in the Northern Pakistan (Baltistan region), sampling campaigns on air particulates and fresh snow at high altitude in Mount Everest an K2 areas (Valsecchi et al., 1999; Rizzio et al., 2000; Giaveri et al., 2001; Marinoni et al., 2001; Rizzio et al., 2001a,b; Rizzio et al., 2002; Balerna et al., 2003a; Balerna et al., 2003b; Bergamaschi et al., 2004; Giaveri et al., 2005; Pecci, 2005) and in the study of more than 60 lake bodies located in Sagarmatha National Park, Nepal (Bertoni et al., 1998; Lami et al., 1998; Manca et al., 1998; Tartari et al., 1998a,b).

The conceptual scheme developed in the SHARE-Asia project (Fig. 8.1) has been born out by these experiences and consolidated since 2001 in participation with the

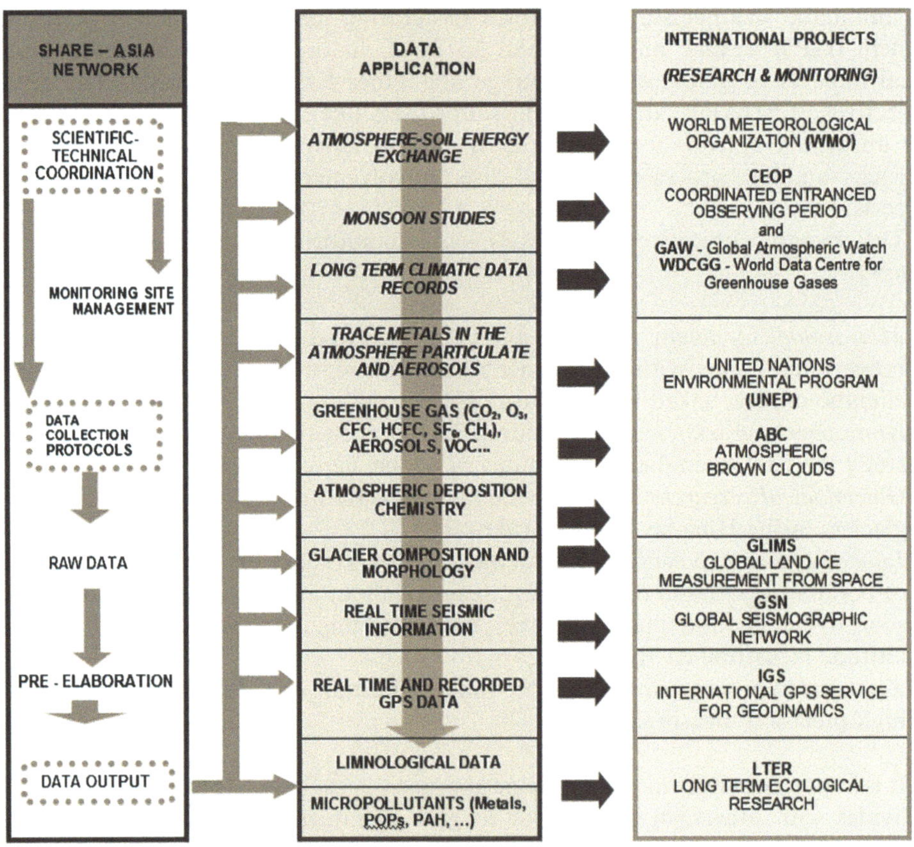

Figure 8.1. Conceptual scheme of data collection and data sharing in the SHARE-Asia project.

framework of the Coordinated Enhanced Observing Period (CEOP) Project (Koike, 2004). The raw data collected at the monitoring sites are managed directly by technical staff of Ev-K^2-CNR and sent to Italy periodically (2–3 time per year) and stored in relational databases. The data are validated and pre-elaborated to give a basic statistical description of the site and periodically published in a summary report (Bertolani et al., 2000a,b). The data collected are used in several scientific fields (atmosphere-soil energy exchange, monsoon modelling, particulate and trace metals atmospheric transport, atmospheric composition and depositions, glaciers morphology, limnological studies, etc.) that are developed in national and international projects.

The fluxes of data are governed by very simple rules that restrict the access to the scientific community only within 2 years from the collection, with the consent of Ev-K^2-CNR to validate and elaborate the data. After this period data are shared with end-users that must agree to acknowledge the Ev-K^2-CNR Committee and accept co-authorship in the case of publication.

3. Interaction with international networks

One of the final aims of SHARE-Asia is to contribute at international networks. The peculiar location of the SHARE-Asia stations (site located at high altitude and in remote areas) give a special meaning to the data collected. Actually the meteorological climate and atmospheric chemistry monitoring stations of SHARE-Asia are part of important international scientific projects:

- CEOP. This project, part of the World Climate Research Programme (WCRP) of the World Meteorological Organization (WMO), is aimed at understanding and modelling the influence of hydro-climatic processes on forecasting of global atmospheric circulation and changes in hydrological resource availability. Interactions between aerosols and the water cycle are also studied (http://www.ceop.net).
- ABC. The project, promoted by UNEP, will create a network of aerosol observatories and pollutant species monitoring stations for studying the impact of aerosols and pollutants on the regional and global physical climate system, as well as on agriculture, the water cycle and human health (http://www.abc-asia.ucsd.edu).

The expertise produced by SHARE-Asia will also imply potential collaboration with other key international programs, such as:

- Global Atmospheric Watch (GAW). This WMO Project provides one of the main contributions to understand the complex mechanisms of natural and anthropogenic atmospheric change through atmospheric physical–chemical measurements on a global and regional scale (http://www.wmo.ch/web/arep/gaw/gaw_home.html).
- The International Global Atmospheric Chemistry Project (IGAC). The project intends to pursue the study of rapid atmospheric changes through the analysis of global distribution and concentration of chemical species and the study of their impacts on global change and air quality (http://www.igac.noaa.gov).

- Global Land Ice Measurement from Space (GLIMS). GLIMS is a project designed to monitor the world's glaciers primarily using data from the Advanced Spaceborne Thermal Emission and reflection Radiometer (ASTER) instrument aboard the EOS Terra spacecraft (http://www.glims.org).

The increasing participation in international projects is mandatory for the future of the SHARE-Asia Project. This objective will be reached depending not only by the quality of data collected and the completeness of the time series, but also by scientific contribution of researchers that are involved in the project.

4. The SHARE-Asia contribution to ABC and future AIMS

Since early 2006 SHARE-Asia is developing an ABC monitoring station at the Pyramid Laboratory-Observatory, located near Lobuche in Khumbu Valley, Nepal. The importance of this ABC site consists in the uniqueness of altitude. As well described in several papers (Ramanathan et al., 2001; Ramanathan and Ramana, 2005; Chung and Ramanathan, 2006), the understandings of the atmospheric brown clouds impacts on South Asian climate and on the hydrological cycle require a description of the atmospheric composition of the whole air column. Generally these studies are performed using aircraft. Recently Ramanathan et al. (2005) and Ramana et al. (2004), however, have emphasized that direct observation of these large aerosol radiative forcings, in particular at high altitude where the effects of the interaction with the Himalayan chain can be put in evidence, are fundamental to description of the persistent, widespread and strongly absorbing haze over the Himalayan region.

In this context the role of ABC-Pyramid Laboratory will be fundamental in the general context of the ABC network. In future the integration of deposition chemistry with aerosol chemistry will close the study of the pollutant circulation at the local level, to obtain complementary information on atmospheric processes connected to the Atmospheric Brown Cloud. The ABC-Pyramid is the first of a network of stations that is planned for installation at altitudes between 2500 and 5000 m along the Himalayan–Karakoram chains. Each monitoring site will include also automatic weather station, and other equipment for monitoring purposes.

5. Conclusion

The SHARE-Asia Project is a challenge for high altitude research. The complexities of instrumentation that will be installed in the monitoring sites, and the need to operate in shared remote sites with limited collaboration at the local level by specialized personnel, require a well-calibrated organization such as that developed in past years by the Ev-K^2-CNR Committee.

These difficulties are in any case secondary aspects with respect to the need for competing at the same level of quality and quantity of data produced by other monitoring station located in more "comfortable" areas. The final target of SHARE-Asia is in fact to give support to international science networks, possibly without any technical limitations, considering the importance of data collection that can describe unique situations.

References

Balerna, A., Bernieri, E., Chiti, M., et al., 2003a. In situ measurements of cesium-137 gamma-ray emission at very high altitudes using a fully portable detector. Nuclear Instruments Methods in Physics Research 512, 631–636.

Balerna, A., Bernieri, E., Pecci, M., et al., 2003b. Chemical and radio-chemical composition of freshsnow samples from northern slopes of Himalayas (Cho Oyu range, Tibet). Atmospheric Environment 37, 1573–1581.

Baudo, R., Tartari, G., and Munawar, M., (Eds), 1998. Top of the World Environmental Research: Mount Everest-Himalayan Ecosystem The Great Ecovision World Monograph Series. Backhuys Publishers, The Netherlands, 293 pp.

Bergamaschi, L., Rizzio, E., Giaveri, G., et al., 2004. Determination of baseline element composition of lichens using samples from high elevations. Chemosphere 55, 933–939.

Bertolani, L., Bollasina, M., and Tartari, G., 2000a. Recent biennial variability of meteorological features in the Eastern Highland Himalayas. Geophysics Research Letters 27, 2185–2188.

Bertolani, L., Bollasina, M., Verza, G.P., and Tartari, G., 2000b. Pyramid Meteorological Station. Sagarmatha National Park 5050 m – Lobuche, Khumbu Valley, Nepal. Summary Report 1994–1998. Ev-K^2-CNR Committee, Milan, Italy, 85 pp.

Bertoni, R., Calmieri, C., and ContesinI, M., 1998. Organic carbon and microorganisms in two Nepalese lakes. In: Lami, A. and Giussani, G. (Eds), Limnology of High Altitude Lakes in the Mt. Everest Regions (Himalaya, Nepal). Mem. Ist. Ital. Idrobiol. 57, 99–106.

Bollasina, M. and Benedict, S., 2004. The role of the Himalayas and the Tibetan plateau within the Asian monsoon system. American Meteorological Society, BAMS 1001–1004.

Bollasina, M., Bertolani, L., and Tartari, G., 1999. Recent interannual variability of local climate in eastern highland Himalayas. UCLA Tropical Meteorology Newsletter, No. 31.

Bollasina, M., Bertolani, L., and Tartari, G., 2002. Meteorological observations at high altitude in the Khumbu Valley, Nepal Himalayas, 1994–1999. Bulletin of Glaciological Research 19, 1–11.

Chung, C.E. and Ramanathan, V., 2006. Weakening of North Indian SST Gradients and the Monsoon Rainfall in India and the Sahel. Journal of Climate, May 15, 2006.

Dubowik, B., Holben, T.F., Eck, A., et al., 2002. Variability of absorption and optical properties of key aerosol types observed in worldwide locations. Journal of Atmospheric Sciences 59, 590–608.

Giaveri, G., Bergamaschi, L., Rizzio, R., et al., 2005. INAA at the top of the world: elemental characterization and analysis of airbone particulate matter collected in Himalayas at 5,100 m high. Journal of Radioanalytical Nuclear Chemistry 263 (3), 725–732.

Giaveri, G., Rizzio, E., and Gallorini, M., 2001. Preconcentration and preseparation procedure for platinum determination at trace levels by neutron activation analysis. Analytical Chemistry 73, 3488–3491.

Koike, T., 2004. The coordinated enhanced observing period-an initial step for integrated global water cycle observation. WMO Bulletin 53 (2), 2–8.

Lami, A., and Giussani, G., (Eds), 1998. Limnology of high altitude lakes in the Mt. Everest Region (Himalayas, Nepal). Mem. Ist. Ital. Idrobiol. 57, 235 pp.

Lami, A., Guilizzoni, P., Marchetto, A. et al., 1998. Palaeolimnological evidence of environmental changes in some high altitude Himalayan lakes (Nepal). In: Lami, A. and Giussani, G. (Eds), Limnology of High Altitude Lakes in the Mt. Everest Regions (Himalaya, Nepal). Mem. Ist. Ital. Idrobiol. 57, 107–130.

Lau, W.K.M., 2005. Aerosol-hydrologic cycle interaction: a new challenge in monsoon climate research. GEWEX News, February 7–9.

Manca, M., Ruggiu, D., Panzani, P. et al., 1998. Report on a collection of aquatic organisms from high mountain lakes in the Khumbu Valley (Nepalese Himalayas). In: Lami, A. and Giussani, G. (Eds),

Limnology of High Altitude Lakes in the Mt. Everest Regions (Himalaya, Nepal). Mem. Ist. Ital. Idrobiol. 57, 77–98.
Marinoni, A., Polesello, S., Smiraglia, C., and Valsecchi, S., 2001. Chemical composition of fresh snow samples from the southern slope of Mt. Everest Region. Atmospheric Environment 35, 3183–3190.
O'Dowd, C.D., Facchini, M.C., Cavalli, F., et al., 2004. Biogenically driven organic contribution to marine aerosol. Nature 431, 676–680.
Pecci, M., 2005. High altitude in situ surveys and researches on the snow cover in high altitude: case studies in Italian and Himalayan mountain ranges. Supplemento Geografia Fisica e Dinamica Quaternaria VII, 253–260.
Ramana, M.V., Ramanathan, V., Podgorny, I.A., et al., 2004. The Direct Observations of Large Aerosol Radiative Forcing in the Himalayan Region. Geophysics Research Letters 31, L05111. doi:10.1029/2003GL018824
Ramanathan, V., Chung, C., Kim, D., et al., 2005. Atmospheric brown clouds: Impacts on south asian climate and hydrological cycle. PNAS 102 (15), 5326–5333.
Ramanathan, V., Crutzen, P.J., Kiehl, J.T., and Rosenfeld, D., 2001. Aerosols, climate and the hydrologic cycle. Science 294, 2119–2124.
Ramanathan, V. and Ramana, M.V., 2005. Persistent, widespread, and strongly absorbing haze over the Himalayan foothills and the Indo-Ganges plains. Pure and Applied Geophysics 162, 1609–1626.
Rizzio, E., Bergamaschi, L., Profumo, A., and Gallorini, M., 2001a. The use of the neutron activation analysis for particles size fractionation and chemical characterization of trace elements in urban air particulate matter. Journal of Radioanalytical Nuclear Chemistry 248, 21–28.
Rizzio, E., Bergamaschi, L., Valcuvia, M.G., et al., 2001b. Trace elements determination in lichens and in the airborne particulate matter for the evaluation of the atmospheric pollution in a region of northern Italy. Environment International 26, 543–549.
Rizzio, E., Bergamaschi, L., Valcuvia, M.G., et al., 2002. Determination of trace elements and evaluation of their enrichment factors in Himalayan lichens. Environmental Pollution 120, 137–144.
Rizzio, E., Giaveri, G., and Gallorini, M., 2000. Some analytical problems encountered for trace elements determination in the airborne particulate matter of urban and rural areas. Science of the Total Environment 256, 11–22.
Shrestha, M.L., 2000. Interannual variation of summer monsoon rainfall over Nepal and its relation to Southern Oscillation Index. Meteorology and Atmospheric Physics 75, 21–28.
Tartari, G., Verza, G.P., and Bertolani, L. 1998a. Meteorological data at the Pyramid Observatory Laboratory (Khumbu Valley, Sagarmatha National Park, Nepal). In: Lami, A. and Giussani, G. (Eds), Limnology of High Altitude Lakes in the Mt. Everest Regions (Himalaya, Nepal). Mem. Ist. Ital. Idrobiol. 57, 23–40.
Tartari, G.A., Tartari, G., and Mosello, R., 1998b. Water chemistry of high altitude lakes in the Khumbu and Imja Kola Valleys (Himalaya, Nepal). In: Lami, A. and Giussani, G. (Eds), Limnology of High Altitude Lakes in the Mt. Everest Regions (Himalaya, Nepal). Mem. Ist. Ital. Idrobiol. 57, 49–76.
Valsecchi, S., Smiraglia, C., Tartari, G., and Polesello, S., 1999. Chemical composition of monsoon depositions in the Everest region. Science of the Total Environment 226, 187–199.
Wania, F. and Mackay, D., 1993. Global fractionation and cold condensation of volatile organochlorine compounds in polar regions. Ambio 22, 10–18.

Mountains Witnesses of Global Changes
R. Baudo, G. Tartari, and E. Vuillermoz (Editors)

9

Merging regional and global chemistry, air quality and global change: SHARE-Asia in the context of the IGAC project

Sandro Fuzzi

Abstract

The International Global Atmospheric Chemistry (IGAC) project was created in the late 1980s to address growing international concerns over rapid changes observed in Earth's atmosphere. Much of IGAC's research effort during its first decade was directed towards assessing the effects of anthropogenic emissions on the background atmosphere. While questions remain concerning the point at which observed global/regional mean trends in component concentrations (signal) unambiguously rise above background natural variability (noise), it is now well recognized that human activities have perturbed the chemical composition of the atmosphere at local, regional, and global scales. Two overarching questions have emerged that constitute the basis for the IGAC action plan over the next decade: (1) What is the role of atmospheric chemistry in amplifying or damping climate change? (2) Within the Earth System, what effects do changing regional emissions and depositions, long-range transport, and chemical transformations have on air quality and the chemical composition of the planetary boundary layer? Within the context of the larger ABC-Asia project, a proposal was put forward for the activation of an ABC-IGAC Task focusing on the monitoring of aerosol and trace gases over the Asia–Pacific region and on estimating their impact on atmospheric chemistry and the radiation budget. The activities currently in progress, to include the Ev-K^2-CNR Pyramid observatory in the Atmospheric Brown Cloud (ABC) monitoring network under the Stations at High Altitude for Research on the Environment in Asia (SHARE-Asia) project, aim at providing an important contribution to ABC-Asia and, at the same time, will undoubtedly be an opportunity for the Italian community of global-change researchers.

1. Global tropospheric chemistry and the role of IGAC

Over the past century, humanity has been altering the chemical composition of the atmosphere in an unprecedented way, in an astonishingly short time. Worldwide

ISSN: 0928-2025 DOI: 10.1016/S0928-2025(06)10009-7

emissions from growing industrial and transportation activities and more intensive agricultural practices have caused widespread increases in atmospheric concentrations of photochemical oxidants, acidic gases, aerosols, and some toxic chemical species. Many of these air pollutants are known to have detrimental impacts on human health and/or natural and managed ecosystem viability. Furthermore, higher fossil fuel consumption, coupled with agriculturally driven increases in biomass burning, fertilizer usage, crop by-product decomposition, and production of animal-based food and fiber have led to increasing emissions of key greenhouse gases, such as carbon dioxide, methane, and nitrous oxide. The net effects of the build-up of radiatively active trace gases and the changing burden of atmospheric particles appear to be responsible for much of the climate trend observed during the 20th century, particularly the warming over the last few decades (IPCC, 2001). Predicted impacts of climate change include disruptions of agricultural productivity, fresh water supplies, ecosystem stability, and disease patterns. Significant increases in sea level and changes in the frequency of severe weather events are also forecast. The resulting effects of all these stresses on biogeochemical cycles could exacerbate changing atmospheric composition and result in further effects on climate. If current trends are unchecked, much more significant warming is predicted, potentially driving a wide range of perturbations in other components of the climate system (Steffen et al., 2004).

Twenty years ago, scientific programs addressing global tropospheric chemistry and the issues described above were in their infancy. Almost no observations of tropospheric composition on a large scale were available, many chemical transformation mechanisms were unknown, and global atmospheric chemistry models were rather crude. The past decade has seen global atmospheric chemistry research blossom. We have learned much about the global cycles (sources, transformations, and sinks) of most important atmospheric chemical species. Existing satellite observations have provided a wealth of data regarding the chemical composition of the stratosphere, and new satellite instruments probing the troposphere have recently been, or are about to be launched. Multi-platform process studies of atmospheric chemical processes have been conducted on an unprecedented scale. Global chemical transport models can now simulate with some success the distribution of key tropospheric chemical species, and are capable of simulating future global atmospheric composition scenarios. Furthermore, short lived, radiatively active substances such as ozone and aerosols are now incorporated as active constituents in most global climate models. As scientific understanding of the elements of atmospheric chemistry has been developed, the necessity of understanding the linkage between atmospheric composition and other components of the earth system has been realized more explicitly (Fig. 9.1). Ten years ago, the concept of having an "earth system" level view was a rather abstract idea. Feedbacks between, for example, changing climate and changing terrestrial emissions, or changing climate and atmospheric chemical composition, were not included in models. Now, we are on the threshold of a more quantitative understanding of the role of atmospheric chemistry in Earth's system processes and of developing strategies to integrate that knowledge into a predictive capability.

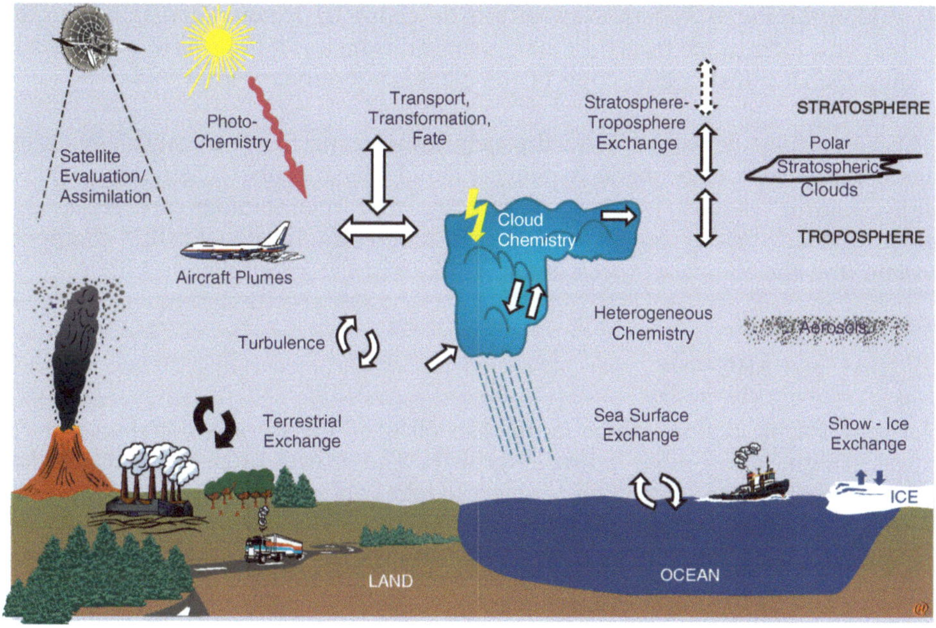

Figure 9.1. Interdisciplinary collaboration is required to adequately understand both the vital interactions among the Earth's component systems and the complex processes taking place in the atmosphere (Bates and Scholes, 2004).

Global tropospheric chemistry and the issues described above have received growing attention, not only from the members of the scientific community, but also from decision makers in governments and industries. With increased recognition in society of the importance and value of the environment, the relation between atmospheric-chemistry research and environmental-policy design has been growing substantially over recent decades. In some cases, international treaties to reduce emissions have been enacted and actions to protect the global environment have been taken. Major challenges remain, however. Although substantial advances have been made in understanding fundamental processes in the chemical system of the atmosphere, our predictive capability remains limited in spite of its importance for informed decision making. The uncertainties in our forecasts of air quality and climate change are still high. In addition, new and challenging problems at the interfaces of weather chemistry, climate chemistry, and ecology chemistry are emerging and will require much attention in the future. Beginning in 1999, as part of IGACs integration and synthesis of a decade of tropospheric chemistry research (Brasseur et al., 2003), a series of discussions and workshops was held to define the research challenges of atmospheric chemistry for the next decade. These discussions are summarized in the following two overarching research questions that will be the basis for the next decade of IGAC research (Bates and Scholes, 2004).

(1) What is the role of atmospheric chemistry in amplifying or damping climate change?

(2) Within the Earth System, what effects do changing regional emissions and depositions, long-range transport, and chemical transformations have on air quality and the chemical composition of the planetary boundary layer?

Collaborations between those with such diverse expertise as meteorology, atmospheric chemistry, atmospheric dynamics, cloud physics, photochemistry, oceanography, marine chemistry, plant physiology, and microbiology are required to understand adequately both the vital interactions among the Earth's component systems and the complex processes taking place in the atmosphere itself.

2. IGAC and ABC-Asia

The Asia–Pacific region, where more than 60% of the world's population live, is experiencing fast and rapid industrial growth. The growth of large urban conglomerates with more than ten million people also takes place in this region at a rapid rate. All this is resulting in higher demands for energy and mobility, at the cost of the degradation of the environment, especially with respect to air pollution at local and regional levels. In many respects, local and regional air quality and global atmospheric changes are different aspects of the same issue. Although air quality deterioration was initially regarded as a local and later as a regional problem, it is now clear that the atmosphere transforms and transports a multitude of gases and aerosols intercontinentally and even globally. Importantly, model-scenario studies have predicted that the rapidly emerging emissions in Asia will increasingly affect the global background levels of tropospheric ozone. Furthermore, this region presently accounts for about 30–50% of the anthropogenic aerosol loading and a positive trend of the pollutant concentration is foreseen in the near future. Consequently, in the coming decades, air quality improvements by emission reductions in the western hemisphere may be overpowered by hemispheric transports from Asia.

The ABC project (Ramanathan and Crutzen, 2001; http://www-abc-asia.ucsd.edu) is an international research effort initiated by the United Nations Environment Program (UNEP) and its first focus is on the Asian and the Pacific regions. The goal is to address the major environmental and climate challenges facing these regions in the coming decades, due to the rising levels of air pollutants. The major components of the ABC project are observations and modeling of regional impacts.

Within the context of the larger ABC-Asia project, a proposal was put forward for the activation of an ABC-IGAC Task focusing on the monitoring of aerosol and trace gases over the Asia–Pacific region and on estimating their impact on atmospheric chemistry and the radiation budget.

The following are the aims of this Task, organised and run by proactive ABC scientists:

- To develop regional air quality and climate observatories in order to establish baseline measurements of pollutant gases, particulates, and column burden of aerosols. Data from these observatories, in combination with those from satellites, will be used to track the pollution plumes on regional scales. In fact, the current lack of high quality data for the ABC-Asia region is the major reason for the uncertainties in environmental impact assessments of this region's atmospheric pollutants.

- Develop an integrated modeling approach that combines data assimilation techniques with predictive models in order to estimate the impact of pollution on climate and atmospheric chemistry.

3. SHARE-Asia and ABC-Asia

SHARE-Asia is a multidisciplinary project of the Ev-K^2-CNR Committee (http://www.evk2cnr.org/en/) integrating research in the fields of Environmental and Earth Sciences. SHARE-Asia focuses, in particular, on mountain areas, given the environmental sensitivity and the importance of these regions for monitoring the Earth's state of health. The Himalayan–Karakoram range, with its elevation and geographic location, represents an ideal place for studying long-range pollutant transport systems on a regional scale and for monitoring changes induced by mechanisms that act on a global scale through monsoon circulation.

SHARE-Asia has recently supported the creation of an Aerosol Observatory program in the Himalayan and Karakoram mountain regions, taking advantage of the Pyramid Station (Fig. 9.2) with the specific aim of contributing to the ABC project. The research team in charge of this project, led by the Italian National Research Council Institute of Atmospheric Sciences and Climate (CNR-ISAC), also includes scientists from the CNRS/Blaise Pascal University, the German Aerospace Center (DLR), and the University of Urbino.

The Aerosol Observatory station will be placed near the Pyramid International Laboratory-Observatory in Nepal at an altitude of 5079 m a.s.l. Implementation of continuous measurements at the Pyramid station will contribute to better quantify

Figure 9.2. The Ev-K^2-CNR Pyramid International Laboratory-Observatory in Nepal at an altitude of 5050 m a.s.l.

Table 9.1. Summary of the measurements to be performed at the Pyramid station in Nepal.

Measurement	Instrumentation	Collaboration with
On-line measurements		
Aerosol size distribution	Scanning Mobility Particle Sizer	CNRS
Aerosol size distribution	Optical Particle Counter	
Aerosol scattering properties	Integrating Nephelometer	
Aerosol absorption properties	Multi Angle Absorption Photometer	DLR
Aerosol optical depth	Sky Radiometer	
Ozone concentration	U.V. Photometric Gas Analyzer	
Off-line measurements		
Aerosol chemical composition	Sampling and Chemical Analysis	
Halocarbons	Flask Sampling and GC-MS Analysis	University of Urbino

the impact of anthropogenic activities over the Himalaya area, in a region between India and China influenced by the monsoon circulation and where little information is presently available.

In order to observe the temporal variation of several atmospheric compounds at a very high altitude in the Himalaya throughout the whole seasonal cycle over several years, continuous in-situ measurements of chemical, physical and optical properties of aerosol and ozone, halocarbons and other greenhouse gases concentration measurements will be carried out. The instrumental set-up (Table 9.1) has been defined in accordance with the ABC project standards. Measurements of the particle properties will provide the basic information required to detect any long-term change in aerosol source emissions and to assess possible climatic effects of aerosols that may result from these changes.

4. Conclusions

The Italian SHARE-Asia program will provide an important contribution to ABC Asia, providing long-term, quality-assured data on atmospheric composition change and on aerosol physical and chemical properties in a region where little information is presently available. Cooperation with the ABC-Asia project will also be implemented in capacity-building activities directed towards Nepalese and Pakistani scientists and technicians to manage atmospheric and environmental monitoring sites and to acquire, analyze, and model the scientific data collected there. The SHARE-Asia/ABC-Asia liaison will, at the same time, be an opportunity for the Italian global change research community, providing an international frame for national activities.

References

Bates, T.S. and Scholes, M. (Eds), 2004. IGAC science plan and implementation strategy, (available at http://www.igac.noaa.gov/).

Brasseur, G.P., R.G. Prinn and A.A.P. Pszenny (Eds), 2003. The Changing Atmosphere. An Integration and Synthesis of a Decade of Tropospheric Chemistry Research. Global Change: The IGBP Series. Springer-Verlag, Heidelberg, Germany.

IPCC, 2001. Climate change 2001: the scientific basis. In: Houghton, J.T., Ding, Y., Griggs, D.J., et al. (Eds), Contribution of Working Group I to the Third Assessment Report of the Intergovernmental Panel on Climate Change. Cambridge University Press, Cambridge, UK, pp. 881.

Ramanathan, V. and Crutzen, P.J., 2001. Concept paper on Asian Brown clouds. Air pollution in the Indo–Asia–Pacific region: impact on climate and environment. United Nations Environment Program/ Environment Assessment Program, (available at http://www.abc-asia.ucsd.edu/ABCconceptFinal23May01.pdf).

Steffen, W., Sanderson, A., Jäger, J., et al., 2004. Global Change and the Earth System. A Planet Under Pressure. Springer, Heidelberg, Germany.

Mountains Witnesses of Global Changes
R. Baudo, G. Tartari, and E. Vuillermoz (Editors)

10

The ABC-Pyramid: a scientific laboratory at 5079 m a.s.l. for the study of atmospheric composition change and climate

Paolo Bonasoni, Paolo Laj, Ubaldo Bonafè, Francescopiero Calzolari, Paolo Cristofanelli, Angela Marinoni, Fabrizio Roccato, Maria Cristina Facchini, Sandro Fuzzi, Gian Paolo Gobbi, Jean-Marc Pichon, Hervè Venzac, Karine Sellegri, Paolo Villani, Michela Maione, Jgor Arduini, Andreas Petzold, Michael Sprenger, Gian Pietro Verza and Elisa Vuillermoz

Abstract

The Himalayan–Karakoram range is located in one of the most densely populated and very rapidly developing world areas. Monitoring of atmospheric composition in this area can play a relevant role in evaluating the background conditions of the free troposphere and quantifying the pollution present at high altitudes, as well as in studying regional and long-range transport phenomena. Due to technical and logistic difficulties in carrying out measurements at high altitude in the Himalaya, no systematic observations of atmospheric constituents are available for this area. Thus, a new measurement station in such a region represents a unique source of data, able to make up for the prior lack of this information. For these reasons, in the framework of the SHARE-Asia and ABC projects, a remote monitoring station, the ABC-Pyramid Laboratory, will be installed in the Khumbu valley near Mt. Everest at 5079 m a.s.l. Continuous in situ measurements of chemical, physical and optical properties of aerosol, surface ozone concentration, as well as non-continuous measurements of halocarbons and other greenhouse-gas concentrations will be carried out. This monitoring station was projected, realised and tested in Bologna at CNR-ISAC Institute during autumn 2005. It was designed to be controlled by remote login and to operate for the long-term in extremely adverse weather conditions. This station represents an ideal place for studying regional and long-range air mass transport, due to natural and human processes. Precious 5-day forecast information about air-masses circulation at the ABC-Pyramid site will be supplied daily by Lagrangian backward trajectories, including suitable forecasts of stratosphere-troposphere exchange phenomena.

ISSN: 0928-2025 DOI: 10.1016/S0928-2025(06)10010-3

1. Introduction

The Himalayan–Karakoram range, because of its elevation and geographic location, represents an ideal place for studying long-range pollutant transport systems on a regional scale, and for monitoring changes induced by mechanisms acting on a global scale through monsoon circulation. This area is placed within two of the most densely populated and very rapidly developing world sites, India and China. In these countries, the increased industrial activities and vehicular traffic lead to a massive growth of anthropogenic pollutant emissions, as pointed out in recent UNEP studies (UNEP and C4, 2002). These emissions especially increase the concentrations of aerosol, greenhouse gases and ozone precursors, promoting intense photochemical smog episodes. In Southeast Asia, one of the most impressive effects of human activities, in particular of heavy biomass burning and fossil fuel emissions, is the brownish haze that covers and envelopes this region (Ramanathan et al., 2001). It has been estimated that aerosols may reduce by up to 10% of the mean solar radiation reaching the Earth's surface, producing a global cooling effect that opposes global warming (IPCC-Climate Change, 2001). Through so-called "direct-effect", aerosols scatter and/or absorb solar radiation, thus cooling the Earth's surface and changing the radiative balance in the atmosphere. Aerosols also affect the water cycle through the so-called "indirect-effect", whereby increasing the number of cloud condensation nuclei, thus inhibiting the growth of cloud drops to raindrops and increasing the life of clouds. This leads to more clouds, increasing reflection of solar radiation, and further cooling of the Earth's surface. The study of aerosol influence on Southeast Asian climate was one of the main goals carried out by the INDOEX experiment (1996–1999); one of the most important findings was that the regional radiative perturbations by anthropogenic aerosols is one order of magnitude greater than that due to anthropogenic greenhouse gases (Ramanathan and Crutzen, 2003). Studies concerning this Asian Brown Cloud (ABC) have shown that this haze blocks up to 15% of solar radiation, causing a possible cooling of the ground and a heating of the atmosphere, which can affect monsoons and other rainfall patterns (Ramanathan et al., 2005). This kind of brown haze has assumed continental-scale proportion; moreover, in tropical areas, the presence of a dry season can increase aerosols and cloud lifetime and thus enhance both the direct and indirect effect (Lau, 2005).

Air quality networks are still developing in Asiatic cities, while no systematic observations of atmospheric compounds are available for high altitude areas, except at Mt. Waliguan in China. Thus, the setting up of new stations for the monitoring of atmospheric compounds represents a step necessary to understand the actual conditions of the Asian background atmosphere and consequently to enable quantification of pollution at high altitudes. Moreover, these experimental activities will also provide precious information for a better understanding of the complex interactions between high mountain ranges and climate processes (including monsoon circulation) as well as to produce input data for atmospheric chemistry and climate modelling.

For these reasons, in the framework of the UNEP-ABC, a high altitude monitoring stations (one in the Himalaya and one in the Karakoram) have been planned within the SHARE-Asia EV-K^2-CNR project. The goal of this project is to increase the knowledge on (1) how the physical, chemical, and optical properties of aerosols change at high Himalayan altitudes, with season and air mass origin; (2) how aerosol size and light scattering change during pollution and dust transport episodes; (3) establish how much stratospheric intrusions and polluted long-range transport episodes contribute to background O_3 concentrations; and, (4) determine the concentrations of greenhouse active and ozone-destroying halocarbons.

The first step of this project consists of installation of a remote monitoring station—the "ABC-Pyramid Laboratory"—in the heart of Himalaya: the Khumbu valley near Mt. Everest at 5079 m a.s.l. This monitoring station will be placed on a ridge 200 m away from, and 30 m above the Italian "Pyramid Laboratory-Observatory", the high-altitude centre for multi-disciplinary researches inaugurated in 1990. This action represents a technological and scientific challenge able to notably improve the scientific knowledge on atmospheric changes by carrying out a continuous monitoring of the atmospheric composition at high altitude in the Himalaya, an activity usually restricted by logistical constraints; lacking infrastructure and power supply, with difficult transport and adverse weather conditions. In particular, the ABC-Pyramid high altitude measurement station will represent a unique source of data, able to make up for the lack of information on atmospheric composition at high altitude in this region. Other than scientific aspects, this project will contribute to implementation on capacity-building processes in very high mountain areas, in the fields of environmental and geophysical monitoring.

2. Experimental

2.1. *Shelter laboratory design*

A laboratory shelter was designed in order to permit the monitoring and sampling of atmospheric compounds at high altitude. The laboratory is composed by 20 aluminium panels for a total weight of 650 kg. The whole structure measures 23 m^3, of which about a third is devoted to holding the batteries for power supply in the back part of the shelter (Fig. 10.1). It is completely dismounted, allowing the transportation at 5000 m by helicopter and Sherpa porters. An aluminium roof in two parts and a protective cover were previewed in order to avoid any water leaks and to have a UV protective sheet.

On the roof five holes for inlet sampling are present. They will host: (1) a PM_1 DIGITEL HVA-80 head for a nephelometer and DMPS instruments (flow rate 1 $m^3 h^{-1}$); (2) a TPS for the OPC that has a specific head with probe with sensor for T and RH (0.3 $L min^{-1}$); (3) a PM_{10} for high volume (500 $L min^{-1}$) discontinuous sampling of aerosol on quartz fibre filters; (4) a second TPS head (15 $m^3 h^{-1}$) for ozone and black carbon measurements, and (5) an additional hole foreseen in order to easily provide eventual improvement of measurement system with other instruments. All the sampling heads for aerosol and gases will be equipped with thermal resistances in order to avoid ice formation in extreme weather conditions. An equipotential copper bar will be installed around the internal part of the shelter in order to avoid any electric charge accumulation by keeping contact between the different components of the shelter and the instruments. A stirrup outside the shelter will

Figure 10.1. The Laboratory-Shelter mounted during the test performed in Bologna.

allow the assembly of the CIMEL radiometer that will provide measurements of aerosol optical depth. The greenhouse gases (GHGs) sampling carried out with steel flasks will be performed once a week by trained local personnel.

2.2. *Power supply*

At the ABC-Pyramid laboratory, the experimental activity will be completely carried out using renewable energy from 96 photovoltaic panels, installed on tilted 35° aluminium supports, which will assure good solar exposure. The energy produced by these photovoltaic panels will be stored in 120 electric storage cells lodged on shelves in the rear of the shelter. The inverters will guarantee current stabilisation, in order to avoid power surges able to damage the instruments. A suitable PC connection will provide real-time information about the power system. In the case of insufficient power supply from the photovoltaic panels, additional energy supplies can be derived from additional photovoltaic panels located in the main Pyramid station.

2.3. *Instrumentation*

In order to provide long-term records of temporal variation of atmospheric compounds at very high altitude in the Himalaya, continuous in situ measurements of chemical, physical and optical properties of aerosol, surface ozone concentration, as well as non-continuous measurements of Halocarbons and other greenhouse gas concentrations will be undertaken. The instrumental setup, defined also in accordance with the Project ABC standards, includes:

(1) Multi-angle absorption photometer (MAAP 5012, Thermo Electron Corporation) measuring aerosol light absorption properties, corresponding to Black Carbon content (firmware, acquisition and inversion procedures pointed out by DLR, German Aerospace Center, Institute of Atmospheric Physics; data acquisition system modified by ISAC-CNR);
(2) Optical particle counter (GRIMM Ambient Aerosol Spectrometer 1.109, GRIMM AEROSOL Technik GmbH & Co. KG) determining size distribution for particles with diameters ranging between 300 nm and 32 μm (data acquisition system developed by ISAC-CNR);
(3) Differential mobility particle sizer (designed and build up by the LAMP-CNRS team) allowing measurement of the complete aerosol physical spectrum with the fine and ultra fine particle size distribution from 10 nm to 500 nm;
(4) Integrating nephelometer (model 3563, TSI Inc.) determining aerosol integral scattering coefficient at three wavelengths (data acquisition system developed by LaMP-CNRSISAC-CNR);
(5) Surface ozone analyser (49C UV photometric gas analyzer, Thermo Electron Corporation, data acquisition system developed by LaMP-CNRS/ISAC-CNR);
(6) Sun Photometer CIMEL CE 318. An automatic sun and sky brightness radiometer, measuring aerosol optical depth and radiative properties.
(7) High Volume ($500\,\mathrm{L\,min^{-1}}$) Sampler for aerosol chemical composition (PM_{10} mass, inorganic, organic, metals and dust). In the first year the HVS procedure (sampling and shipping protocols included) will be tested. Positive test results will allow the chemical analysis to be performed at the ISAC-CNR.
(8) Halocarbon sampling (25 Halocarbons relevant for climate issues will be analysed in grab samples collected weekly); the steel flasks sampled in Nepal will be analysed by GC-MS (HP 6890 GC system, Agilent) at Mt. Cimone Station (Italy) by the Urbino University team. Concentrations will be referred to the SIO and UB scales.
(9) Weather station (WXT5100, Vaisala) providing six weather parameters: air temperature, atmospheric pressure, relative humidity, precipitation, wind intensity and direction.

This experimental activity relies on the expertise in scientific activities carried out at Research Stations located in high mountain and remote areas such as Mt. Cimone (Italy), Puy de Dôme (France) and Dome C (Antarctica), where the institutions noted above have been directly engaged from a long time.

3. Backward-trajectory and stratosphere-troposphere exchange forecasts

An ensemble of 5-day, three-dimensional, backward trajectories are started daily for the ABC-Pyramid site at three different pressure levels (600, 500 and 400 hPa) thanks to ETHZ. The wind fields for this Lagrangian forecast are based upon the operational daily ECMWF forecast, and the LAGRANTO model (Wernli and Davies, 1997) is used to calculate the back-trajectories. The output allows the identification of the regions from where the air masses at the receptor site originated 5 days before. In addition to the 3D path, potential vorticity and relative humidity are also traced along each trajectory. The former directly separates stratospheric from tropospheric air masses (considering the dynamical tropopause as the 2 pvu), whereas the latter is a valuable quantity to identify dry stratospheric intrusions.

4. Remote control and data acquisition

One of the major challenges in designing the ABC-Pyramid laboratory has been represented by both the autonomy and the remote control of on-line instrumentation. Dedicated, integrated state-of-the-art data acquisition and satellite communication systems have been designed. The project of the network connection is shown in Fig. 10.2. All the acquisition systems are working on two industrial computers (K2-1 and K2-2) placed in the shelter laboratory. The power supply of each

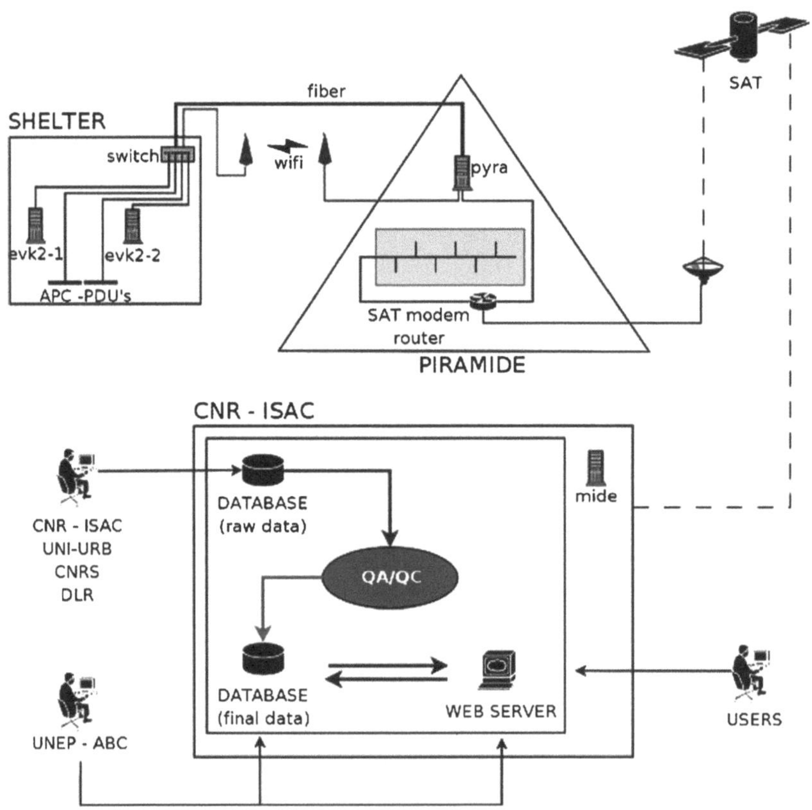

Figure 10.2. Network outline for remote control and data transmission.

instrument will be remotely controlled by two APC Switched Rack Power Distribution Units.

The two industrial computers in the shelter will be connected with the server "PYRA", located in Pyramid Laboratory, via coupled optical fibre buried 50 cm underground. An additional coupled optical fibre was planned for a backup to replace the primary connection in case of emergency. A third connection, based on wireless technology, is also present. The switch procedure between these connection systems is completely automatic with priority on the fibre cable.

All the online instruments are completely autonomous and remotely controlled, while manual interventions are necessary only for the off-line measurements (aerosol sampling on filters, greenhouse gases sampling on flasks) and for the instrumentation maintenance. The server "PYRA" will be linked through satellite connection to the server "MIDE", located at ISAC-CNR in Bologna. All the instruments will be continuously accessed directly from ISAC-CNR (Bologna) and from OPGC-CNRS (Clermont-Ferrand) for remote action. Obtained data will be stored on "MIDE" server and, after control and validation checks, will be included in a database available on web server. Quality control for instrumentation and data reduction will be performed according to EUSAAR/GAW/AGAGE procedures for aerosols and gases (GAW, 1992; WMO/GAW, 2003).

5. Test performed in Bologna

The setup of the ABC-Pyramid shelter-laboratory started at the beginning of September 2005 in Bologna. All the experimental activities that will be performed in the Himalaya were simulated at ISAC-CNR from 12 October to 5 November 2005, except for the CIMEL sun photometer and GHGs sampling. At first, the shelter laboratory was tested for water tightness. Then the racks with the instruments were installed inside the shelter laboratory. The network connection and the data acquisition systems were also tested and improved.

The whole instrumentation was working simultaneously during the period 29 October to 5 November. An example of the dataset collected during this test period is shown in Fig. 10.3, reporting only a part of the measurements carried out. In the urban area, the regional networks for air quality monitoring are usually not equipped with this kind of high technology instrumentation. Even if the main goal of our test concerned the technical setup of the Himalayan laboratory, the scientific meaning of the dataset collected during this "Bologna field campaign" represents a unique and interesting opportunity to better study the air quality in one of the main urban areas in the Po basin.

6. Conclusions

In order to observe continuous aerosol loading, chemical composition, optical and physical characteristics of aerosols, as well as greenhouse trace-gas concentration at a remote high altitude site in the Himalayan range (at 5079 m a.s.l., in the Khumbu

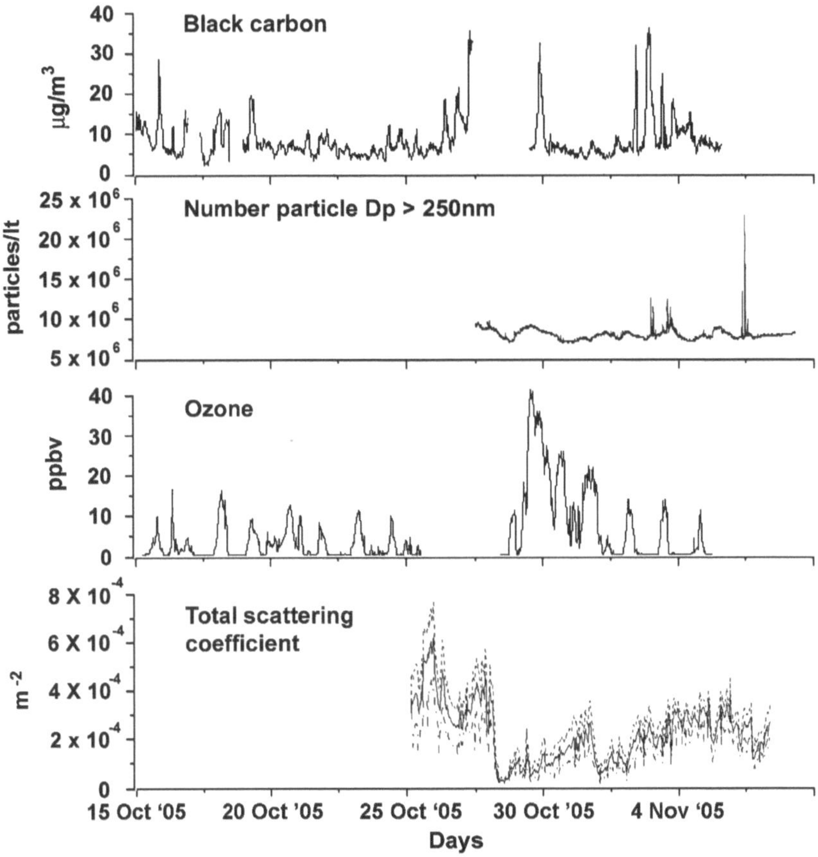

Figure 10.3. Overview of the dataset collected during the instrumentation test performed in Bologna (Italy).

Valley), a new monitoring station (the ABC-Pyramid Laboratory) has been planned to be operational since March 2006.

A quite autonomous, long-term operational monitoring station was designed to run at extremely adverse condition. Realised and tested in Bologna (Italy) in autumn 2005, this station can be controlled by remote login for online measurements while routine maintenance and offline measurements (high volume aerosol and air sampling) require the intervention of trained local people. This project leads the creation of a unique laboratory for investigating atmospheric composition changes. A team of skilled scientists and engineers from different scientific institutions was engaged to develop and manage the ABC-Pyramid Laboratory. During the test performed in Bologna the whole instrumentation and remote communication system worked properly. Actually, the open challenge is to succeed at very high altitude site, with new technical difficulties to face. By providing long-term, quality-assured data on aerosol physical and chemical properties, tropospheric ozone and other greenhouse gases in a region where little information is presently available, the Italian

Table 10.1. Parameters measured during the test performed in Bologna.

Measured parameter	Instrument	Unit
Black carbon	MAAP	$ng\,m^{-3}$
Total number of particles ($d>250$ nm)	GRIMM#190	L^{-1}
Total number of particles $10<d<500$ nm	DMPS/SMPS	cm^{-3}
Ozone	Thermo Electron 49C	ppbv
Air-temperature	Vaisala capacitive ceramic	°C
Atmospheric pressure	Vaisala capacitive silicon	hPa
Relative humidity	Vaisala capacitive thin film polymer	%
Wind speed	Vaisala ultrasonic transducer	$m\,s^{-1}$
Wind direction	Vaisala ultrasonic transducer	deg
Precipitation (30 min cumulative)	Vaisala piézoélectrique transducer	mm
Total scattering coefficient	TSI integrating nephelometer 3563	m^{-1}
Back scattering coefficient	TSI integrating nephelometer 3563	m^{-1}

SHARE-Asia Project will provide an important contribution to the ABC Asia and GAW research programmes (Table 10.1).

References

GAW, 1992. Report of the WMO Meeting of Experts on the Quality Assurance Plan for the GAW, Garmisch- Partenkirchen, Germany, 26–30 March 1992 (WMO TD No. 513).

IPCC, 2001. Contribution of working group I to the third assessment report of the intergovernmental panel on climate change. In: Houghton, J.T., Ding, Y., Griggs, D.J., Noguer, M., van der Linden, P.J., Dai, X., Maskell, K., and Johnson, C.A. (Eds), Cambridge University Press, Cambridge, UK and New York, USA, p. 881.

Lau, W.K.M., 2005. Aerosol-Hydrologic cycle interaction: a new challenge in monsoon climate research. Gewex Project Office.

Ramanathan, V., Chung, C., Kim, D., et al., 2005. Atmospheric Brown Clouds: Impact on South Asian climate and hydrological cycle. Proceedings of the National Academy of Sciences (PNAS) 102, 5325–5333.

Ramanathan, V. and Crutzen, P.J., 2003. New Directions: Atmospheric Brown "Clouds". Atmospheric Environment 37, 4033–4035.

Ramanathan, V., Crutzen, P.J., Kiehl, J.T., and Rosenfeld, D., 2001. Aerosol, Climate and the hydrological cycle. Science 294, 2119–2123.

UNEP and C4, 2002. The Asian Brown Cloud: Climate and Other Environmental Impacts UNEP, Nairobi.

Wernli, H. and Davies, H.C., 1997. A Lagrangian-based analysis of extra-tropical cyclones. I: The method and some applications. Quarterly Journal of the Royal Meteorological Society 123, 467–489.

WMO/GAW, 2003. Aerosol Measurement procedures guidelines and recommendations (WMO TD No. 1178).

Mountains Witnesses of Global Changes
R. Baudo, G. Tartari, and E. Vuillermoz (Editors)

11

The EV-K^2-CNR Pyramid and the AERONET network (Himalayan atmospheric brown cloud characterization via sunphotometer observations)

Gian Paolo Gobbi, Federico Angelini, Francesca Barnaba and Paolo Bonasoni

Abstract

A Cimel sunphotometer operating in the framework of the AERONET project has been installed at the Himalayan Ev-K^2-CNR Pyramid (5079 m a.s.l.) in the year 2006, as site Ev-K^2-CNR. The observational activity will provide a characterization of the optical and microphysical properties of atmospheric aerosols, in particular of the atmospheric brown cloud (ABC) in the Himalayan region. This paper will describe the Cimel sunphotometer measurement technique, will introduce to the AERONET programme and will evaluate the contribution of the proposed Ev-K^2-CNR AERONET site to the study of the ABC.

1. Introduction

Atmospheric aerosols are micrometer-sized particles mostly made of sulphates, nitrates, mineral dust, black carbon, ash, and hundreds of organic compounds. Their origin can be either natural or anthropogenic (e.g., d'Almeida et al., 1991). Because of their small size these particles float in the atmosphere until precipitation, or impaction removes them. Their typical lifetime is of the order of one week. When sunlight hits, absorbing aerosols, it reveals a brown-coloured haze. In the case of anthropogenic emissions, this haze has been defined as the atmospheric brown cloud (ABC) (Ramanathan and Crutzen, 2003). All clouds prevent some sunlight from reaching the Earth's surface, mainly by reflecting it back to space (e.g., Kinne and Pueschel, 2001). In addition, brown clouds can absorb the solar radiation they intercept, heating up the atmosphere (e.g., Markowicz et al., 2002; Ramana et al., 2004). Furthermore, high aerosol concentrations reduce a water cloud's ability to dissipate by turning into rain. Overall, ABC aerosols prevent sunlight from reaching the surface, warm up the air layers they dwell in, and reduce the amount of rainfall of a region. Through these mechanisms, ABCs can have significant impacts on both regional and global climate, such as temperature changes, rainfall changes, and crop

ISSN: 0928-2025 DOI: 10.1016/S0928-2025(06)10011-5

growing season alterations (e.g., Ramanathan et al., 2001; Menon et al., 2002). Last but not least, air pollution (and ABC is pollution) is a cause of premature death (e.g., Krewski et al., 2000).

Satellite data reveal thick, polluted haze layers scattered all over the globe (e.g., Kaufman et al., 2002). Mostly originating from populated regions, these emissions quickly extend to the most remote areas and oceans. Findings from the INDOEX experiment (http://www-indoex.ucsd.edu/publications/) revealed that the so-called "brown cloud phenomenon" in Asia spreads to the Himalayas and over the North Indian Ocean region, spanning an entire continent and an ocean basin (e.g., Lelieveld et al., 2001). Figure 11.1, from Kaufman et al. (2002), provides a snapshot of the way the aerosol load of densely populated regions is increased with respect to less-populated areas. Assessment of the exact level to which anthropogenic aerosols are affecting our climate is still incomplete. However, the overall (direct plus indirect) forcing might be close to balance the warming effects of anthropogenic greenhouse gases (e.g., Bellouin et al., 2005; Lohmann and Feichter, 2005).

In the year 2006, within the framework of the SHARE-Asia Project, we installed at the Himalayan Ev-K^2-CNR Pyramid a Cimel sunphotometer operating as part of the NASA Aerosol Robotic Network (AERONET). The aim of this project is to provide a remote-sensing characterization of the optical and microphysical properties of the ABC reaching the Himalayan region. In fact, such phenomena are expected to be rather important, particularly in the pre-monsoon season (Shrestha et al., 2000). The Pyramid site laboratories are also part of the ABC Asia research initiative (http://www-ABC-ASIA.ucsd.edu). The photometric observations will allow following the time evolution of the aerosol optical depth (AOD) and other properties of the cloud

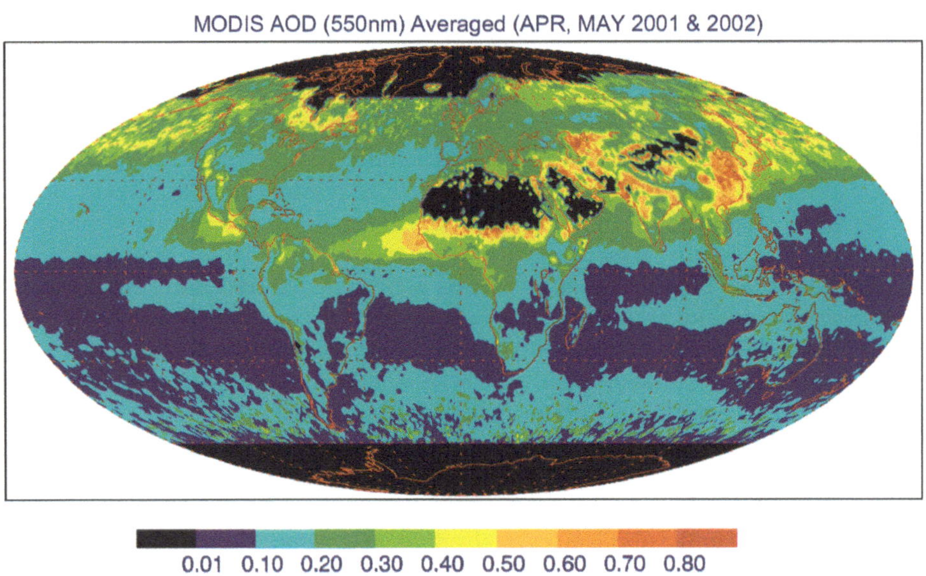

Figure 11.1. Global distribution of mean aerosol optical depth (AOD) for April and May during 2001 and 2002. The black area represents no data. The data are from MODIS instrument onboard Terra, from Kaufman et al. (2002).

as single scattering albedo (SSA, fundamental in computing the ABC effects on solar radiation), particle size distribution (SD, relevant to determination of the ABC effects on formation and lifetime of water vapor clouds) and refractive index.

2. AERONET and the Ev-K2-CNR pyramid

The AERONET programme (http://aeronet.gsfc.nasa.gov) is a federation of the ground-based remote sensing aerosol stations. The goal is to assess aerosol optical properties and validate satellite aerosol retrievals. The network imposes standardization of instruments, calibration and data processing. The standard AERONET sunphotometer is made by Cimel (http://www.cimel.fr, shown in Fig. 11.2). The instrument performs automated observations of direct solar irradiance and of sky radiance at several scattering angles (in almucantar and principal plane configurations). In fact, sky-scanning can provide extended information on aerosol optical properties (Nakajima et al., 1983; Dubovik and King, 2000; Dubovik et al., 2000). All observations are made at a number of wavelengths defined by interferential filters. The calibration of each AERONET instrument is carried out at NASA Goddard with periodicity of about one year.

AERONET observations are transmitted to NASA Goddard for processing, either via satellite link or Internet. Analysis products are then reported and freely accessible on the network website in quasi-real time. AERONET data provide observations of spectral AODs, inversion products, and rainfall in about 150 globally distributed, geographically diverse aerosol regimes. Three levels of data are available from this

Figure 11.2. A Cimel CE 318-N sun-tracking and sky-scanning sunphotometer, similar to the one installed at the Ev-K^2-CNR Pyramid. The optical head has two collimators with 1.2° FOV, one for direct sun and another for sky radiance measurement. Sun tracking is performed by a 4-quadrant detector, with accuracy better than 0.1°. Wavelength selection is made by interferential filters, in our case seven, centered at 340, 380, 440, 500, 675, 870, and 1020 nm (bandwidth 10 nm FWHM). Accuracy of calibrated instruments AOD: ± 0.01 for wavelenghts $\lambda > 440$ nm. Operating temperature: -30 to $+60$ °C. Data transfer via satellite link to NASA GSFC. Power requirements: sunphotometer 12 V–10 Ah battery; transmitter 12 V–30 Ah battery. All rechargeable by solar panels.

website: Level 1.0 (unscreened), Level 1.5 (cloud screened), and Level 2.0 (cloud-screened and quality-assured). An example of the products available on the AERONET site is reported in Fig. 11.3. In particular, observations related to the site of Kampur in northern India are shown to partly illustrate the behaviour of the aerosol in this region.

The Ev-K^2-CNR Pyramid Laboratory is located at 5079 m a.s.l. (27°57′31″ N–86°48′48″’ E), at the foot of the Nepali side of Mt. Everest. Figure 11.4 shows the location of the Pyramid on an Indian–Nepali map. On the same map are reported the Kampur station (approximately 600 km W-SW) and the new sites planned to constitute by the year 2007, the AERONET Indo–Gangetic transect. This project is intended to provide a thorough characterization of the major ABC affecting this region. The Ev-K^2-CNR AERONET site, together with the aerosol laboratory that has been installed at the same location in 2006 (Bonasoni et al., 2006) are expected to provide an original characterization of the impact of pollution on this remote region of the planet and contribute answers to some of the issues related to the effects of anthropic activities on the climate and on the environment of our planet.

In conclusion, the photometric observations performed at the Ev-K^2-CNR station in the framework of the SHARE-Asia Project are expected to answer the following questions:

- What is the climatology of aerosol physical, chemical, and optical properties at this remote, high altitude station?
- How do the radiative properties of aerosols change as a consequence of transport of ABC or other type of air masses?
- What are the ABC effects on formation and lifetime of clouds?
- How good are satellite retrievals of aerosol AOD over this region?

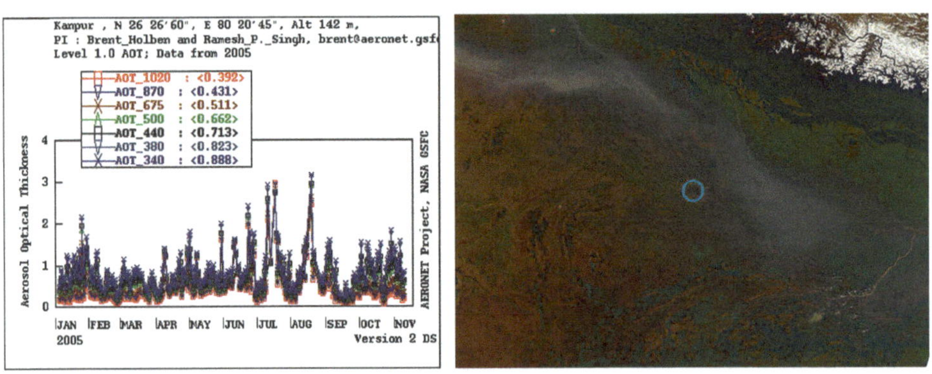

Figure 11.3. Example of three products available on the AERONET web site: (1) Spectral optical thickness of the year 2005 at the site of Kampur, India, just south of the Himalayas; (2) Modis image for the site of Kampur showing the Northern India ABC. (3) Inverted aerosol size distribution, refractive index, single scattering albedo, and asimmetry factor from Cimel observations made on November 7, 2005 at the site of Rome Tor Vergata.

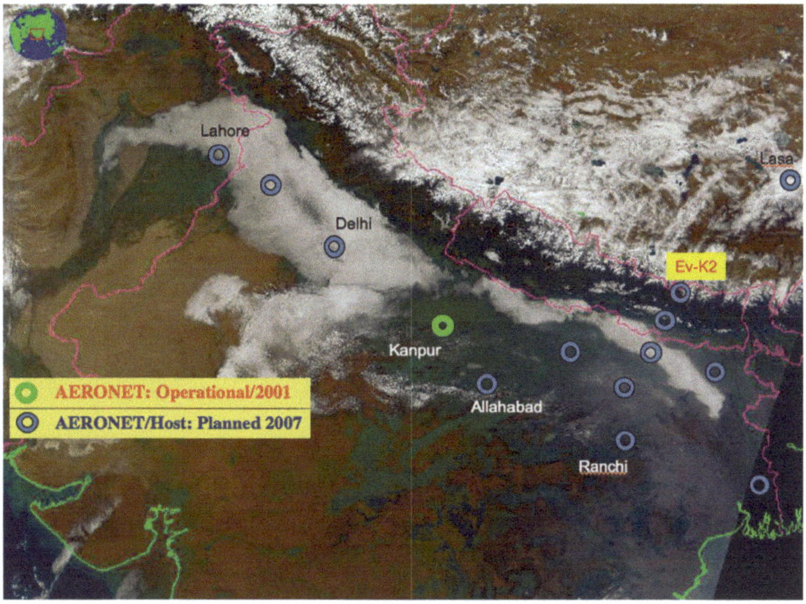

Figure 11.4. The Indo–Gangetic transects of AERONET stations planned to be operating by the year 2007 plotted on MODIS image of the winter 2004 ABC pollution affecting this region.

Acknowledgements

We are grateful to Brent Holben and Yoram Kaufman of NASA–GSFC for constructive collaboration in setting up this experiment and for providing some of the images presented in this paper.

References

d'Almeida, G.A., Koepke, P., and Shettle, E.P., 1991. Atmospheric Aerosol – Global Climatology and Radiative Characteristics. A. Deepack, Hampton, VA.

Bellouin, N., Boucher, O., Haywood, J., and Reddy, M.S., 2005. Global estimate of aerosol direct radiative forcing from satellite measurements. Nature 438, 1138–1141.

Bonasoni, P., Laj, P., Bonafè, U., et al., 2006. The ABC-Pyramid: a scientific laboratory at 5079 m a.s.l. for the study of atmospheric composition change and climate; Proceedings Conference "Mountains witnesses of global changes. Research in the Himalaya and Karakorum: SHARE-Asia Project," Rome, November 16–17, 2005.

Dubovik, O. and King, M.D., 2000. A flexible inversion algorithm for retrieval of aerosol optical properties from Sun and sky radiance measurements. Journal of Geophysics Research 105, 20,673–20,696.

Dubovik, O., Smirnov, A., Holben, B.N., et al., 2000. Accuracy assessment of aerosol optical properties retrieval from AERONET sun and sky radiance measurements. Journal of Geophysics Research 105, 9791–9806.

Kaufman, Y.J., Tanré, D., and Boucher, O., 2002. A satellite view of aerosols in the climate system. Nature 419, 215–223.

Kinne, S. and Pueschel, R., 2001. Aerosol radiative forcing for Asian continental outflow. Atmospheric Environment 35, 5019–5028.

Krewski, D., Burnett, R.T., and Goldberg, M.S., 2000. Reanalysis of the Harvard Six Cities Study and the American Cancer Society Study of particulate air pollution and mortality. Health Effects Institute Special Report, Boston, MA (http://www.healtheffects.org/Pubs/Rean-ExecSumm.pdf).

Lelieveld, J., Crutzen, P.J., and Ramanathan, V., 2001. The Indian Ocean experiment: widespread air pollution from South and Southeast Asia. Science 291, 1031–1036.

Lohmann, U. and Feichter, J., 2005. Global indirect aerosol effects: a review. Atmospheric Chemistry and Physics 5, 715–737.

Markowicz, K.M., Flatau, P.J., Ramana, M.V., et al., 2002. Absorbing Mediterranean aerosols lead to a large reduction in the solar radiation at the surface. Geophysical Research Letters 29 (20), 1968, doi:10.1029/2002GL015767.

Menon, S., Hansen, J., Nazarenko, L., and Luo, Y., 2002. Climate effects of black carbon aerosols in China and India. Science 297, 2250–2253.

Nakajima, T., Tanaka, M., and Yamauchi, T., 1983. Retrieval of the optical properties of aerosols from aureole and extinction data. Applied Optics 22, 2951–2959.

Ramana, M.V., Ramanathan, V., Podgorny, I.A., et al., 2004. The direct observations of large aerosol radiative forcing in the Himalayan region. Geophysical Research Letters, 31, L05111, doi:10.1029/2003GL018824.

Ramanathan, V. and Crutzen, P., 2003. Atmospheric brown clouds. Atmospheric Environment 37, 4033–4035.

Ramanathan, V., et al., 2001. The Indian Ocean experiment: an integrated analysis of the climate forcing and effects of the Great Indo-Asian Haze. Journal of Geophysics Research 106, 28371–28398.

Shrestha, A.B., Wake, C.P., Dibb, J.E., et al., 2000. Seasonal variations in aerosol concentrations and compositions in the Nepal Himalaya. Atmospheric Environment 34, 3349–3363.

SHARE-Asia Scientific Fields: Atmospheric Physics & Chemistry, & Global Change

Mountains Witnesses of Global Changes
R. Baudo, G. Tartari, and E. Vuillermoz (Editors)

12

Global earth observation system of systems and the coordinated enhanced observing period high altitude observatories

Toshio Koike

Abstract

The coordinated enhanced observing period (CEOP) is one of the foundations of the World Climate Research Program with affiliations to many of the world's environmental agencies, including the Global Earth Observation System of Systems (GEOSS) of the U.S. Environmental Protection Agency. The CEOP Phase 2 science framework, based on new analytical tools, is constructed for making maximum use of opportunities and for addressing CEOP guiding goals, by modifying and adding to the Phase 1 overall science objectives established in the Water and Energy Simulations and Predictions (WESP) and the CEOP Inter-Monsoons Model Study (CIMS). In addition to the monsoonal region study that CIMS is undertaking, the water and energy cycle in semi-arid and cold regions, both of which are vulnerable and sensitive to climate change and global warming, are now being targeted in the framework of WESP. To address the natural and human-induced effects on the water cycle, aerosol-water cycle interaction in the monsoonal regions will now be investigated in the framework of CIMS. It is important to aggregate from information at a reference site scale and downscale from global and regional scales to a watershed scale for making usable information. To address these issues, a watershed hydrology study, including a downscaling study, is being established in Phase 2 as a focused activity that spans WESP and CIMS and provides linkage to water resources studies. Two cross-cutting activities, namely a CEOP analysis intercomparison project, and a project for impact analysis of extreme events for increased understanding of hydroclimate processes and improving model predictability, are being introduced to address certain basic aspects and to synthesize other elements that are common to the CEOP objectives. During the first two years of CEOP Phase 2, 2005–2006, CEOP has made efforts for accomplishing data collection and science targets for the CEOP Phase 1 and has prepared for establishment of a reference basin network in addition to reference site upgrading, improvement of the CEOP data system, and preliminary studies on the additional science targets. During the following four years, 2007–2010, CEOP plans to implement

ISSN: 0928-2025 DOI: 10.1016/S0928-2025(06)10012-7

its Phase 2 enhanced observation period and address the CEOP Phase 2 science targets, simultaneously. CEOP will also take steps toward the establishment of the GEOSS in situ observation network, as well as data integration and an information fusion system for the water cycle, in an experimental way. At the same time, CEOP will receive benefits from GEOSS for accomplishing its own science objectives. In this manner, the CEOP Phase 2 observation and data integration system will make the transition into an element of the GEOSS operational framework. This holistic approach will generate shared scientific targets with existing scientific projects and programs in due course. CEOP will cooperate with and support these activities by contributing the two unique functions established in Phase 1 and working jointly to exploit the opportunity these tools provide.

Mountains Witnesses of Global Changes
R. Baudo, G. Tartari, and E. Vuillermoz (Editors)

13

The coordinated enhanced observing period (CEOP) report: integrated data systems in the study of the water cycle in Asia

Sam Benedict

Abstract

CEOP has made important progress towards the realization of its long-term guiding scientific goal: "To understand and model the influence of continental hydroclimate processes on the predictability of global atmospheric circulation and changes in water resources, with a particular focus on the heat source and sink regions that drive and modify the climate system and anomalies." The scientific objective for the CEOP Water and Energy Simulation and Prediction (WESP) working group is to use enhanced observations to diagnose, simulate and predict water and energy fluxes and reservoirs over land on diurnal to annual temporal scales as well as apply these predictions for water resource applications. The CEOP Monsoon Studies Working Group aims to achieve another CEOP science objective: to document the seasonal march of the monsoon systems, assess their driving mechanisms, and investigate their possible physical connections. The CEOP Inter-monsoon Model Study (CIMS) has been undertaken to assess, validate and improve the capabilities of climate models in simulating physical processes in monsoon regions around the world. CIMS and WESP are demonstrating the utility of CEOP data for better understanding of the regional and global water cycle and for model physics improvement. The CEOP Data Management, Satellite Data Integration and Model Output Development and Management Working groups are attaining the CEOP goal to provide a database of common measurements from both in situ and satellite remote sensing measurements, as well as matching model output that includes Model Output Location Time Series (MOLTS) data along with four-dimensional data analyses (4DDA; including global/regional reanalyzes) for a specified period. By setting these goals, CEOP is responding to the challenges and priorities that relate to variations in the Earth's water and energy budgets and the cycling rate of the hydrological cycle as posed by the International Panel on Climate Change (IPCC) and will contribute to the development of WCRP COPES and the GEO, GEOSS.

ISSN: 0928-2025 DOI: 10.1016/S0928-2025(06)10013-9

1. Overview

Objective(s): CEOP's guiding goal is:

> To understand and model the influence of continental hydroclimate processes on the predictability of global atmospheric circulation and changes in water resources, with a particular focus on the heat source and sink regions that drive and modify the climate system and anomalies.

The key science objectives of CEOP are associated with its Water and Energy Simulation and Prediction (WESP) and Monsoon Systems Studies activities. These include respectively: (1) to use enhanced observations to diagnose, simulate and predict water and energy fluxes and reservoirs over land on diurnal to annual temporal scales as well as apply these predictions for water resource applications; and (2) to document the seasonal march of the monsoon systems, assess their driving mechanisms and investigate their possible physical connections.

2. Status

CEOP has made progress in the establishment of two sets of unique functional components:

- components to integrate observations based on coordination among field science groups, space agencies and numerical weather prediction (NWP) centers in the local, regional and global scales;
- components required to exchange and disseminate observational data and information including data management that encompasses functions such as Quality Assessment/Quality Control, access to data and archiving of data, data integration and visualization, and information fusion.

In this context, one of the key achievements of CEOP, in this reporting period, has been to greatly refine the establishment of an integrated observation system by combining different types of observations, in situ and satellite. In addition, the outputs from the NWP model are merged with the observed data to provide spatially and temporally continuous coverage in a complementary way. The coordinated enhanced observation and model output generation were carried out during the first Enhanced Observing Period (EOP-1) (July–September 2001), the EOP-3 (October 2002–September 2003), and the EOP-4 (October 2003–December 2004).

2.1. In situ data issues

Key agreements were reinitiated to maintain the provision of in situ data from 35 selected globally distributed "reference" stations mainly involved in the Global Energy and Water Cycle Experiment (GEWEX), Hydrometeorological Panel (GHP) and Continental Scale Experiments (CSEs). These reference sites provide enhanced observations of sub-surface (soil profiles), surface (standard meteorological and radiation), near surface (flux tower), atmospheric profiles (rawinsonde and profiler) and ancillary data sets (radar, special observations) in a common format.

Collection of the data from the CEOP reference sites for the initial CEOP period has shown that adherence by the sites to a consistent format is especially important to ensure an efficient continuation of the CEOP data set development and delivery process. The CEOP Reference Site Data Management Internet page is at: http://www.joss.ucar.edu/ghp/ceopdm/. Data are deposited at this site through the CEOP Central Data Archive (CDA) at NCAR.

2.2. *Satellite observations*

Almost all components of the water cycle among atmosphere, land and ocean can be observed by currently available satellite sensors. CEOP has, therefore, taken the essential step of integrating the satellite products for generating new data sets of the overall water cycle. The work associated with satellite data set development and integration that was undertaken by CEOP during this reporting period has progressed as planned. Data set documentation and background information has been made available at the following Internet page: http://monsoon.t.u-tokyo.ac.jp/camp-i/doc/sat_info/index.htm. A contribution by Japan Aerospace Exploration Agency (JAXA) in coordination with the University of Tokyo has been established, which applies to an on-going effort to provide CEOP satellite data sets for integration with the CEOP in situ and model output data. The data set consists of the main water cycle parameters necessary to accomplish CEOP scientific goals; these data are geo-coded (i.e. re-sampled to a regular LAT/LON Grid). They are generated at three scales, 250 km rectangular, monsoon regional and global scales, associated with product levels 1b, 2 and 3. The processing levels have also been defined to ensure a clear understanding of the nomenclature and reduce ambiguity in the statement of requirements. The levels of processing have been established to be:

[Level-1b] – Radiance product with full resolution at reference sites.
[Level-2] – Geophysical product at the same resolution at reference sites and monsoon regions.
[Level-3] – Statistical geophysical product in space and/or time at reference sites, monsoon regions and global. (e.g., Monthly mean rain rate at reference sites, etc.)

They consist of an image element and a metadata part element that are compliant with the ISO-19115 standard.

2.3. *NWP model outputs*

The basis of the NWP model output component of CEOP, which was established through a letter sent in September 2001 to NWP and data assimilation centers worldwide, was expanded and extended during this reporting period. The initial letter requested that the NWP centers archive specific model output data for CEOP during the CEOP EOP-3, -4 time period from 1 October 2001 through 30 September 2004 (later extended through 31 December 2004 to match the extension of EOP-4). The letter also requested both analysis/assimilation and forecast model products from both global and regional NWP suites, and from both operational and reanalysis systems.

The new letter, sent in 2005, asked that this work be extended through the end of CEOP Phase 2. The response to the request from WCRP and CEOP for this extension to the support of the CEOP model output component was a special success for the start of CEOP Phase 2 implementation. All nine operational NWP and two data assimilation centers that are currently contributing to this component of CEOP are, therefore, expected to continue to do so at least at the current level of commitment through 2010. To assist with the organization of this activity, a CEOP Model Output Management Document was drafted as a guide for the participating centers to use in setting up their processes for meeting their commitments to CEOP.

The Max-Planck-Institute for Meteorology with the International Council for Science (ICSU) World Data Center for Climate (WDCC) in Hamburg, Germany, which earlier undertook to serve the CEOP model output archive center, has also agreed to continue in this capacity through CEOP Phase 2. Results up to this point in the CEOP model output generation effort make it clear that the transfer aspect of the data handling effort has been progressing well. Data from all 11 Centers (NCEP, UKMO, NASA–GMAO, NASA–GLDAS, JMA, BMRC, ECMWF, NCMRWF, ECPC, CPTEC–INPE and CMC) participating in CEOP have been received at the data archive center and have either been placed into the database at the Hamburg facility, or are in the process of being entered into the database. The gateway to the CEOP Model database can be found at: http://cera-www.dkrz.de/CERA/cera2browser_-CEOP/index.html.

2.4. CEOP data integration activities

The total amount of the CEOP Phase 1 data is estimated to be around 300 Terabytes. As originally produced by the various sources, the data were in a wide variety of formats and structures and CEOP began addressing possible solutions to this important issue. In response to this situation there was recognition of need for data management systems for the collection, sharing and provision of data from which users can obtain precisely the data they need, whenever they want it and in formats familiar to the science community. It is essential to transform observation data into scientifically and socially relevant information through the systematic collection and integration of data, merging of essential related information and building of systems for sharing this knowledge on an international basis.

Recently, the University of Tokyo, JAXA and the Committee on Earth Observation Satellites (CEOS) have, therefore, begun working together to create a distributed "data mining" system for the CEOP data archive. CEOS membership encompasses the world's government agencies responsible for civil Earth Observation Satellite (EOS) programs. Within CEOS, the Working Group on Information Systems and Services (WGISS) is responsible for systems and services that manage and supply the data and information from participating agencies' missions. The purpose of the Distributed Data Integration System, known as the WGISS Test

Facility for CEOP (WTF–CEOP), is to support development of data services associated with data integration. WTF–CEOP provides metadata and documentation on CEOP data and supports browse and review of the data, including various data integration functions and services. In addition, CEOP scientists are invited to submit requests for specialized processing of CEOP data to support their research. The system is based on server technology from the Distributed Oceanographic Data System (DODS) (now known as OPeNDAP). WTF–CEOP uses Live Access Server (LAS) software to provide menus to select the data variable, location, time range and other factors. After menu selections are complete and the data type has been chosen, LAS will access the data via the DODS server at the appropriate data archive center. The Ferret client software then performs data processing functions such as data plots, on screen display of the data values, and downloading the data in ascii, spreadsheet and NetCDF formats. The LAS software was designed so that it was possible to add other analysis software.

In addition to the Distributed System, a Centralized Data Integration System has been developed at the University of Tokyo that utilizes the centralized CEOP data archive sited at the Institute of Industrial Sciences, the University of Tokyo and allows handling of significantly larger amounts of heterogeneous observation data. The browse and analysis interface is performed by dedicated clients, which provides the users with menus, integrated access to the data and analysis tools. The connection between the clients and the server is based on HTTP. Users can access all types of data through a single interface and can view the retrieved data as graphic charts or bitmap images, depending on their dimensionality. Some analysis operations such as average, difference, correlation and visualization can be applied to single or multiple data types through the interface.

Prototypes of both the Centralized Data Integration System and the Distributed Data Integration System were opened to the CEOP community on 1 June 2005. They are accessible, along with more information about what services they provide, at the websites: http://jaxa.ceos.org/wtf_ceop/ and http://monsoon.t.u-tokyo.ac.jp/ceop-dc/ceop-dc_top.htm for Distributed and Centralized Systems respectively.

3. New directions (longer term vision)

During its deliberations at the Fourth CEOP International Implementation and Planning Meeting (March 2005) the CEOP Advisory and Oversight Committee (AOC) and the CEOP Science Steering Committee (SSC) endorsed a number of CEOP goals and objectives for the future that includes:

- A CEOP Phase 2 plan and schedule (as presented at the meeting) that proceed in two stages that run from 1 January 2005 to 31 December 2010 extending existing data and observation processes and adding greater emphasis on the research and analysis components of CEOP, providing for CEOP to meet its commitments to CEOS/IGOS–P Water Theme, WCRP/COPES and GEOSS.
- Selection of the diurnal cycle as a unifying scientific theme in CEOP Phase 2.

- Phase 2 plans to continue to select a small set of hydrological reference sites, from the existing CEOP reference sites that will have dual roles as validation sites for the land-surface parameterizations in coupled land–atmosphere–ocean models, essentially at a point or small area scale and as "tie points" or ground validation reference sites for remote sensing products.
- Phase 2 plans to exploit progress in WESP on the recycling of moisture associated with local evaporation in affecting regional water budgets as a means for CEOP to contribute towards, what may be a larger goal associated with examining some inter-connectivity between land areas including the middle and higher latitudes, and in general towards better understanding of some of the means through which land plays a role in the larger climate system.
- Phase 2 plans to entrain a team of researchers embarking on a Worldwide Integrated Study of Extremes (WISE) that focuses in a collective manner on extremes during the CEOP period as an extension of the work currently under way in CEOP WESP.
- CEOP Phase 2 plans to undertake a joint initiative with the WCRP Climate and Cryoshpere CEOP/CliC project and the International Polar Year (IPY) initiative that will exploit a number of CEOP reference sites in regions where snow occurs to provide the required data to begin and complete this project in Part 2 of CEOP Phase 2 (2007–2010).
- Phase 2 plans to exploit CEOP reference sites located in semi-arid regions to produce data that can be applied using capabilities now existing in WESP to better understand variations in water and energy budgets in such regions.
- Plans to augment and formalize the CEOP international coordination function to provide more consistent collaboration with national and international funding agencies and relevant research groups related to organizing and managing CEOP.

4. Future activities

Data from the carefully selected CEOP reference stations will continue to be delivered to the CEOP CDA at UCAR/JOSS. The CEOP sites will continue to be closely involved with the existing network of observing sites in the GEWEX Continental Scale Experiments (CSEs) distributed around the world.

As planned, model products will be received from the major National and Multi-National Numerical Weather Prediction Centers around the globe, and a CEOP model products archive and distribution center operated by the Max Planck Institute (MPI) will continue handling the data. Also, during the next reporting period the CEOP Satellite Data Integration Center at the University of Tokyo (UT) will receive and store satellite data. Data sets from ESA will be added to the current database in 2006.

The network that links the CEOP reference site, model and satellite data archives developed with joint JAXA and NASA support will continue to evolve in 2006. Because of a NASA funded proposal that was accepted in 2005 the CEOS, WGISS, will continue to develop the CEOP WGISS Test Facility for the purpose of integrating the CEOP data centers and making the data available to the broader user community. The collection, archive and distribution of coordinated, quality checked, in situ, satellite and model data sets are expected to lead to success in meeting CEOP science goals. With these new data sets and integration/visualization tools, CEOP WESP will examine the vertical structure in the atmosphere and the impact of land processes on closing and simulating vertically integrated water and energy budgets with observations and analyses at global scales.

In the CEOP Inter-monsoon Study (CIMS), a hierarchy of models including general circulation models (GCM), regional climate models (RCM) and cloud resolving models (CRM) will be examined to run numerical experiments that target simulation of fundamental physical processes and lead to identification of basic errors and biases in model physics. Because of the funding by NASA of a specific CEOP related proposal in 2005, the CEOP Model Output Working Group plans to apply CEOP data sets/tools to specifically focus on the ability of current global data assimilation systems, individually and in ensemble, to reproduce all of the components of the water and energy cycles (precipitation, evaporation, transports, water and energy content, and radiation). They will investigate processes related to the diurnal cycle and seasonal progression (e.g. monsoons).

CEOP will increase its interaction with the elements of WCRP and other international organizations and efforts that are focused on the measurement, understanding and modeling of water and energy cycles within the climate system. This includes holding joint implementation and planning meetings with the Integrated Global Water Cycle Observations (IGWCO) theme within the framework of the International Global Observing Strategy Partnership (IGOS–P) in future years. Since the ad hoc Group on Earth Observations (GEO), has prepared a 10-year Implementation Plan for a comprehensive, coordinated and sustained Global Earth Observation System of Systems (GEOSS), in 2006 CEOP will attempt to undertake a number of activities that are related closely to work that is being defined in GEOSS so that CEOP can also be viewed as an example of a coordinated activity in support of GEOSS.

5. Key results

There has been a great deal of discussion and work on using CEOP data for science applications particularly in three areas where specific examples can be provided:

(1) The CEOP initiative on focusing model "validation" against the reference site data has a web page where some of the initial work in this area was presented for reference by the contributors to the model output component of CEOP. The page is at: http://monsoon.t.u-tokyo.ac.jp/ceop/model/telecon/. This page has been set up by the CEOP Secretariat at Tokyo. Several of the outcomes from these studies have already been published in CEOP Newsletters and presented at international meetings such as the American Geophysical Union spring and fall meetings last year.

(2) On a broader scale there is work that is just now getting started that includes multiple models run alone and in ensembles. This effort is being spearheaded by

CEOP Working Group Chairs/Co-Chairs. The results have already shown the ability of several current global data assimilation systems, individually and in ensemble, to reproduce all of the components of the water and energy cycles (precipitation, evaporation, transports, water and energy content, and radiation). The analysis data have been collected in conjunction with CEOP. An article in the CEOP Newsletter has shown the results and has increased international involvement so that the effort is now being expanded into a CEOP-wide approach to model improvement.

(3) On the science objectives CEOP has now had four workshops on Monsoon Systems studies. Several articles have been published in the CEOP Newsletter on the application of CEOP data to improved understanding of monsoon characteristics.

(4) From the WESP group an interesting science effort is under way called the CEOP Inter-CSE Transferability Study (ICTS). The home page for that is at: http://w3.gkss.de/ICTS/index.html. In the ICTS, regional atmospheric climate models (RCM) from different Continental Scale Experiments (CSEs) are being transferred from their "home" CSE to other CSEs involved in GEWEX. The models are being initialized and forced at their boundaries by several state-of-the-art Global Circulation Models (GCMs). At http://www.joss.ucar.edu/ghp/ceopdm/model/model.html one can find a list of global analyses data and associated data centers. For evaluation CEOP data from the CEOP reference site data archive and the CEOP satellite data archive are being considered. Main emphasis is on the energy and water cycle components.

(5) CEOP has submitted a pre-proposal on a CEOP Polar Observations Project to be undertaken in Part 2 of CEOP Phase 2 that will be undertaken jointly with CliC and IPY in 2007–2008. Help and inputs will be sought on data sites in the higher latitudes to form the basis for meeting the science objectives.

(6) Another new thrust will be related to "Extremes" that has begun and which will be expanded in the coming year.

6. List of key publications

- CEOP Brochure March 2002;
- CEOP Newsletters (Total of eight from January 2002–August 2005);
- CEOP article of the WMO Bulletin in the April 2004;
- Thirty-eight technical Papers presented at the CEOP 4th International Implementation and Planning Meeting March 2005. Fifteen of these and related technical materials are being readied for publication in a special CEOP issue of the JMSJ in 2006.

Mountains Witnesses of Global Changes
R. Baudo, G. Tartari, and E. Vuillermoz (Editors)

14

Verification of numerical model forecasts of precipitation and satellite-derived rainfall estimates over the Indian region: monsoon 2004

Laura Bertolani and Raffaele Salerno

Abstract

This work describes the preliminary results of a study aimed at: (1) assessing the ability of a general circulation model routinely run at the Epson Meteo Centre (CEM) in predicting daily rainfall; (2) evaluating the performance of satellite-derived precipitation estimates (namely, NOAA CPC CMORPH) over the same domain and during the same period. The CPC daily rain gauge analysis is used as reference for validation. The study focused on the Indian Monsoon during summer 2004, and comparison with a similar analysis at the mid-latitudes is also shown.

1. Introduction

Precipitation is one of the most difficult weather elements to predict. It depends on many physical and dynamical processes, such as large-scale motion of moist air, orographic lifting, convection. Since all these processes are represented in Numerical Weather Prediction (NWP) models, the quality of the model predicted precipitation is often used as a critical indicator of the overall model skill (Ebert et al., 2003).

Near real-time satellite-derived estimates of precipitation, which are becoming increasingly available in recent years, provide a valuable tool to the scientific community for analysing and investigating the physical processes associated to the water cycle, due to their high spatial and temporal resolution over large areas. There are many fields of application for these measurements, (e.g. flood warning, nowcasting techniques, water resources monitoring, data assimilation and verification of NWP forecasts, especially in regions where other observing systems are not available), and many validation/intercomparison efforts are trying to properly assess the accuracy of the algorithms and to compare the products with numerical model forecasts over various domains and at different spatial resolution (e.g. Ebert, 2004; Janowiak, 2004; Kidd, 2004). These studies showed that quantitative precipitation forecasts (QPFs) usually perform best for mid-latitudes, during the cool season and

ISSN: 0928-2025 DOI: 10.1016/S0928-2025(06)10014-0

in case of large-scale rainfall; at the same time, QPFs overestimate light and moderate precipitation, while underestimate the heaviest amounts. On the opposite, satellite-derived precipitation estimates perform best for tropical, warm season and convective rainfall, underestimate light precipitation, and usually overestimate heavy rainfall. Moreover, the satellite-based techniques show difficulties in detecting solid precipitation, drizzle, orographic rain and precipitation over cold surfaces. Thus, NWP models outperform the satellite algorithms in the cool season and at high latitudes.

The motivation of this work was to study and to evaluate the capabilities of a global circulation model (CEM-GCM) in predicting precipitation on a daily basis (up to 3 days), when and where important thermal and dynamical processes are involved over a large domain, in comparison with satellite-derived precipitation estimates (an alternative source of rainfall information for verification purposes over sea, land and remote areas), and during the coordinated enhanced observing period (CEOP) Phase I Project. Thus, this study focuses on the Indian Monsoon during summer 2004.

2. Data and methods

The CEM-GCM is based on the 1997 National Centers for Environmental Prediction (NCEP)/Experimental Climate Prediction Center (ECPC) Global Spectral Model (GSM, Roads et al., 1999) and runs at T126 (horizontal resolution of about $1^\circ \times 1^\circ$) and with 28 unevenly spaced sigma vertical levels. The model physics includes the Simplified Arakawa-Shubert convective scheme. The two-layer Oregon State University Land Surface Model (LSM) is employed, and heterogeneous boundary conditions are used at the surface (14 vegetation classes and 12 soil types). The initial conditions are provided by the NCEP $1^\circ \times 1^\circ$ global gridded analysis (GDAS).

The satellite precipitation estimates considered for validation are $0.5^\circ \times 0.5^\circ$ lat/lon gridded data available every 3 h, derived from blended passive microwave (PMW) and Infrared (IR) data combined with the National Oceanic and Atmospheric Administration (NOAA) Climate Prediction Center (CPC) Morphing method, the so-called CMORPH (Joyce et al., 2004), which uses cloud motion and evolution derived from IR imagery to "morph" the PMW rain rates between successive scans. Both model forecasts and satellite-derived estimates were validated against the NOAA CPC daily rain gauge analysis available at $0.5^\circ \times 0.5^\circ$ lat/lon resolution.

The analysis was carried out interpolating all data to a common $1^\circ \times 1^\circ$ grid, and daily 06-06 UTC accumulations (the daily time interval used by the reference for verification over the Indian region) were used. The model was initialised at 12 UTC, with output every 6 h and forecast length of 90 h; accumulated daily precipitation up to 3 days was then computed (thus, the first day of forecast (day 1) included precipitation accumulated from +18 to +42 h, the second day (day 2) included precipitation accumulated from +42 to +66 h, and the third day (day 3) included rainfall accumulated from +66 to +90 h).

The main recommendations for the verification and intercomparison of QPFs from operational NWP models, provided in December 2004 by the World Weather

Research Program/Working Group on Numerical Experimentation (WWRP/WGNE) Joint Working Group on Verification (JWGV), were strongly taken into account to carry out this first verification effort over the Indian region. The WWRP/WGNE JWGV highly recommended statistics (continuous and categorical statistics) were computed for the domain 5°–27°N/70°–90°E, land points only, and considering only grid boxes with at least one rain gauge inside. Unfortunately, more than 50% of the CPC daily rain gauge analysis data were not available for June; therefore this month, which represents the onset phase of the Asian Summer Monsoon, was not included in the study. The period examined was then July–September 2004.

3. Verification of results

The seasonal mean precipitation rate over India from CEM-GCM, CMORPH and CPC daily rain gauge analysis is depicted in Fig. 14.1. The region with heavy precipitation along the West coast, due to topographic forcing, the rain shadow leeward of the Western Ghats, and the eastern precipitation area, associated mainly with dynamical forcing, are quite well reproduced by the model. Anyway, a severe overestimation in rainfall amount and a spreading in rainfall distribution are evident, especially at day 1, and over the Arabian Sea (comparing with CMORPH), along the Western Ghats, the Bay of Bengal and the southern slope of the Himalaya. On the other hand, CMORPH shows a better agreement with the reference for validation, both in terms of amount and pattern, except for the Western Ghats, where the satellite-derived product tended to underestimate rainfall.

The visual evaluation was confirmed and quantified by the time series of precipitation rate, RMSE (Fig. 14.2a and b, respectively), and mean error (not shown) computed for the investigated area.

The model reproduced quite well the intraseasonal variability (active and break phases) of the monsoon, but systematically overestimated precipitation amounts, especially in day 1, when rainfall amounts are almost double to the observed one (a seasonal predicted mean of 12.6 mm d^{-1}, against an observed mean of 6.2 mm d^{-1}). Day 2 and day 3 scored very similar values (10.2 and 10.5 mm d^{-1}, respectively). CMORPH, on the contrary, sometimes showed a light underestimation in daily precipitation (seasonal mean of 6.1 mm d^{-1}) and achieved a better correspondence with the gauge analysis. As expected, satellite-derived rainfall estimates exhibit the best performance (lowest RMSE), followed by the second day of model forecast. Figure 14.3 shows categorical statistics as a function of an increasing rainfall threshold.

The Bias Score (BS) gives the ratio of the predicted/estimated precipitation frequency to the observed rainfall frequency, without any regard to the correspondence between the rainfall patterns, and it is used to assess the tendency to over- or under-predict/estimate rain occurrence. Clearly, the model overpredicted ($BS > 1$) rainfall frequency for all the thresholds and forecast lead times, especially in the day 1 and for moderate precipitation (more than double with respect to observation), while the heaviest amounts did not show any evident tendency. Day 2 and day 3 had the same trend, but with lower values. CMORPH showed a very light underestimation in the

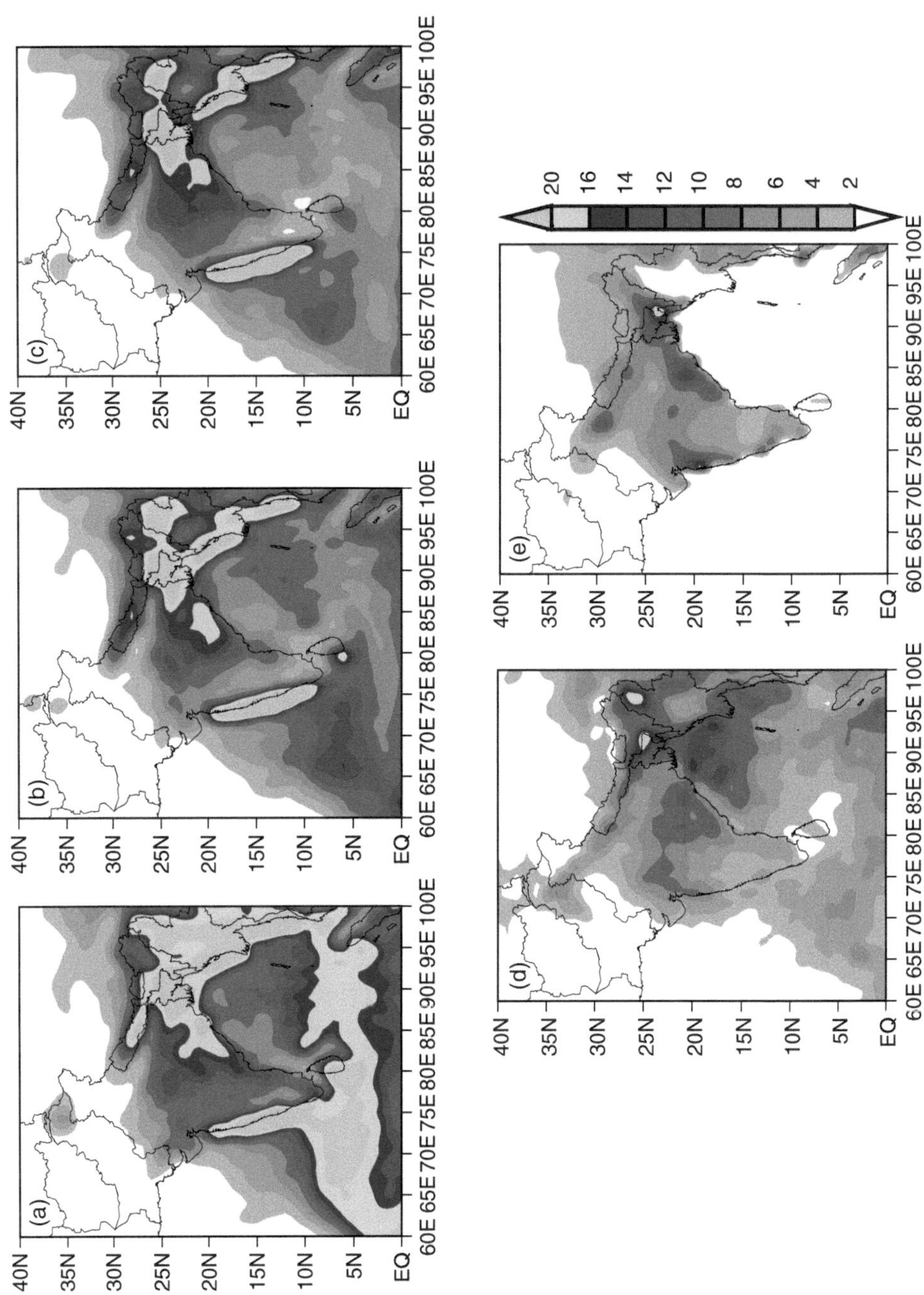

Figure 14.1. Seasonal (JAS) averaged precipitation rate (mm d^{-1}) for CEM-GCM (a: day 1; b: day 2; c: day 3), CMORPH (d) and gauge analysis (e).

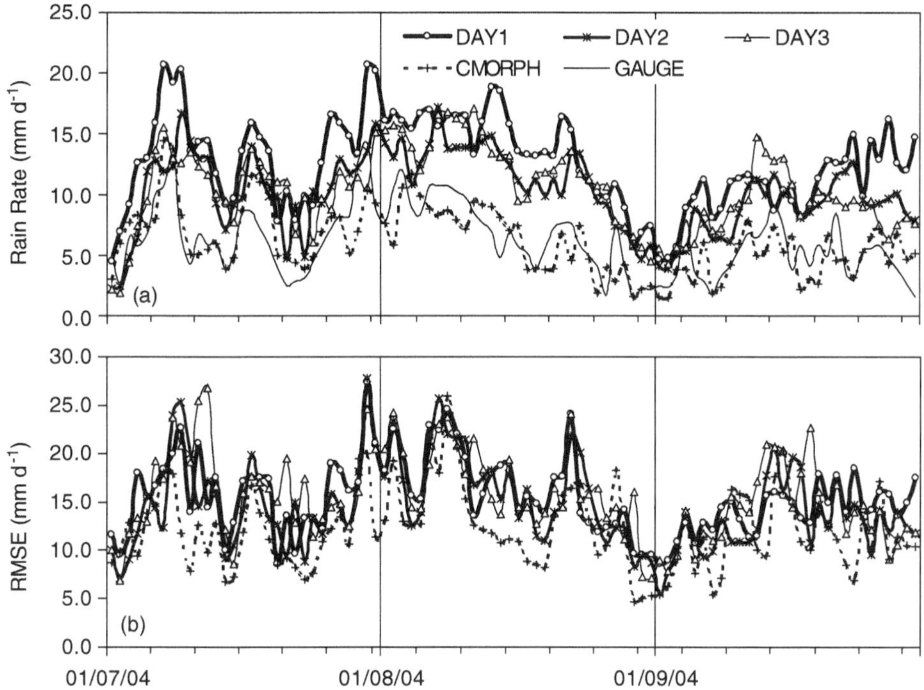

Figure 14.2. Time series of averaged precipitation rate (mm d^{-1}) for CEM-GCM, CMORPH and gauge analysis, and root mean square error (mm d^{-1}) for the model and the satellite-derived estimates.

frequency of light precipitation, and a very light overestimation for moderate precipitation.

The probability of detection (POD) is the fraction of observed rain events that are correctly predicted/estimated and a high value indicates a good ability to capture the occurrence of the rainfall events. In this case, POD indicated that the model correctly forecasted more than 50% of the observed precipitation for amounts lower than 20 mm d^{-1}, while CMORPH captured more than 50% of the observed rainfall for amounts below 10 mm d^{-1}.

The false alarm ratio (FAR) measures the fraction of the incorrect predicted/estimated rain detections. For perfect prediction/estimation, this score should be 0.0. In the studied case, more than 50% of all model predicted rain above 5 mm d^{-1} was incorrect (due to false alarms), while more than 50% of all satellite-derived rainfall estimates above 10 mm d^{-1} were incorrect.

The Equitable Threat Score (ETS) measures the fraction of all the rain events that were correctly predicted/estimated taking into account the hits that would occur purely due to random chance. Its "equitability" allows scores to be compared across different weather regimes, and for this reason ETS is one of the most used scores in the verification of QPFs. This score ranges from –1/3 to 1, with the limit of 1.0 indicating perfect correspondence between predicted/estimated and observed precipitation occurrence. The ETS clearly showed that the highest detection skill

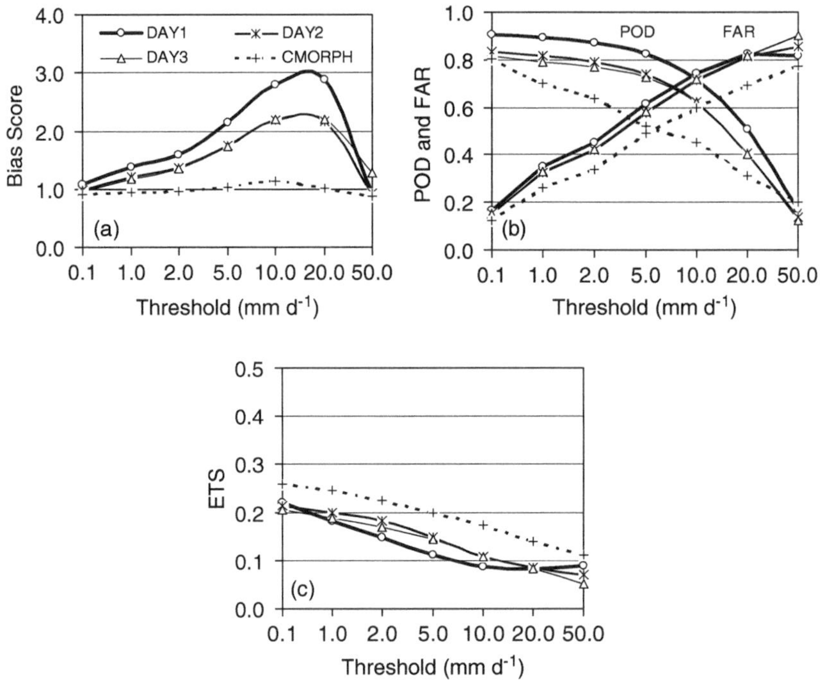

Figure 14.3. Bias Score (a), POD and FAR (b) and ETS (c), as function of precipitation threshold, for CEM-GCM and CMORPH, during the monsoon season (JAS) 2004 over India.

was achieved for the lighter amounts in both CEM-GCM and CMORPH, with higher values by the satellite-derived product, then skill decreased with increasing threshold.

Figure 14.4 shows the same analysis illustrated in Fig. 14.3, but depicts a completely different scenario: mid-latitudes (Europe 36°–56°N, 10°W–25°E) during Fall (SON) 2004, so when "stratiform" large-scale precipitation dominates and snowfall events can occur. Also in this case the CEM-GCM overpredicted the rainfall frequency, especially for moderate amounts, but with lower values then the Indian summer monsoon case, and without differences among the three forecast lead times. CMORPH, on the contrary, strongly underestimated precipitation, especially at the higher latitudes (above 50°N), for all the amounts, mainly due to snowfall events occurring in late November over Central and Northeastern Europe. In this case CEM-GCM outperformed CMORPH for all the thresholds, especially with light precipitation and with the first day of forecast.

4. Summary and conclusions

The CEM-GCM reproduced well the rain distribution over the Indian regions, but overestimated precipitation amount over the whole domain, particularly over the Arabian Sea, windward of the monsoon flow, and, interestingly, with the first day of

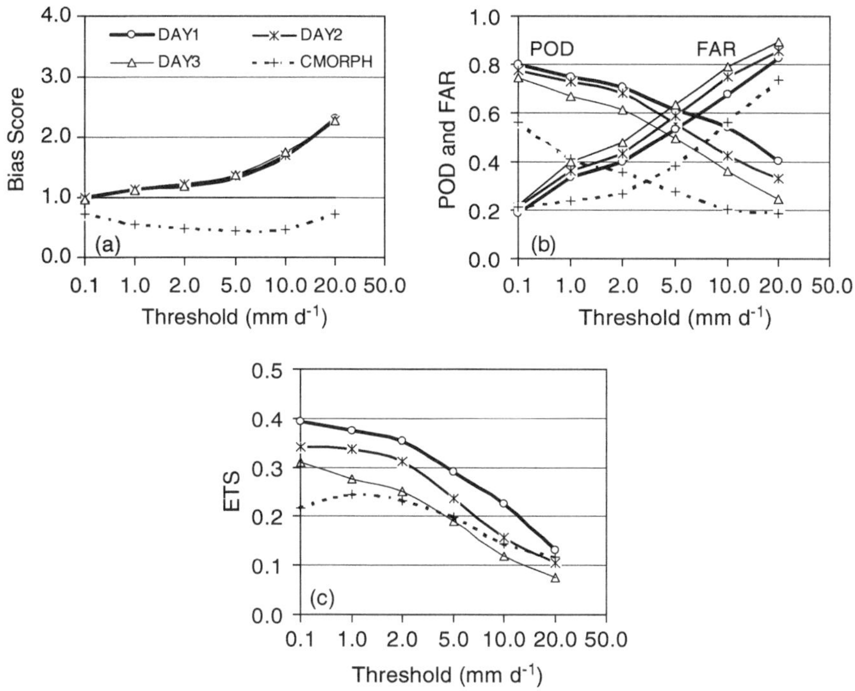

Figure 14.4. Bias Score (a), POD and FAR (b) and ETS (c), as function of precipitation threshold, for CEM-GCM and CMORPH, during Fall (OND) 2004, over Europe.

forecast. This last evidence was not found in mid-latitudes during Fall (when "stratiform" large-scale precipitation dominates), and was probably due to a prolonged "spin-up" time, associated to a widespread area with strong instability and deep convective activity, particularly enhanced at the initial time used to run the model (12 UTC, late afternoon in the Indian Monsoon regions). The duration of the spin-up depends on the model and on the domain for which precipitation is calculated (Krishnamurti et al., 1988); about this topic, Arpe (1991) demonstrated that the northern hemispheric extratropics ECMWF short-range forecasts of precipitation were accurate estimates of the truth for the day to day variability, while the tropics and southern hemisphere showed a realistic distribution, but with a strong spin-up in the data. Moreover, a well-known problem found in this kind of model (especially the NCEP/GFS T170L42, but also with other resolutions), consists of producing the so-called "precipitation bombs", which are erroneous grid-scale convective systems occurring in presence of enhanced low-level moisture convergence, vertical shear and conditional instability (see http://meted.ucar.edu/nwp/pcu3/cases/DVN_082202_flood/frameset.htm). Besides, further investigations are needed to clarify this point, and improvements to the convective scheme, the LSM, as well as data assimilation could reduce this problem.

CEM-GCM overpredicted the frequency of all rain amounts, particularly those exceeding 10–20 mm d^{-1}, and at the first day of forecast, except the heaviest ones (exceeding 50 mm d^{-1}), which did not show any evident trend. Moreover, the more the rain amount increased, the more the ability of the model ability decreased, due to the increasing amount of false alarms and to the greater difficulty in correctly predicting the location of the heaviest rain. The second day of forecast was overall better than the first and third ones, except for the heaviest amounts, as it probably provided the best compromise between spin-up and forecast error. CMORPH showed a better agreement with the reference for validation, both in terms of amount and occurrence, except along the Western Ghats, where the satellite product seemed to underestimate precipitation. In conclusion, as expected, the satellite-derived precipitation estimates provided by CMORPH outperformed CEM-GCM during monsoon 2004.

5. Future works

Since a full evaluation of the global model output and satellite-derived estimates can be done studying their performance during different seasons (at least 1 year) and regions (e.g. tropics, extra tropics), the analysis presented here will be extended to other months and to areas where reliable gridded reference data are available. As suggested by the WWRP/WGNE JWGV recommendations, a comparison between forecasts and persistence (or climatology) will be carried out to put the verification results into perspective and to highlight the usefulness of the forecast system itself. Other satellite-derived products will be validated, and the skill of a regional climate model (the Regional Spectral Model), set up for the European and Indian/Himalayan domains, will be assessed. In doing this, the highly valuable source of data provided by CEOP will be used for intercomparison/verification purposes on a daily and sub-daily basis.

Acknowledgement

The authors thank Massimo Bollasina (Department of Atmospheric and Oceanic Science, University of Maryland) for reviewing this work and for his useful suggestions.

References

Arpe, K., 1991. The hydrological cycle in the ECMWF short range forecasts. Dynamics of Atmosphere and Oceans 16, 33–59.

Ebert, E., 2004. Monitoring the quality of operational and semi-operational satellite precipitation estimates—The IPWG validation/intercomparison study. Online Proceedings of 2nd International Precipitation Working Group Workshop. 25–28 October 2004, Monterey. http://www.isac.cnr.it/%7Eipwg/meetings/monterey/mry2004-proc-cont.html

Ebert, E., Damrath, E.U., Wergen, E., and Baldwin, M.E., 2003. The WGNE assessment of short-term quantitative precipitation forecasts. Bulletin of the American Meteorological Society 84, 481–492.

Janowiak, J.E., 2004. Validation of satellite-derived rainfall estimates and numerical model forecasts of precipitation over the United States. Online Proceedings of 2nd International Precipitation Working

Group Workshop. 25–28 October 2004, Monterey. http://www.isac.cnr.it/%7Eipwg/meetings/monterey/mry2004-proc-cont.html

Joyce, R.J., Janowiak, J.E., Arkin, P.A., and Xie, P., 2004. CMORPH: a method that produces global precipitation estimates from passive microwave and infrared data at high spatial and temporal resolution. Journal of Hydrometeorology 5, 487–503.

Kidd, C., 2004. Validation of satellite rainfall estimates over the mid-latitudes. Online Proceedings of 2nd International Precipitation Working Group Workshop. 25–28 October 2004, Monterey. http://www.isac.cnr.it/%7Eipwg/meetings/monterey/mry2004-proc-cont.html

Krishnamurti, T.N., Bedi, H.S., Heckley, W., and Ingles, K., 1988. Reduction of the spin-up time for evaporation and precipitation in a spectral model. Monthly Weather Review 116, 907–920.

Roads, J., Chen, S., Kanamitsu, M., and Juang, H., 1999. Surface water characteristics in NCEP global spectral model reanalysis. Journal of Geophysical Research 104, 19307–19327.

Mountains Witnesses of Global Changes
R. Baudo, G. Tartari, and E. Vuillermoz (Editors)

15

Circulation and relationship between pollutant sources and atmospheric composition in the Himalayan region

Giuseppe Calori, Gregory R. Carmichael, Domenico Anfossi, Pietro Malguzzi and Silvia Trini Castelli

Abstract

Long-term source–receptor relationships in Asia have been studied by means of a multilayer Lagrangian model. Here a summary of results for sulfur concerning the Himalayan region is presented, with a reconstruction of the past trends during the 1975–2000 period, an estimation of the role played by sources inside the region and adjacent areas and an outline of intra- and inter-annual variations. The remarkable increase of emissions in the past decades, with a corresponding rise of deposition levels in mountains is also noted, together with the relatively moderate contribution of local sources to regional budgets, as opposed to the ones of the Indo–Gangetic plain and the rest of the Indian subcontinent.

1. Introduction

Estimating past, present and future levels of deposition, ambient levels of concentration and aerosol column loading are central to the evaluation of risks to ecosystems and human health, and net changes in radiative forcing and resulting changes in climate. This is particularly true in Asia and in the Indian subcontinent, where the pressing environmental problems of urban pollution, acid deposition and climate change are already extensively documented and are expected to intensify. Population increase accompanied by expanding economies and change of lifestyles during the last two decades has in fact boosted energy demand throughout the area. The primary energy demand in Asia is currently increasing at a pace much more rapid than the world average. Presently ~80% of the demand is satisfied by fossil fuels, with coal being the primary energy source; but the contribution of biofuels is largest in South Asia and the Himalayan regions. Most of the energy scenarios up to 2020 are characterized by a further increase in energy use, with fossil fuels of growing importance (WEC–IIASA, 1995; Nakicenovic and Swart, 2000). There is also increased scientific interest and political concern regarding the long-range transport and fate of these

ISSN: 0928-2025 DOI: 10.1016/S0928-2025(06)10015-2

species because countries are receiving increasing amounts of pollutants from neighboring and even distant countries. This growth in emissions has stimulated various atmospheric modeling studies, investigating different aspects of transport, transformation and deposition mechanisms of pollutants compounds (mostly sulfur and particulate matter) throughout the region. Integrated assessment models can help to better understand the consequences of future scenarios and emission abatement strategies, by coupling atmospheric models with consistent energy projections, related control technologies information and environmental impacts. This has been the main purpose of the RAINS-Asia Project (Foell et al., 1995; Downing et al.; 1997). Within its framework, long-term source–receptor relationships have been developed by mean of an atmospheric dispersion model, covering most of Asia.

In the following the modeling approach and the data employed in these studies are summarized, followed by the results on the Himalayan region, concerning the reconstruction of the past trends, the relationships with sources, seasonality and inter-annual variability. The focus will be on sulfur, on which most of the work has been done. Sulfur in fact, is one of the most important trace constituents for the atmosphere in the area, with implications for regional environmental pollution and radiation budget; moreover, some of the points revealed by its study could be applied also to other constituents.

2. Methodology

Within the RAINS-Asia project, atmospheric calculations were carried out by means of the ATMOS model (Arndt and Carmichael, 1995; Arndt et al., 1997, 1998), a multilayer source-oriented Lagrangian trajectory model. ATMOS simulates emission, transport, chemical transformation and deposition of sulfur. It assumes that the atmosphere is subdivided in two or three vertical layers (a time-varying planetary boundary layer and the free troposphere during daytime; a surface layer, a residual boundary and the free troposphere during nighttime), in which source emissions are injected as puffs and among which mass transfers happen according to layers variation. Puffs are followed for up to 5 days forward in time, and the conversion from sulfur dioxide to sulfate is modeled using a rate depending on latitude and day of the year, accounting for gas-phase as well as in-cloud transformations. Both species are removed via dry and wet deposition processes, with parameters depending on land-sea terrain type, season and precipitation intensity. Details about the model formulation can be found in Arndt and Carmichael (1995); Arndt et al. (1997), while model parameters are based on a series of reviews and studies (Huang et al., 1995; Ichikawa and Fujita, 1995; Ichikawa and Hayami, 1998; Phadnis and Carmichael, 1998; Xu and Carmichael, 1998).

ATMOS was run for multiple years on a domain covering Asia from 60° to 150° E and from 20° S to 55° N, computing concentrations and deposition on a reference grid at one degree resolution (the same resolution as the emissions data set). Two set of runs were performed, one covering individual years 1985 and 1990 through 1997, and one every fifth year from 1975 to 2000. Year-specific meteorology and emissions were used.

The NCEP Reanalysis data, consisting of four-times daily values at ~2.5° resolution provided by the NOAA–CIRES Climate Diagnostics Center (Boulder, Colorado), were used to prepare the meteorological input: wind at standard pressure levels to drive mass transport in model layers, precipitation fields for deposition, local heat flux, temperature profile and surface wind speed (as well surface properties assigned according to a global land cover database) for PBL height estimation.

Considered emission fluxes included different types of sources. Anthropogenic fluxes, including ground-based sources and maritime traffic, were based on the work of Streets et al. (1999). The 1990 base year inventory Streets et al. (1995) uses detailed data on energy and fuel use by sector for 99 administrative entities (countries, regions, megacities) in which the domain was subdivided. Inventories for all other years were then derived from the baseline inventory using country-based energy-use statistics and emission-control levels. For modeling purposes emissions were subdivided into area sources, arranged on a 1° resolution grid, and ~250 large point sources. The emissions from the portion of the former Soviet Union falling inside the simulation domain were taken from Ryaboshanko et al. (1996). Volcanic sources were assigned according to Andres and Kasgnoc (1998), integrated for Japan with data from Fujita (1992). Model calculations were compared with observations in the region (Arndt and Carmichael, 1995; Arndt et al., 1997, 1998; Ueda and Kang, 1999; Guttikunda et al., 2001) as well as against six other long-range transport models, including detailed Eulerian codes, in MICS-Asia inter-comparison exercise (Carmichael et al., 2002a). These results indicate that ATMOS is able to capture many of the features and the long-term variability of deposition levels in different areas of Asia and that its performance in predicting sulfur deposition was shown to be very similar to these more detailed models.

3. Modeled historical trends

The modeling exercise spanning from year 1975 to 2000 allowed reconstruction of the main features of the past historical trends of emissions, concentrations and deposition in the area considered. Estimated emission totals in year 2000 for the countries most relevant for the Himalayan region are as follows (in kton yr^{-1}): Bangladesh 156, Bhutan 2, India 7487, Nepal 35 and Pakistan 1234. Such figures are the result of the demographic and economic growth in the previous decades, increasing energy consumptions at a high pace. Variations of yearly country emissions, with respect to year 1975, are presented in Fig. 15.1 for every fifth year from 1975 to 2000. According to the estimates, all countries over the whole period considered experienced an increase in emissions by a factor between two and five, with a noticeable acceleration during last years in the relevant cases of India and Pakistan. Such trends were the highest in Asia, when compared to countries experiencing substantial but less dramatic increases (e.g. southeast Asia) or even starting to decelerate or decrease (as in the case of Japan and the Republic of Korea) as the result of the adoption of control measures (for a discussion, see Carmichael et al., 2002b).

The resulting trends of total deposition on a selected country, or regions of the Himalaya, or adjacent to it, are also presented in Fig. 15.1 (Indian "western Himalaya"

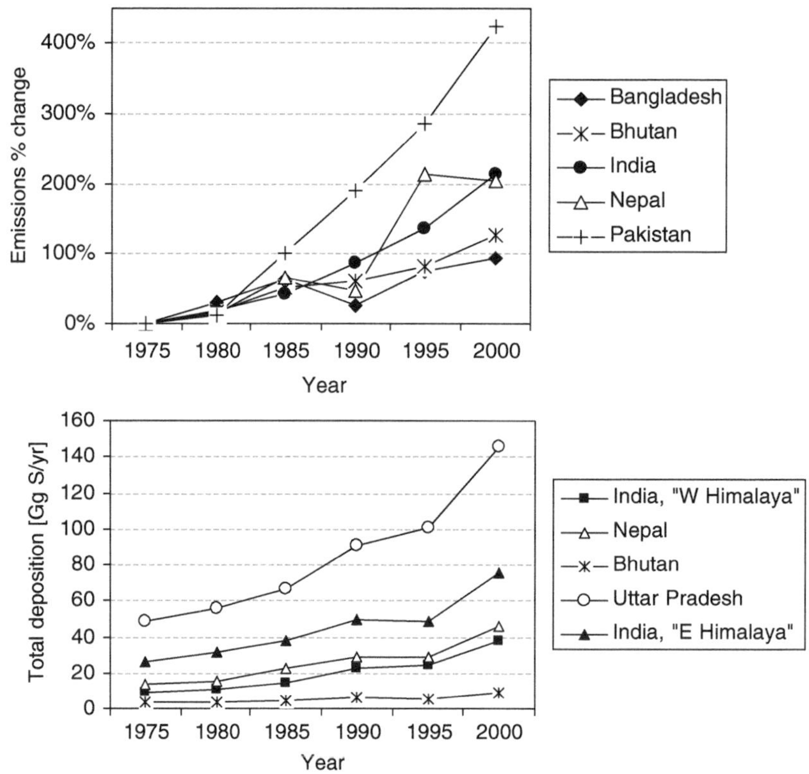

Figure 15.1. Variation (relative to year 1975) of yearly emissions of countries in the Indian subcontinent (above); trends of yearly total sulfur deposition in selected countries and regions (below).

corresponds to Jammu and Kashmir and Himachal Pradesh; Uttar Pradesh also includes Uttaranchal; Indian "eastern Himalaya" corresponds to Assam, Arunachal Pradesh and all other states south of them). Consistent with emission trends, depositions over the 25-year period increased by factor between three and four, with the highest increase in the Indian western Himalaya and the lowest in Indian eastern Himalaya. From absolute amounts, the contrast between mountain regions and the Indo–Gangetic plain is evident, a combined result of the emission distributions and the barrier effect of the orography; e.g., the deposition in Uttar Pradesh is three to five times higher than the ones in Nepal or Indian W Himalaya, despite the fact that the areas are comparable. Further elements in this sense emerge when we look closer at the contributions given by sources in different areas.

4. Relationships with sources

Figure 15.2 shows the main contributions to sulfur budgets for five different target countries/regions in the Himalayan area: Indian "western Himalaya", Bhutan, Nepal, Indian "eastern Himalaya" and Uttar Pradesh. In each map, the percentage

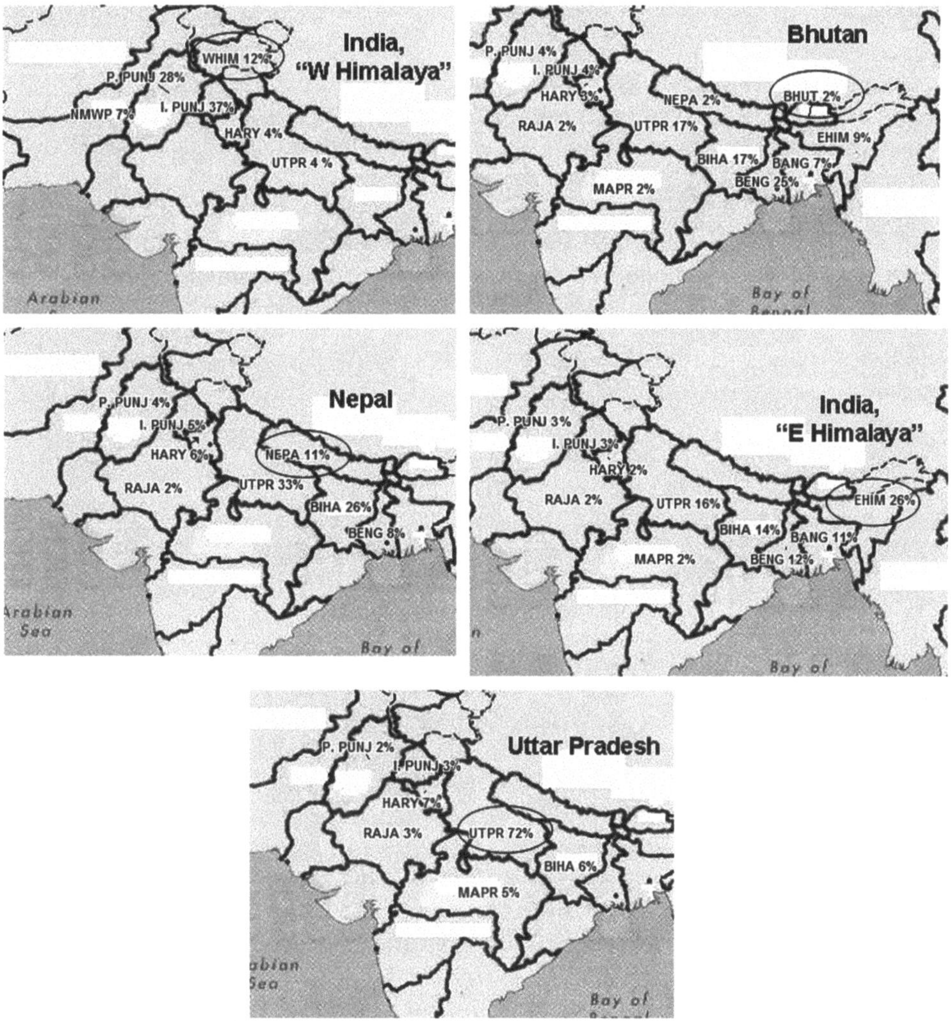

Figure 15.2. Average relative contributions of different source regions to total sulfur deposition in different regions in or near the Himalaya. NMWP: Pakistan NW Frontier Province; P.PUNJ: Pakistani Punjab; WHIM: Indian "western Himalaya"; for example, the map labeled Nepal shows where the sulfur deposited in Nepal was emitted (i.e. 11% from sulfur emitted in Nepal and 33% from emissions in UTPR). I.PUNJ: Indian Punjab; HARY: Haryana (including Delhi); UTPR: Uttar Pradesh and Uttaranchal; MAPR: Madhya Pradesh and Chhattisgarh; NEPA: Nepal; BIHA: Bihar and Jharkhand; BENG: West Bengal; BHUT: Bhutan; BANG: Bangladesh; EHIM: Indian "eastern Himalaya".

over a given region corresponds to the contribution of the sources of that region to the total deposition on the target country/region; contributions represent an average of the calculations performed for the years from 1985 to 1997.

In all mountain regions, local sources are not the main contributors to regional budgets: most of the deposition is the result of the depletion of air masses carrying pollutants from the plains. In the case of Indian western Himalaya and Nepal, the local contribution is around one-tenth of the total. The largest contributors are instead the populated regions at their south–southwest: Pakistani and Indian Punjab in the first case, and Uttar Pradesh and Bihar in the second, are collectively responsible for around 60% of the deposition. The remaining part comes from the long-range transport from more distant areas, some of which are also reported in the maps. The opposite situation happens in Uttar Pradesh, where more than 70% of the deposition is of local origin, and the remaining part comes mostly from the Indo–Gangetic plain and central India. Indian eastern Himalaya, being in this study a mix of mountain and plain areas, shows an intermediate situation, with around one-fourth of the sulfur budget given by local sources and the rest coming mostly from the west. An extreme situation is Bhutan, where due to the very low local emissions and geographic location of the country almost all the sulfur is imported from the surrounding areas. It should be noted that these arguments apply to regional budgets over large areas, on which the studies have been conducted: local situations can differ substantially, due to specific emission conditions or local circulation features.

5. Seasonality and inter-annual variability

In the Himalayan region most of the source contributions exhibit a clear seasonal pattern. This is shown in Fig. 15.3, where monthly total depositions and monthly contributions of selected source countries to yearly depositions in Nepal and Bhutan are reported, for the years 1985 and 1990 to 1997. In most cases, the different timing of the peaks can be explained according to differences in the beginning of the rain season associated with the monsoon. In Nepal, deposition from local sources is concentrated during the summer monsoon, when most of the rain occurs. In 1994 and 1997 (El Niño years) the peak deposition occurred slightly earlier. In the same part of the year the Indian contribution also peaked, as a result of sulfur advected by the southern monsoonal flow. The advected contributions from the west, as is the case from Pakistan, peak in the transitional season (spring), coincident with the prevailing westerlies (*andhis* in Pakistan and NW India). In Bhutan, most of the deposition occurs in the April–June period. The earlier peak can probably be attributed to a combination of westerly disturbances in the Ganges delta during spring and the earlier arrival of the monsoon in the upwind source areas. Also in this case the contribution from the west occurs earlier. Moving in the opposite direction along the Himalayan–Karakoram range, in northern India and Pakistan we have the opposite phenomenon, with deposition peaking in July–August due to the late arrival of the monsoon (not shown here).

The multiyear simulations also provide insight on the role played by the inter-annual meteorological variability on deposition levels. To estimate its magnitude, year-by-year depositions for the 1985–1997 period were also computed using as a reference the fixed emission levels of year 1990 (the baseline year in RAINS-Asia framework). Results for different parts of Asia are discussed in Calori et al. (2001). Figure 15.4 (left) summarizes the results for the Indian subcontinent, with a map of the normalized variation of yearly total S depositions, computed using year 1990

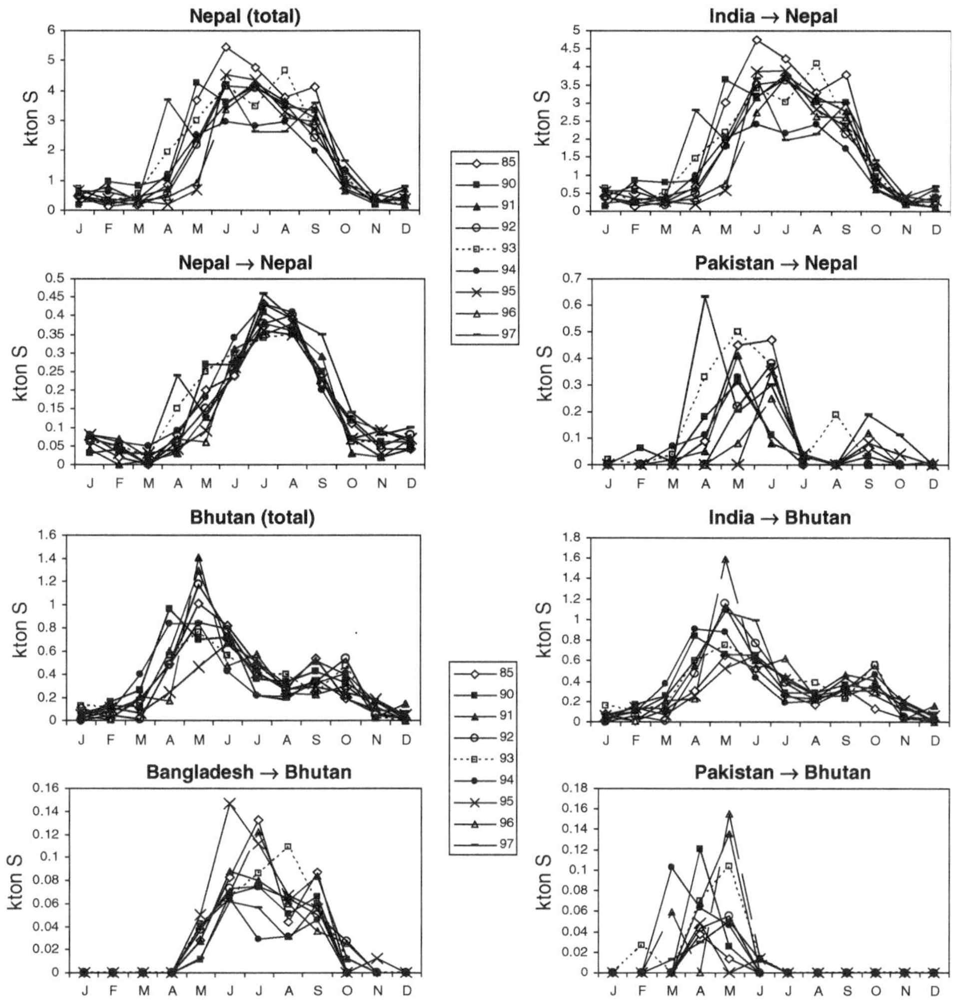

Figure 15.3. Monthly contributions of selected source regions to yearly depositions in Nepal and Bhutan, for years 1985 and 1990–1997. "Nepal (total)" shows seasonal variation in total sulfur deposition in Nepal, while "India → Nepal", shows the seasonal variation in emissions in India to the deposition in Nepal.

emission levels. According to the calculations, the lowest values in the area (5–15%) are located in the plains, in correspondence to the highest emitting areas, while highest values (20–40%) are located in central India. The Karakoram–Himalayan area and South India exhibit intermediate values (10–20%). A comparison with the corresponding map of the normalized variation of the yearly precipitation (Fig. 15.4, right) reveals only a broad correspondence. This indicates that inter-annual variations cannot be easily explained only by changes in precipitation, but are probably due to a combined effect of changes in circulation patterns, a subject of further studies.

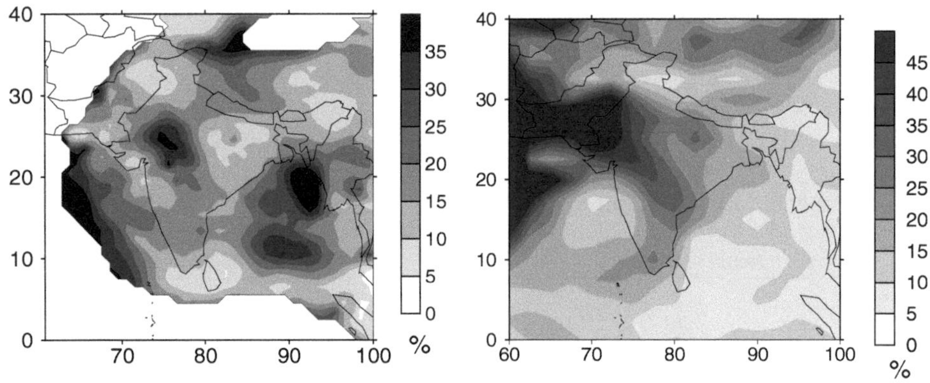

Figure 15.4. Left: normalized variation of yearly total S depositions (1990 emission levels); right: normalized variation of yearly precipitation; calculation period: 1985–1997.

References

Andres, R.J. and Kasgnoc, A.D., 1998. A time-averaged inventory of subaerial volcanic sulfur emissions. Journal of Geophysics Research 103, 25,251–25,261.

Arndt, R.L. and Carmichael, G.R., 1995. Long-range transport and deposition of sulfur in Asia. Water, Air and Soil Pollution 85, 2283–2288.

Arndt, R.L., Carmichael, G.R., and Roorda, J.M., 1998. Seasonal source-receptor relationships in Asia. Atmospheric Environment 32, 1397–1406.

Arndt, R.L., Carmichael, G.R., Streets, D.G., and Bhatti, N., 1997. Sulfur dioxide emissions and sectorial contributions to sulfur deposition in Asia. Atmospheric Environment 31, 1553–1572.

Calori, G., Carmichael, G.R., Street, D., et al., 2001. Interannual variability in sulfur deposition in Asia. Journal of Global Environmental Engineering 7, 1–6.

Carmichael, G.R., Calori, G., Hayami, H., et al., 2002a. The MICS-Asia study: model intercomparison of long-range transport and sulfur deposition in East Asia. Atmospheric Environment 36 (2), 175–199.

Carmichael, G.R., Streets, D.G., Calori, G., et al., 2002b. Changing trends in sulfur emissions in Asia: implications for acid deposition, air pollution, and climate. Environmental Science and Technology 36 (22), 4707–4713.

Downing, R.J., Ramankutty, R., and Shah, J.J., 1997. RAINS-Asia: an Assessment Model for Acid Deposition in Asia. Directions in Development. The World Bank, Washington DC, USA.

Foell, W., Amann, M., Carmichael, G., et al. (Eds), 1995. RAINS-ASIA: An Assessment Model for Air Pollution in Asia. Phase-I Final Report. The World Bank, Washington, DC.

Fujita, S., 1992. Acid deposition in Japan. Technical Report ET91005, Central Research Institute of Electrical Power Industry.

Guttikunda, S.K., Thongboonchoo, N., Arndt, R.L., et al., 2001. Sulfur deposition in Asia: seasonal behavior and contributions from various energy sectors. Water Air and Soil Pollution 131 (1/4), 383–406.

Huang, M., Wang, Z., He, D., et al., 1995. Modeling studies on sulfur deposition and transport in East Asia. Water, Air and Soil Pollution 85, 1921–1926.

Ichikawa, Y. and Fujita, S., 1995. An analysis of wet deposition of sulfate using a trajectory model for East Asia. Water, Air and Soil Pollution 85, 1927–1932.

Ichikawa, Y. and Hayami, H., 1998. Comparison of long-range transport models of sulfur oxides for East Asia. In: Workshop on Transport of Air Pollutants in Asia, 27–29 July 1998, Interim Report, International Institute for Applied System Analysis, Laxenburg, Austria.

Nakicenovic, N. and Swart, R. (Eds.), 2000, Emissions scenarios: a special report of Working Group III of the Intergovernmental Panel on Climate Change. Cambridge University Press, Cambridge, UK.

Phadnis, M.J. and Carmichael, G.R., 1998. Evaluation of long-range transport models for acidic deposition in East Asia. Journal of Applied Meteorology 37, 1127–1142.
Ryaboshanko, A.G., Brukhanov, P.A., Gromov, S.A., et al., 1996. Anthropogenic emissions of oxidized sulfur and nitrogen into the atmosphere of the former Soviet Union in 1985 and 1990. Report CM-89, Department of Meteorology, Stockholm University.
Streets, D.G., Amann, M., Bhatti, N., et al., 1995. Chapter 4 of RAINS-ASIA: An Assessment Model for Air Pollution in Asia. Phase-I Final Report. The World Bank, Washington DC, USA.
Streets, D.G., Tsai, N.Y., Waldhoff, S.T., et al., 1999. Sulfur dioxide emission trends for Asian countries, 1985–1995. Workshop on the Transport of Air Pollutants in Asia, Interim Report, International Institute for Applied System Analysis, Laxenburg, Austria, July 22–23, 1999.
Ueda, I. and Kang, S.J., 1999. Evaluation of the performance of the ATMOS model. In: RAINS-Asia Phase II, Report submitted to the World Bank, The Overseas Environmental Cooperation Center.
WEC-IIASA (World Energy Council – International Institute for Applied System Analysis), 1995. Global Energy Perspectives to 2050 and Beyond. World Energy Council, London, UK.
Xu, Y. and Carmichael, G.R., 1998. Modeling the dry deposition velocity of sulfur dioxide and sulfate in Asia. Journal Applied Meteorology 37, 1084–1099.

Mountains Witnesses of Global Changes
R. Baudo, G. Tartari, and E. Vuillermoz (Editors)

16

Italian air force observatory network for environmental and meteorological monitoring: from data control to quality assurance

Fabio Malaspina, Francesco Foti and Emanuele Vuerich

Abstract

The main purpose of a meteorological observatory network is to completely describe natural phenomena so to record their characteristics forever. Such recordings increase in importance through time, becoming part of the patrimony of unique and unrepeatable knowledge. These records are useful to all humanity for scientific research for making choices that will have effects on us and our planet. Especially, recently, some of this information can verify and initialize mathematical models. Natural phenomena can have different global or local characteristics, and, according to their typology, the right network for measurements is necessary to fix instrumentation specifics and establish appropriate procedures. Study of global-scale natural phenomena requires fundamental performance measures for long periods in places far from human pollution, and high mountains are particularly suitable, for such purposes. In fact in 1654 the world's first meteorological network was created in Italy by the Grand Duke of Tuscany, who was already worried about climate change. The Meteorological Service of the Italian Air Force continues this job with its own national network that performs special measures and even takes part in Antarctica's surveys. The answers to problems discovered in the management of such meteorological stations are mainly due to technology changes, measures of small quantities, the necessity of highly specialized personnel, remote and often uncomfortable measurement locations, and heterogeneity of instrumentation, calibration, methods of measure, and exposure. In this epoch of telecommunications observation systems are ever more distant from those that analyze the data. Such systems analyses are commonly received only as encoded messages or numerical files, without anyone knowing the history, the limits, or the different characteristics of instrumentation used in the monitoring stations. The highest quality data are those that guarantee the requisites dictated by the purpose for which they were produced. The complex climatic system is mainly characterized by incessant variability of its configurations that are irreplaceable. A subsequently assessed "quality control" based on past statistical descriptions is not enough, but it is also necessary a "quality-assurance"

ISSN: 0928-2025 DOI: 10.1016/S0928-2025(06)10016-4

system, in order to give to consumers enough information about measurements that were obtained, and in order to establish limits that must be accounted for in data analysis. Only "well defined" information changes uncertainty into measurable risk that can be emphasized in decision-making.

1. Introduction: Department for aeronautical experimentations (Re.S.M.A.) and observation networks of italian air force meteorological service

This Italian Air Force Department is located in Vigna di Valle next to Bracciano Lake, in an extinct volcano. This constitutes an historic headquarters that was the first aerologic station in Italy since 1909. At the present moment it is in charge of doing standard meteorological observations, UV and ozone measurements, and meteorological soundings. It works 24 h a day and is equipped with instrumental emplacements where it is possible to make comparisons among different instruments for a long time and in different weather conditions. Lots of tests or special measurement campaigns are carried out in order to assess the performances of various observation devices (Fig. 16.1). Department personnel are able to state the quality of measurements and observations coming from the Service's network.

The Italian Air Force Meteorological Service Network has two different observation sections: the 84 human-controlled stations part and the 110 automatic weather stations part. Some of these Italian stations are situated in uncomfortable but meteorologically interesting places. For example: in the Alps, the Paganella station is at 2129 m a.s.l. and the Plateau Rosà station at 3480 m a.s.l.; in the Apennine the Cimone station is at 2165 m a.s.l.; and on some occasions the Italian Meteorological Service takes part in important measurement campaigns, as is the case with the Antarctic National Program.

Figure 16.1. ReSMA experimental site that includes six equipped areas under both automatic and manual control every day. ReSMA has been designated as a site for World Meteorology Organization field intercomparisons of rainfall intensity (RI) gauges to be held in Vigna di Valle since August 2007 for two years.

The Italian meteorological network includes also three special measurement networks: the chemical analysis of the precipitation network (7 stations) activated in 1975, the total ozone network (3 stations) activated in 1957, and the solar radiation network (37 stations) activated in 1958. Cimone station also started measures of carbon dioxide concentration in 1979.

2. What would we like to measure? Two different monitoring networks for two different purposes

Interest in natural phenomena is an innate part of human nature. Greek thinkers first began the long tradition of stars-meteorology and empiric-meteorology. They collected and ordered in an organic way knowledge of the atmosphere acquired in the past and raised it to a natural philosophy. Almost 2000 years after that, meteorology was changed from qualitative to quantitative. In the 17th century, after the invention of the first principal meteorological instruments by the Italian scientists, meteorology acquired the characteristics of a quantitative science. It was recognized that natural phenomena should be described and recorded in the most complete way for the present and future studies. Galileo Galilei wrote about this (in Italian) in *Il Saggiatore* in 1623:

> Natural phenomena are written in this big book which is always opened in front of our eyes. It is written in a mathematical language, and the characters are triangles, circles, and other geometric figures; without these it is a vain wandering in a dark labyrinth.

Later Lord Kelvin (1824–1890) more clearly stated:

> I affirm that when you can measure and express in numbers what about you are speaking, you know indeed something.

A single station is not enough to study meteorological phenomena, so a well-organized network is needed. The purpose of observation networks is to produce quality data of atmospheric parameters, which must be selected in the right way to describe the phenomenon in which we are interested.

The importance of good recordings increases more and more through the years to make a part of the patrimony of unique and irreplaceable knowledge, useful to the whole of humanity for scientific research, useful to us and to our children for making choices that will have effects on human beings and planet Earth. It is important to understand that every new measure modifies or creates new knowledge. Only "quality data" from the past and the present acquire additional value over time, so only these kinds of long historical series represent a national and human patrimony.

Natural phenomena can have different global or local characteristics. According to their typology, it is necessary to design the appropriate network for measurement of them. It is also necessary to fix instrumentation specifications and establish appropriate procedures. In the case of global pollution, which has a climatic impact and which is not "directly" related to human activities, special and meteorological measurements must be performed in stations settled far from polluting sources. These "global measurements" generally monitor very small changes and very small concentrations. They also must be carried out for long periods and in apparently uncontaminated areas such as high mountains. In the case of local pollution, which has mainly an impact on human health, measurements must be collected near the polluting sources. These "local measurements" generally monitor high concentrations and local pollution has a short-period impact.

In this epoch of advanced telecommunications, observation systems are more and more distant from those that analyze the data. System analysis can often involve receipt of only encoded messages or numerical files, without knowledge of history, limits, or different characteristics of instrumentation used in monitoring stations. It sometimes happens that the measurement accuracy (i.e., the closeness of the agreement between the result of measurement and the true value of the object to be measured – the *measurand*) is replaced by instrumental resolution (which is a quantitative expression of the ability of an indicating device to distinguish meaningfully between closely adjacent values of the quantity indicated, for display presentation in tenths, hundredths, etc.). Sometimes global and local measurements are confused, and the instrumental uncertainty is not considered.

Therefore, it is our experience that it is sometimes possible to see the risk that the network's goal is no longer describing as completely as possible the natural phenomena of our interest, but it is, for example, becoming more for the initialization of mathematical models or the monitoring of legal threshold concentrations in pollution areas. In this way, instruments are not used at the maximum of their performance and not all possible data are recorded, preventing the best description of nature to leave to future generations for their future strategic decisions and research.

3. Problems in performing measurements

In the fields of observation and analysis, it may be possible to move toward a progressive automation; but in meteorology, machines cannot yet completely replace human observation, human intuition and human critical mind. About this argument, it has been written (Anonymous, 1993)

> Although this guide provides a framework for assessing uncertainty, it cannot substitute for critical thinking, intellectual honesty and professional skill. The evaluation of uncertainty is neither a routine task nor a purely mathematical one; it depends on detailed knowledge of the nature of the *measurand* and of the measurement. The quality and the utility of the uncertainty quoted for the result of the measurement therefore ultimately depends on the understanding, critical analysis, and integrity of those who contribute to the assignment of its value.

The problems occurring in the management of mountain stations, whose purpose is monitoring global changes, are mainly due to: (1) change of technology; (2) measure of small quantities; (3) necessity of high specialized personnel; (4) remote and often uncomfortable location of the site of measurement; (5) no net homogeneity (instrumentation, calibration, methods of measure, exposure). The following examples can show some of these problems.

A comparison campaign among several thermometer screens and shields can show what happens when a technological change occurs, or the presence of a non-homogeneous network. During a continuous daily sampling, temperature differences

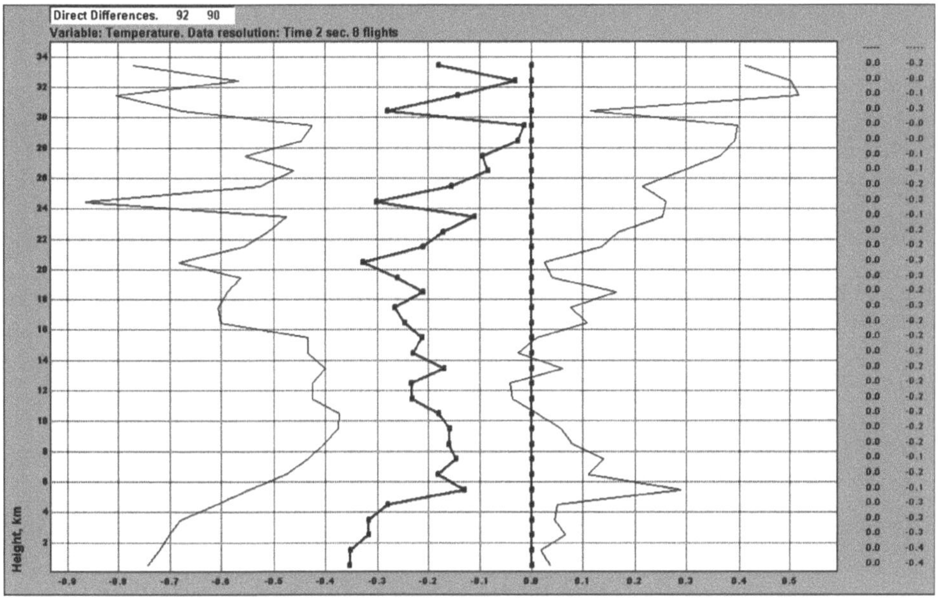

Figure 16.2. The new Radiosonde Vaisala RS92 has been recently introduced to substitute for the previous model RS90. Temperature average direct differences were calculated with RS92-SGP as a reference. Also a "standard deviation" line is presented. The impact of future instrumentation change, considering the measurement accuracy required by current working standards, will be negligible whenever atmospheric sounding is used for local monitoring or for model initial analysis. In the climatic field, however, since statistical analysis is performed at precision of tenths to even hundredths of Celsius degrees, the consequences could be relevant should instrument change not be taken into account during analysis. In fact, as new equipment replaces old, indications of gradual and progressive atmospheric heating will result.

of more than 1.5°C have occurred. Maximum temperature differences correspond to sunrise, when generally the absolute minimum of temperature is recorded, and the sunset (van der Meulen, 1998). In a comparison campaign among different radiosondes of the same manufacturer, possible changes in aerological data due to instrument change were discovered (Malaspina et al., 2004). Eight launches were performed, and a systematic error of 0.2°C was noted (Fig. 16.2).

In another case, inhomogeneity among instruments has produced great problems in UV-A radiation measures (Anav et al., 1994). As shown in the Fig. 16.3, different spectral responses for similar instruments produce very different results and no comparable measures.

Because of these problems, it has long been recognized that it is important not only to record the meteorological and special-measurement values, but also the circumstances in which the measurements are made, i.e. the metadata. Metadata are particularly important in climate studies. For example, temperature measurements are affected by the state of the surrounding, by vegetation, by the presence of buildings and other objects, by ground cover, by conditions and changes in design of the radiation shield or screen, and by other changes in equipment. Temperature is one of the meteorological quantities whose measurement is particularly sensitive to

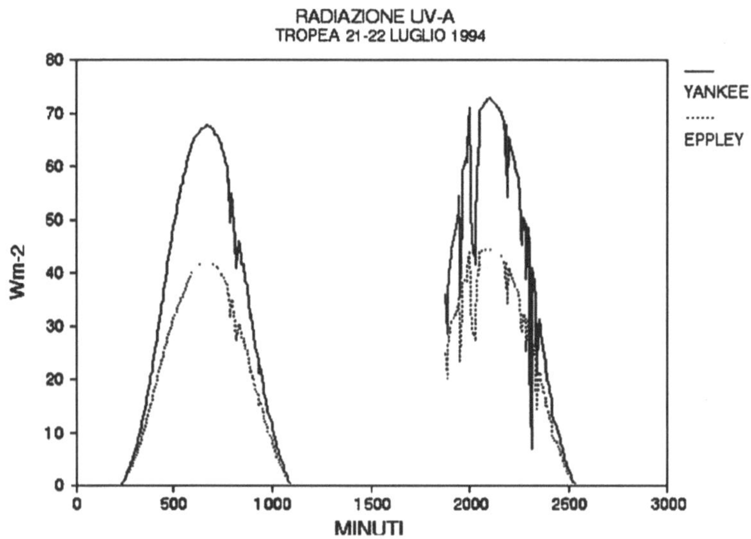

Figure 16.3. Different spectral responses involving similar instruments can produce very different results.

exposure (Anonymous, 1996). In addition, the statistical error in this kind of measure has to be consistent with the measurement accuracy; the same is true for the average and standard deviation. In addition, data analysis must respect reality, at least: RESOLUTION$\geq$ACCURACY (Fig. 16.4).

4. From data control to quality assurance

Only "well defined" information changes uncertainty into measurable risk and allows assessment of decisional processes. When data satisfy the requisites dictated by the purpose for which they have been produced, they are considered of "good quality". These data are not necessarily excellent, but it is essential that their quality is known and demonstrable. Without a quality system, data must be regarded as being of uncertain or unknown quality, and their usefulness is diminished (Anonymous, 1996).

The fundamental difference between "quality control" and "quality assurance" must be noted.

Quality control is the best-known component of a quality management system (qms), and it is the irreducible minimum of any qms. It consists of data examination in stations or/and in data-analysis centers in order to detect possible measurement errors, so that unreliable data can be either corrected or deleted (Anonymous, 1996). This statistical control is perfect if one believes that a climate system is stationary, that is, when the statistical description of the past is the probability evaluation of the future. But some doubts can rise, as in the case of David Hume (1711–1776) who stated:

> Being determined by custom to transfer the past to the future, in all our interferences; where the past has been entirely regular and uniform, we expect the event with the greatest assurance, and leave no room for any contrary supposition. But where different effects have been found to follow from causes, which are to appearance exactly similar, all these various effects must occur to the mind in transferring

SPECIFICATIONS

MULTI-PLATE RADIATION SHIELD

Sensor Types:
Accomodates temperature and humidity sensors up to 26 mm (1 in) diameter

Radiation Error: @ 1080 W/m² intensity
Dependent on wind speed
0.4° C (0.7°F) RMS @ 3 m/s (6.7 mph)
0.7° C (1.3°F) RMS @ 2 m/s (4.5 mph)
1.5° C (2.7°F) RMS @ 1 m/s (2.2 mph)

Construction:
UV stabilized white thermoplastic plates
Aluminum mounting bracket, white
Stainless steel U-bolt clamp

Dimensions:
12 cm (4.7in) Dia. x 27 cm (10.6 in) H.
Mounting fits vertical pipe 25-50mm (1-2in) Dia.

Weight:
Net weight- 0.7 kg (1.5 lb)
Shipping weight- 1.4 kg (3 lb)

PLATINUM TEMPERATURE PROBE

1000Ω Platinum RTD
Accuracy:
±0.3°C at 0°C ....(standard)
±0.1°C at 0°C ....(optional)
Available Outputs:
4 wire RTD ....(standard)
4-20 mA ....(41342L)
0-1 VDC ....(41342V)

Figure 16.4. The choice of significant digits in measurement must be consistent with the accuracy of the measuring system (not only of the sensor itself). In this case, uncertainty in measurement is greater or equal to the "radiation error" plus the sensor accuracy.

> the past to the future, and enter into our consideration, when we determine the probability of the event. Though we give the preference to that which has been found most usual, and believe that this effect will exist, we must not overlook the other effects, but must assignee to each of them a particular weight and authority, in proportion as we have found it to be more or less frequent.

In addition to this important philosophic aspect, which can cause one to delete or to not consider the most meaningful data, other problems are associated with quality control, such as irrecoverable cost of lost data, control sometimes performed after a long time, and an unrepresentative historical series caused by lost data.

Quality assurance operates continuously at all points in the whole observation system, from network planning and training, through installation and station operations to data transmission and archiving. The provision of good-quality meteorological data is not a simple matter, and it is impossible without a quality management system.

5. Conclusion: "Not all numbers are data"

Not all numbers are data and not all stations provide data to study the meteorological phenomena at all space-time scales. The calculators and the new electronic acquisition systems can deceive, but measurement systems have limits. Commonly the enormous consistency of data can reduce statistical error, but, for example, it cannot be less than instrumental error. If measurement-system limits are not considered, a meteorological network does not describe natural phenomena that happen around us, but instead it produces only a computer-mathematical virtual reality. For climatic studies, it is necessary to store both meteorological quantity and metadata. Instruments and networks should be under the continuous control of specific quality management systems. Therefore, according to our experience, we must not only increase the quantity and kind of measurements, but it is also necessary to increase their quality too.

References

Anav, A., DiMenno, M., and Moriconi, M. L., 1994. Misure di Radiazione UV a Tropea, I.F.A.R.I. 94/25.

Anonymous, 1993. Guide to the expression of uncertainty in measurement. International Organization for Standardization (ISO).

Anonymous, 1996. Guide n.8 to meteorological instruments and methods of observation. 1996. World Meteorological Organization.

Malaspina, F., Foti, F., Vuerich, E., and Casu, G., 2004. Radiosounding: possible change in aerological data due to instrument change. Il Nuovo Cimento 27C (5), 503–513.

van der Meulen, J.P., 1998. A thermometer screen intercomparison. WMO/TD n .877: pp. 319. http://www.dwd.de/EUMETNET/Berichte/TECO98temp.pdf

Mountains Witnesses of Global Changes
R. Baudo, G. Tartari, and E. Vuillermoz (Editors)

17

Climate change in Italy: an assessment of data and reanalysis models

Raffaele Salerno, Mario Giuliacci and Laura Bertolani

Abstract

Every year climatic problems occur around the globe. It has been recognized that some of these widely dispersed climatic extremes might have common origins, e.g. the general effects of global warming, the periodic changes of sea surface water temperature in the central and eastern equatorial Pacific Ocean and other modifications in sea surface temperature that influence the sensible and latent heat fluxes across the air-sea interface and the atmospheric circulation at all scales. In the Mediterranean region, some evidences of these changes are almost evident: the modification in temperature distributions and extremes, the droughts in some places and the floods in others, the alteration of alpine glaciers. The models, among them the reanalysis models, can help us to understand and assess some of these changes and how they are related to the others on the whole Earth.

1. Past climate

During past ages movements of the Earth's plates have altered the distribution of land surfaces. This in turn has changed the temperatures of some places, and the regional and global climate is linked to land and ocean distribution, as well as its transformation during the ages. For instance, the waters around New Zealand's South Island cooled from 20°C, about 53 million years ago, to 12°C present day due to poleward movement, the Antarctic cooling and the changing of ocean currents.

Most of the Earth's climate is driven by astronomical factors, such as the axial tilt and the variations of the Sun. The solar constant, i.e. the amount of radiation that the Earth receives outside the atmosphere, has changed: 4.7 billion years ago sun radiation was 70 of what it is now and during the Carboniferous, about 300 million years ago when so much of the coal resources of Earth were deposited, it was the 2.5 less than today.

The Earth's history shows fluctuation in the climate during the ages. Analysis of ice cores from Greenland and Antarctica shows that warm interglacial periods arise

ISSN: 0928-2025 DOI: 10.1016/S0928-2025(06)10017-6

within a few hundred years, but a glacial period takes much longer to form. In the most recent ice age there were 22 warm events, at the beginning and the end of which there could be regional mean temperature changes to up to 7°C within a decade. At each flip from one regime to another or flop back again, temperatures in Greenland would alter by 6–7°C, sometimes within a year, and snowfall varied by a factor of two. Also, huge changes of wind strength occurred. So, large changes of climate can naturally occur with rapid swings between two or more weak states of equilibrium. This may implicate that also fairly modest changes may be enough to trigger another swing.

In the last 10,000 years climate has been more stable and less variable than during the glacial times. There are sensitive regions where the climate can change quickly and where changes have a regional if not more widespread effect. One of these regions is the North Atlantic Ocean, particularly the area near southern Greenland. The variation on this region has a fairly great impact on the Europe and Mediterranean climate.

A lot of research has been dedicated to the discovery of cycles in the climate data series. Any record can be interpreted as the summation of simultaneous cycles of change, each cycle having its own individual period and amplitude. Amongst the many alleged cycles of climate, only a few are credible, usually linked to astronomical factors. For example, the number of sunspots affects the climate. Sunspots have a periodicity of about 11 years; a time interval found analyzing the available data. The solar day and the effects due to the Earth's orbit are noticeable elements of climate cycles.

Ice-core analysis has also shown that the main greenhouse gases, carbon dioxide (CO_2) and methane (CH_4) occur in higher concentration during warm periods. Also, the two variables, temperature and greenhouse gas concentration are clearly consistent, yet it is not clear what drives what.

Also sea level has changed little since 7000 years ago. Sea-level fluctuations during the Pleistocene were larger and they were largely controlled by the amount of continental ice in the Northern Hemisphere. At the time of glacial maxima, the sea level was about 100 m lower than today. The examination of submerged layers of sediments near the mouths of major low-altitude rivers, such as the Indus and Nile, reveals the strength of the river currents at different times in the past. All indicate increased monsoonal rainfall onto catchments of northern Africa, the Middle East, southern Asia, India, at around 9000 years ago. This is confirmed by the results of climate models. The higher rainfall was mainly due to the migration of the summer rain-belt to higher latitude after the end of the last glacial age.

2. Climate models

Models have become basic tools for any atmospheric study. In particular, models in atmospheric physics are mainly used to forecast the weather (Numerical Weather Prediction models, NWP) as well as to simulate changes in climate (General Circulation Models, GCM). They both use the same equations representing natural laws of physics, but the purpose of any climate model is to simulate changes in climate as a result of slow changes in some boundary conditions (such as the solar constant) or physical parameters (such as the greenhouse gas concentration). NWP models are used to predict the weather in the short (1–3 days) and medium-range (4–10 days) future. A good NWP model is able to accurately predict the movements and evolution of disturbances such as frontal systems and tropical cyclones. Climate models are run much longer, for years on end, long enough to learn about the mean climate and its variability.

All types of models err so much after some time (2 weeks) that they become useless from a perspective of weather forecasting. For instance, an error in the (SST) by a few degrees or a small but systematic bias in cloudiness throughout the model matters little to a NWP model. In the perspective of a GCM climate simulation, these factors are important, because they matter over a long term. On the other hand, climate simulations ignore fluctuations when considering long-term changes, whereas NWP models take no notice of very slow processes.

A state-of-the-art GCM is a coupled atmosphere-ocean model, i.e. a model simulating surface and deep ocean circulation is "coupled" to an atmospheric GCM. The interface between the two models is the sea surface. This coupling is essential to simulate some major phenomena like El Niño.

Reanalysis models (Kalnay et al., 1996) are generally fixed-parameter GCMs that are performed to get a description of the atmosphere free of inhomogeneities caused by a change in the analysis system. Reanalysis is effectively a model run constrained by observations. If only a few observations are available, the constraint is weak and the model essentially produces its own intrinsic variability. Due to the increased amount of observations, however, the model is more and more forced to follow the observed variability rather than its own intrinsic one, becoming a powerful tool to investigate climate changes and compare the results of climate simulations.

At the end, a model of the atmosphere numerically simulates the state of the atmosphere itself, using an approximation of the equations of motion. Any numerical simulation uses a time step to advance in time. The time step should be short enough to avoid the fact that the fastest-traveling disturbances have time to traverse the distance of the grid spacing, i.e. the distance among each subdivision of the model domain. Higher resolution models require shorter time steps so that more calculations are needed to simulate the same period. But, as computer processing units become less expensive, models are refined to allow for closer spatial resolutions, more accurate parameterizations and more runs, tweaking parameters or initial conditions as in the ensemble forecasting.

3. Present climate and the global warming

At present, the global mean temperature over the last 140 years shows two periods of rapid change. Global surface temperature has increased by 0.6°C in the last century and the SST has also warmed. Less than a third of the global warm-up can be explained by changes in the Sun's intensity while this trend is largely explained by observed changes in greenhouse gases and aerosol concentrations. On the other side, the upper troposphere seems to have cooled a little. This apparent inconsistency can be explained using models to give an answer or help to understand the climate; the simulations show that volcanic aerosols (such as those from eruptions) warm the stratosphere (10–50 km) and cool the surface. Greenhouse gases cause a large surface warming that offsets the cooling by other forcing. Some studies seem to demonstrate that during the next 10 years, warming of the upper troposphere by well-mixed greenhouse gases should become apparent.

To summarize, the 20th century climate changes include:

- the global surface warming of the air and increase of the mean SST; most of the warming has occurred in late winter and spring and at night;
- a cooling of the lower stratosphere by about 1 °C since 1950 (consistent with model simulations);
- a global increase of rainfall over land of 1% implying greater evaporation caused by warming; the increase in rainfall (and temperature) is larger in winter at high northern latitudes and a slight decrease has occurred in some regions, such as in northern Africa and on the West coast of South America;
- increased atmospheric vapor pressure over the tropics, at least since 1973;
- more clouds over the oceans, especially convective clouds and cirrus clouds; this, in turn, enhances warming;
- a rise of the sea level by 10–20 cm;
- a general retreat of glaciers;
- tropical cyclones are fewer and weaker in the North Atlantic (compared to the middle of the 20th century) but more numerous in the northwest Pacific;
- hurricanes' intensity and rainfall increase due to the rise in ocean's temperatures and more available atmospheric water vapor;
- fewer frost in many places, e.g. northwestern Europe;
- more severe, more frequent, longer lasting heat waves.

4. Climate change in the Mediterranean and Italy

The effects of climate changes can be found in the Mediterranean region with the same signals listed above; global surface warming of the air, increase of mean SST, general retreat of alpine glaciers, more extreme events and more frequent, longer lasting heat waves. The variations are connected to those phenomena with larger amplitude. For example, both frequency spectra of maximum temperature in the Mediterranean region and the Multivariate ENSO (El Niño southern Oscillation) Index (MEI) show a maximum at about 5 years, suggesting a possible connection between the variations of the Pacific equatorial ocean SST and the Mediterranean climate.

In Italy, an analysis of the temperatures from 1979 to 2004 for each season shows an evident difference in the frequency of them which show higher temperatures respect to the mean values, especially in winter and spring (Fig. 17.1). At the same time, precipitation has decreased over the past 20 years. Considering the

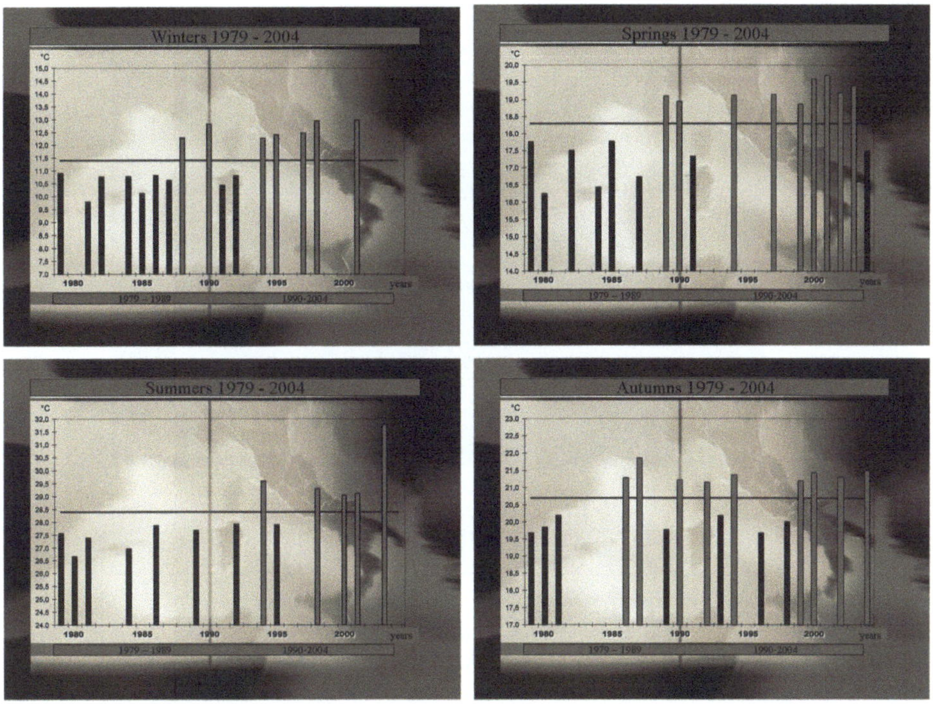

Figure 17.1. The seasonal warmest (light bars) and coldest (dark bars) years in the period 1979–2004. The horizontal line represents the mean value of the whole period. The variation compared to mean value has been chosen to be at least $\pm\sigma$ (about 0.5°C) to be represented in the picture.

winter–spring–summer, 9-month period, 6 years with a deficit of rain greater than 40% occurred from 1960 to the present day, but five of them occurred in the last 15 years. If the number of wet days is considered (wet day = day with a precipitation $\geq$1 mm) we have a decrease of trend from the beginning of 20th century to the present, with a reduction of 20% in one century.

Glacier modifications also occurred all over the alpine area. The general retreat of glaciers is more evident in the central and eastern part of the Italian Alps (Fig. 17.2). Moreover, the mean level of 0°C has raised about 150 m in the last 50 years. Many data in the alpine region show an increase of about 1°C at 3000 m in the last half century during summer, with a loss of the ice mass of about 15% from 1960; the decrease is more evident from 1980, with a loss of the 3% in the year 2003, when the hottest summer of the century occurred.

The occurrence of extreme events has increased in the last decades; floods in autumn, heat waves, extreme rain events, droughts, wind storms, and tornados. Concerning summers with long-lasting or frequent periods of heat waves, from 1950 to present, half of the total cases are found in the last 20 years. An increase of 1°C in the summer mean maximum temperature increases the probability to have hot temperatures (32–35°C) by a mean value of 100% and up to 250%.

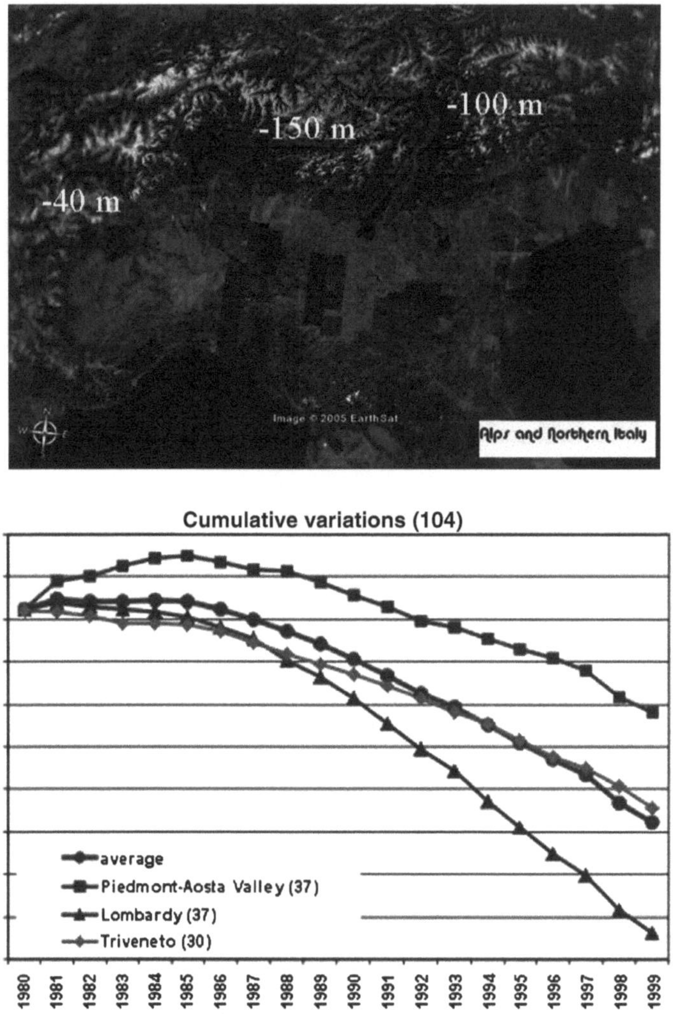

Figure 17.2. The retreat of glaciers in the Italian Alps region. The retreat is more evident in the central and eastern Alps and more pronounced after 1985.

On the other hand, in the latest 60 years, four of six flood events in Italy occurred in the last 12 years. Increase of extreme rain events have occurred all over Italy: +380% in Milan, +250% in Bari, +220% in Naples, +200% in Rome, +190% in Bologna, and +180% in Turin. However, the NCEP reanalysis surface precipitation rate, from December to August, shows a decrease in the period 1990–2004 compared to period 1950–1989, especially in the northern and central part of Italy. This trend is generally found for most of the Europe. On the other hand, in autumn, the surface precipitation rate has an increase in the period 1990–2004 compared to 1950–1989, especially for Italy, but also for Portugal, southern England, and the eastern Mediterranean Sea.

An explanation of some of these behaviors can be found in the changes of circulation occurred in the Mediterranean region. If each season is considered, looking at the sea level pressure composite mean from the NCEP reanalysis, for the period 1990–2004 minus 1950–1989 (Fig. 17.3), it may be evident that in winter and spring most of the frontal systems have to pass at higher latitudes than the Mediterranean ones. But, in summer and especially in autumns, the frontal systems may pass more easily over southern Europe and Mediterranean Sea. This is also confirmed by the general increase of low-level vorticity in autumn during the period 1990–2004 compared to 1950–1989.

This phenomenon interacts with the increase of SST. The comparison among the re-analyzed SSTs of the period 1990–2004 minus the 1950–1989 SSTs shows an increase of about 0.4–0.5°C for the surface temperature of seas around Italy. But, in autumn, this increase is more evident, with SSTs higher than 0.7–0.8°C in the last 15 years compared to the previous forty ones (Fig. 17.4).

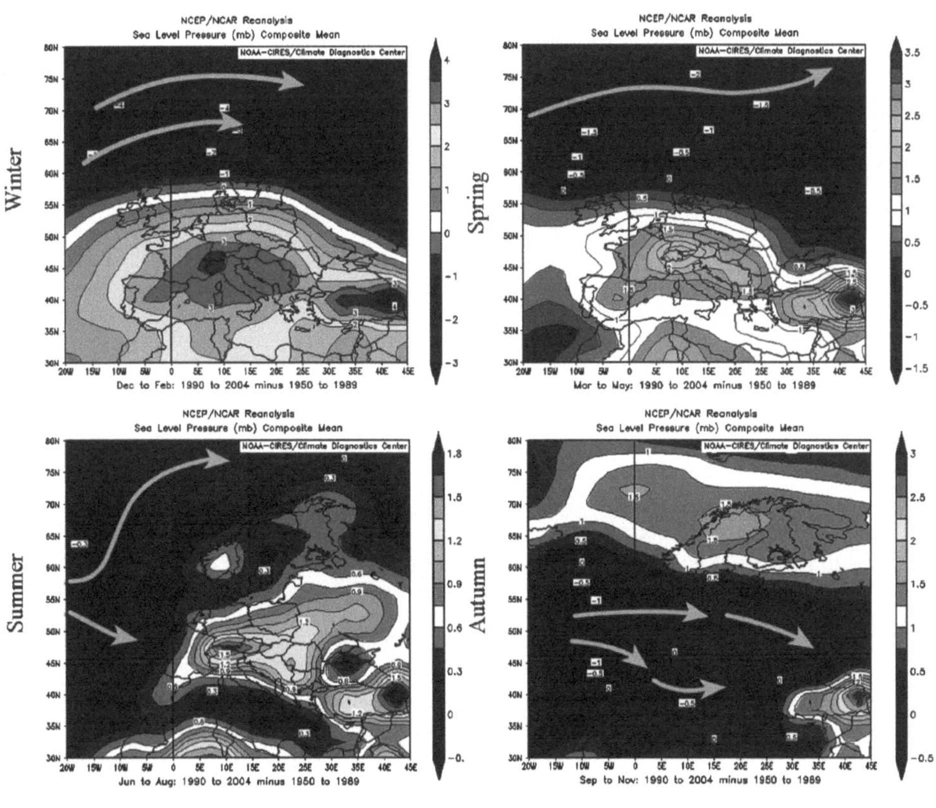

Figure 17.3. The composite mean of sea level pressure difference for the period 1990–2004 minus 1950–1989 in winter, spring, summer, and autumn. The arrows represent the mean circulation patterns.

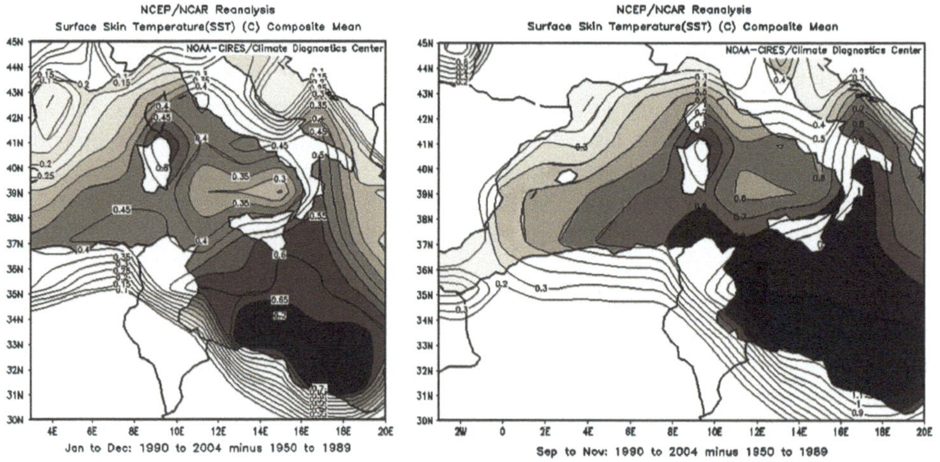

Figure 17.4. The composite mean of SST difference for the period 1990–2004 minus 1950–1989, January to December, on the left and for September to November, same difference, on the right.

5. Future perspective

The key issue at present is whether global models provide a sufficiently reliable picture of future climate. Unfortunately, there is not a simple answer to the question. Modern GCMs contain detailed representations of the atmosphere, oceans, ice, land surface, and vegetation, but:

- they are incomplete; processes known to be important are not yet or not fully incorporated into them;
- they reflect our understanding of the way the climate system works; any gaps will be mirrored within the model;
- not all processes known to be important can be described explicitly and models include parameterizations that are approximate descriptions of particular physical processes;
- some important features of climate are not resolved explicitly (due to finite resolutions);
- the atmospheric and oceanic models tend to drift and artificial adjustments need to be made even if there should be no need for them at all.

On the other end:

- climate models convincingly simulate the most important large-scale processes of climate, such as pressure patterns, jet streams, monsoons, blocking patterns, the seasons, ocean currents and day–night temperature differences;
- they produce climate changes consistent with expectations derived from an understanding of the physical processes governing climate;
- they are an essential tool for improving our understanding of climate, e.g. the reanalysis models as a tool for a deep insight to weather and climate patterns;

- they can reproduce global temperature change from ice age to ice age when the changing orbital parameters of the Earth and greenhouse gases are taken into account;
- they enable experiments to be conducted, which are inconceivable in the real world.

How much uncertainty could be taken into account? In regards to calculations, it should always include an estimate of the possible error. This applies both to weather forecasting and considerations of future climates. For example, in the case of weather forecasting the chance of rain can be mentioned as a percentage. Operationally, a percentage chance of rain means that in the long run more than a trace amount of rain falls sometimes during the forecast period (usually a day) at an arbitrary location within the forecast area, on about a percentage of the occasions when that chance is quoted. The allotted percentage depends on the degree to which all the evidence is mutually consistent. For example, prognoses of weather patterns from several numerical models can be compared with each other, and the lower the various predicted rainfall amounts, the lower the percentage probability will be.

The range of model forecasts is constrained due to the limited predictability of the behavior of the atmosphere. Even a perfect numerical weather prediction (based on available observational input) eventually becomes meaningless, because the atmosphere is an inherently chaotic fluid. It is well known that chaos is due to non-linear processes, and since the smallest (molecular) scale cannot be resolved in macro-scale weather prediction models, the non-linear feedback of unresolved motions on macro-scale motions remains unknown. A non-linear process forces a change that is not simply proportional to a variable. Attempts have been made to parameterize non-linear feedbacks in terms of macro-scale (resolved) variables, but errors remain which eventually overwhelm the prediction. Chaos in fluids such as the atmosphere is a permanent characteristic of reality; nevertheless, numerical weather forecasts have improved over the last few decades for three reasons:

- more detailed, more frequent, and more accurate observations have led to better model initializations. Especially the amount of remotely-sensed data, such as satellite and radar data, have dramatically increased;
- the rapid increase in size and speed of available computers has allowed higher resolution simulations and the use of computationally more demanding parameterizations;
- our knowledge of atmospheric processes has improved, in particular the parameterization of clouds, radiation, sub-grid scale processes, and surface interaction.

In regards to estimating climates some decades hence, there is again considerable uncertainty. Clearly the concentration of greenhouse gases is increasing, and all models, from the simplest one to the most complex climate model, agree that this results in higher global mean surface temperatures. On the other hand, interactions between the atmosphere and other "spheres" (mainly the biosphere and the hydrosphere), which are of generally negligible relevance to weather forecasting, are essential to climate prediction and are poorly or less understood. This uncertainty allows for a diversity of opinions in interpreting climate-modeling results and demands the request of a global "Earth System Model" to understand past variations and predict the future climate.

6. Conclusions

It has been recognized that climatic extremes might have common origins, as the general effects of global warming, the periodic changes of sea surface water temperature in the central and eastern equatorial Pacific Ocean and other modifications in SST which influences the sensible and latent heat fluxes across the air-sea interface and the atmospheric circulation at all scales. In the Mediterranean region, some evidences of these changes are almost evident: the modification in temperature distributions and extremes, the droughts and the general increase of dry days, the increase of autumn floods and the extreme rain events, the alteration of alpine glaciers. The models, particularly the reanalysis models, can help us to understand and assess some of these changes and how they are related to the others on the whole Earth; but we need to quantify the earth system as a whole to understand the climate evolution and the interactions among the atmosphere, the biosphere, and the hydrosphere. This "Earth System Model" has to be compared with the best possible observing system to provide a deep understanding of the past climate and a reliable perspective of the future.

Acknowledgement

The authors thank Dr. A. Perotto for reviewing this manuscript.

Reference

Kalnay, E., Kanamitsu, M., Kistler, R., et al., 1996. The NCEP/NCAR 40-year reanalysis project. Bulletin American Meteorological Society 77, 437–471.

Mountains Witnesses of Global Changes
R. Baudo, G. Tartari, and E. Vuillermoz (Editors)

18

Climate changes and mountains

Giovanni Kappenberger

Abstract

Climate changes are of several types and differ from one region to another. Global warming concerns most of the mountains areas. Glaciers are one of the best climate indicators of the ongoing changes. In this paper a few personal glaciological experiences are presented from three different places on the world: the mountains of the Canadian Arctic, the Swiss Alps and the Nepal Himalaya. For a better understanding of the ice loss due to the atmospheric warming, comparisons with the temperature records of the closest available recordings near the height of their equilibrium line are presented. In the case of the Himalaya, a simple evaluation has been done using the summer 0°C level in the free atmosphere, based on upper air data. With help from the radiosond data of New Delhi and assuming a constant precipitation amount and intensity during the monsoon months (June to September), it has been found that along the southern slopes of the Himalaya the 0°C and snowfall line have risen at least by 100–200 m in recent decades. A similar change has also been observed for the Swiss Alps, using the radiosond data of Milan. Some changes in the alpine flora are occurring along with the general glacier retreat. Two recent exceptional precipitation events, a flood in the Swiss Alps and an avalanche tragedy due to a severe and rather unusual late October storm in Nepal, are consistent with a warmer atmosphere showing a greater energy potential.

1. Introduction

Changes in the mountains are of several and different kinds; they concern the soil, flora, fauna, water, ice and the atmosphere in general and are of both quantitative or qualitative types. In this contribution a few, mainly personal experiences are presented. They are mainly related to the rising temperature of the atmosphere. The loss of ice of the retreating glaciers is the most visible change in the mountains around the world. Observations and measurements of glacier lengths and mass balance describe the general retreat.

ISSN: 0928-2025 DOI: 10.1016/S0928-2025(06)10018-8

In the first part of this contribution we take a quick look at three glaciers of three different regions of the globe: the Canadian Arctic, the Swiss Alps and the Nepal Himalaya. In each region we have a short overview of the temperature evolution in the last decades at a nearby place. Of course the summer temperature is only one parameter that describes glacier changes, though an important one.

In the second part of the contribution, again very short, mostly personal observations of the flora in Swiss Alps are presented. In the third part, two very recent extreme precipitation events, with loss of lives, are mentioned – one in the Swiss Alps and the other in the central Nepal mountains. They also could be related to rising temperatures and higher energy potential in the atmosphere.

2. Glacier changes and temperature evolution in three regions of the northern hemisphere

Summer temperature is only one, but the most important of the parameters determining glacier behaviour. Temperature is easier to measure than precipitation. Summer temperatures of the atmosphere at the height of the melting zone of the glaciers, at a reasonable distance from them, are a good indicator of a possible change in the behaviour of the nearby glacier. For this purpose, temperature records of measurements that have been done at a height close to the equilibrium line of the glaciers have been selected.

2.1. Canadian Arctic

2.1.1. Laika Glacier, Coburg Island

Laika Glacier is situated at 76°N in the Canadian Arctic and was studied in the 1970s (Blatter and Kappenberger, 1988). Its mass balance and movement were the subject of diploma thesis of the author in 1975. The glacier flowing down from a little Ice Cap, with a top of 500 m a.s.l. and reaching the plane close to the sea, has been retreating already in the past, according to an aerial picture of 1949. The glacier has not been investigated in recent times. Navigating on the new instrument of Google-Earth and Image (2005), a recent picture of the glacier was found covered by a thin layer of *Altocumulus* clouds, but with still quite visible ice edges. Compared to pictures from 1947 and 1971, the tongue shows a very important, even dramatic retreat.

The equilibrium line of Laika Glacier was at about 300 m a.s.l., some 30 years ago; according to the satellite photograph mentioned above, it has probably risen in the last decades (Kappenberger, 1976).

To have some atmospheric data from the surroundings, the station of Pond Inlet in the northern part of Baffin Island, about 350 km to the south of the little Coburg Island, was chosen. Pond Inlet is close to sea level, but can be considered quite representative for the climatic conditions of Laika Glacier at Coburg Island.

2.1.2. *Summer temperature, Pond Inlet, Baffin Island*

Meteorological data are available on the homepage of the Canadian weather service, some 30 years back (Environment Canada, 2005). The average of the monthly mean of maximum, mean and minimum summer temperatures of June, July and August have been selected and plotted on a graph.

The result is an increasing value of all three parameters by about 2°C from 1976 to 2004 (Fig. 18.1). Important to note, not only the temperature levels have gone up but also the average of the temperature minimum, which is now +2°, was about 0°C 30 years ago; this means that the length of the melting period has increased considerably too.

A short analysis of the summer precipitation – May–September – of the same time period during the last 3 decades shows an increasing amount, from 100 mm to about 130 mm, in the last years. It should be noted that more and more of the precipitation is falling as rain, flowing off immediately from the ice surface, so that only a part can be used for ice accumulation.

2.1.3. *Changes in the arctic*

The sharp rise of summer temperatures is well known in the Artic. The consequences are of different type and probably not yet all known. It is clear that the environmental conditions undergo a change. It is well known that the decreasing sea-ice thickness and the reduction of its extension will have grave consequences. One of them is the increasing difficulty of the polar bear to get enough food because the longer summer seasons reduce the pack ice, the hunting ground for seals, for the "King of the Arctic".

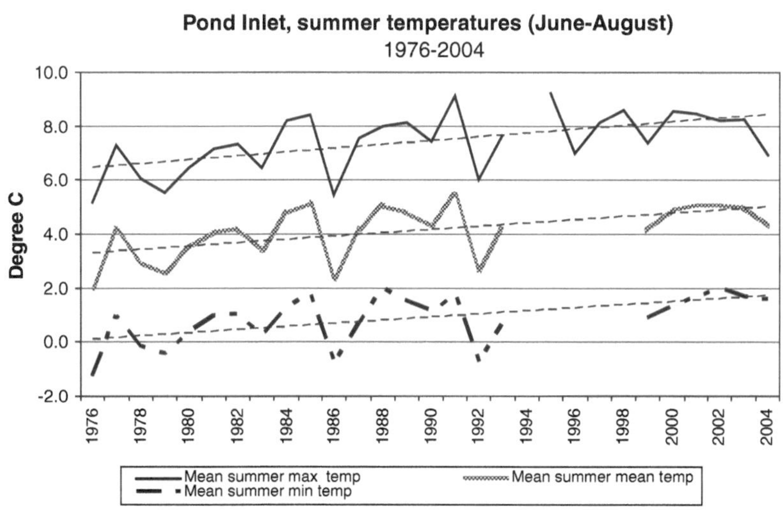

Figure 18.1. The picture illustrates the warming of the summer atmosphere in the Canadian Arctic. Average of the monthly mean of maximum, mean and minimum summer temperatures of June, July and August, Pond Inlet, northern Baffin Island.

2.2. Southern Swiss Alps

2.2.1. The Basodino Glacier

The Basodino Glacier is a small alpine ice body of about 2.3 km^2 surface, situated south of Gotthard Pass in the central–southern Swiss Alps. Mass balance, using the direct glaciological method, has been measured by the author since 1992 (Kappenberger et al., 1995; Aellen et al., 1996). In recent years an increasing ice loss has been observed. The equilibrium line was at about 2800 m a.s.l. earlier, but has risen considerably in the last years. This glacier can be considered representative for the other glaciers of the region, though most of them show an even more dramatic loss of ice (ETHZ, VAW, 2005).

Looking at an atmospheric parameter to explain the retreating glaciers, the summer temperatures of the nearby upper air station of Milan have been chosen.

2.2.2. Summer temperatures at 700 hPa, Milan

The upper air station of Milan is located at a distance of about 140 km from the Basodino Glacier and can be taken as representative for the air mass affecting the southern alpine slope. The value chosen to show a change in the atmospheric conditions is the sum of the positive °C-day at the 700 hPa level, being close to an altitude of 3000 m a.s.l.

Between 1954 and 2005, the sum of the positive °C-days during May through September has risen from about 300 to 450. This is equivalent to 30°C a month or to 1°C a day. Assuming a mean temperature decrease with hight of 0.6–0.8°C each 100 m, we can assume a rise of the 0°C height during the last half century of 150 to 200 m. If we analyse the graph, we can note that the rise was not gradual, but rather abrupt, with a step that took place between 1976 and 1980, probably due to a change in the atmospheric circulation (Fig. 18.2).

2.2.3. Changes in the Swiss Alps

Some examples of changes of the flora on top of mountains are presented in Section 3.

2.3. Central Himalaya, Nepal

2.3.1. Lirung Glacier

In the Langtang of Central Nepal, the Lirung Glacier ranges from more than 7000 m a.s.l. down to about 4200 m a.s.l., with an equilibrium line between 5000 and 5500 m a.s.l. During a stage in Langtang, in 1991 and 1992, some very simple glacier tongue measurements using simple methods have been done (Kappenberger et al., 1993). On three profiles we could find out that the tongue of the Lirung Glacier was practically without movement in the lower part. Also, we measured at one point a lowering of the debris-covered glacier surface of about 5 m. More recent measurements propose that there is no more ice mass flowing into the tongue (Yamada et al., 1992).

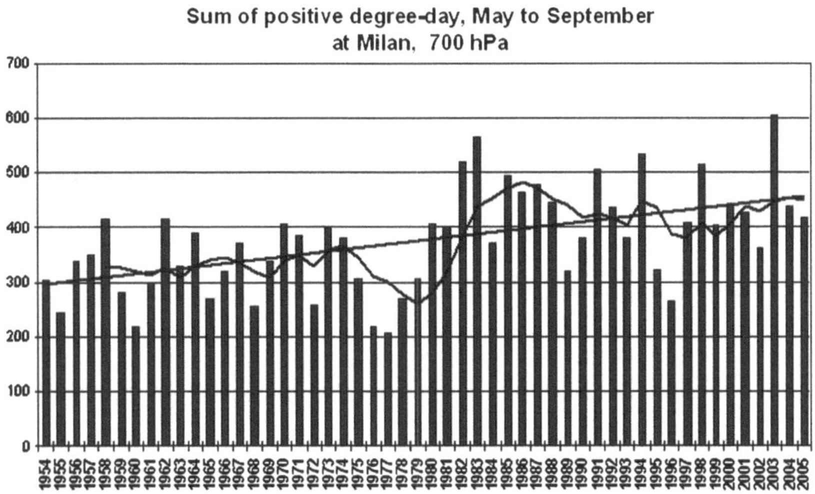

Figure 18.2. The picture illustrates the warming of the summer atmosphere in the Swiss Alps. Sum of the positive °C-day at the 700 hPa level of the upper air station of Milan, May to September. A major change was observed around 1980.

It is very difficult to get atmospheric values that explain the glacier behaviour. The meteo-hydrological station of DHM of Kyangjing at 3900 m a.s.l., close to the Lirung Glacier, shows a yearly temperature increase from 1.5 to about 4.0°C between 1998 and 2000 (DHM, 2001; Shresta et al., 2003). Looking for a temperature of the free atmosphere somehow representing the summer air mass flowing into Himalaya, the upper air station of New Delhi, located in the southwest of the mountain range, was chosen. Assuming that there has been only minimum humidity and precipitation change during the recent decades, we could assume that changes in the 0°C level are quite parallel to changes in the snow line of precipitation during the monsoon season.

2.3.2. Summer temperatures of the atmosphere, New Delhi

To minimize the influence of direct sun radiation on the radiosond, the 00 UTC ascent has been chosen to calculate the mean monthly height of the 0°C level. The information was taken from the University of Wyoming (2005) database, where it is possible to get every ascent back to 1973. Each daily 00 UTC ascent was requested online, and the height was taken directly or by doing an interpolation.

The fact of looking over all the ascent data helped a lot to find errors and missing data, or was useful just to get a feeling of the data quality. It turned out that especially in the 1970s several ascents were missing and some others showed mistakes of every kind. If more than about 10 days were missing, the month was not considered. The results of the monthly mean of the 0°C height for the main monsoon months – June, July, August and September – show a mean rise of about 100–200 m of the 0°C level in the last 30 years, with an even more important value of 300 m in the month of June. Considering a lapse rate of 0.6°C for 100 m, this means that the

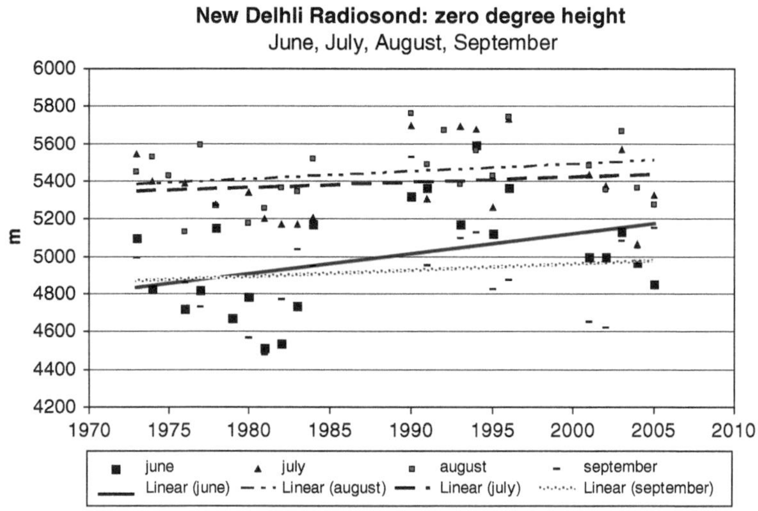

Figure 18.3. The picture illustrates the warming of the summer atmosphere south of the Himalaya. Monthly mean of the 0°C height for the main monsoon months June, July, August and September at the upper air station New Delhi.

temperature increase, close to the 5000 m a.s.l. height, is about 1–2°C for June to September and 2–3°C in June. According to New Delhi data, we can clearly see that the warmest period was recorded in the 1990s. The stronger warming of the month of June means that the melting season has become longer.

One of the main sources of error of the radio sounding is radiation warming. If we assume that in recent years the quality of the instrument and of the sensors has become better, this could indicate that today the measured values could be cooler rather than warmer (Joss, pers. com.). This means that the atmospheric warming would be even stronger (Fig. 18.3).

2.3.3. Changes in the Himalaya

Besides glacier changes (Oerlemans, 2001; Hoelzle et al., 2003; WGMS, 2005; WWF Nepal Program, 2005), we do not consider other changes in the Himalaya. A glaciological study in the Langtang region in the early 1990s proposed a scenario of different glacier extensions in a warmer atmosphere due a change in circulation, a stronger convection and the asymmetric consequences of an east–west trending valley (Kappenberger and Kerkmann, 1997).

3. Changes of the Alpine flora

3.1. Flowering date of Anemone nemorosa in the Southern Swiss Alps

In Prato Sornico, not far away from the Basodino Glacier in the southern Swiss Alps, the date of the flowering of *A. nemorosa* has been recorded since 1957. Of course, not

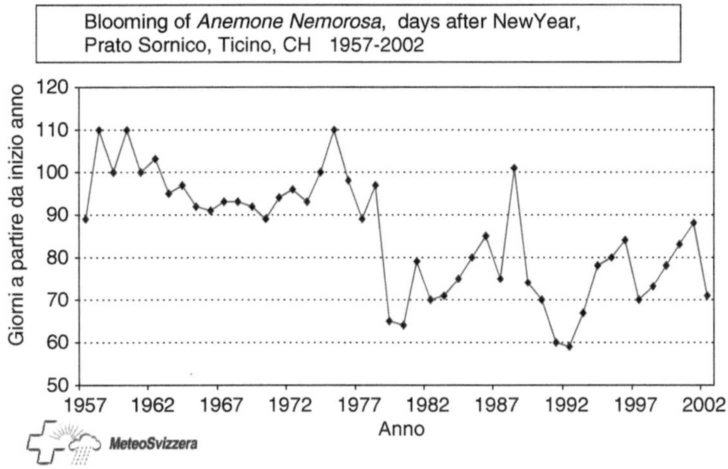

Figure 18.4. Date of the blooming of *Anemone Nemorosa*, in days after the beginning of the year, at Prato Sornico in the southern Swiss Alps. Since about 1980 the blooming date is about 2–3 weeks earlier.

only temperature but also sunshine, humidity and precipitation influence the moment of the blooming of the flower (MeteoSwiss, 2005). Before 1977, the blooming date was still about in the first days of April; after that year the mean date shifted to the middle of March, with a big variability. Again we can observe a change, probably due to a new circulation pattern, such as shown by the graph of the Milan radiosond (Fig. 18.4).

3.2. Change of number of species of alpine flora on top of mountains in Engadin, Swiss Alps

The number of species of the flora on the top of ten mountains, ranging from a height of 2959 to 3262 m a.s.l. in the Bernina region of Grison, has been investigated by the University of Zürich and the University of Hannover. Since the beginning of the last century, it was found to be more than a doubling of the number of species, with a steeper trend during the last years (Burga et al., 2004).

3.3. Tree line observations in the southern Swiss Alps

Two personal observations confirm a tendency of a rising tree line in the southern Swiss Alps; the first one is from the region of Pizzo Cadregh, close to the Lucomagno Pass, in the central Swiss Alps. At four positions above the local forest, at a height of 2000 – 2400 m a.s.l., several dozens of small trees of *Pinus cembrus* and *Larix deciduas* were planted in the beginning of the 1990s. After 12 years, almost all of them had been eaten by the animals. In summer 2005 a few small *Pinus cembrus* have been found, growing in a natural way.

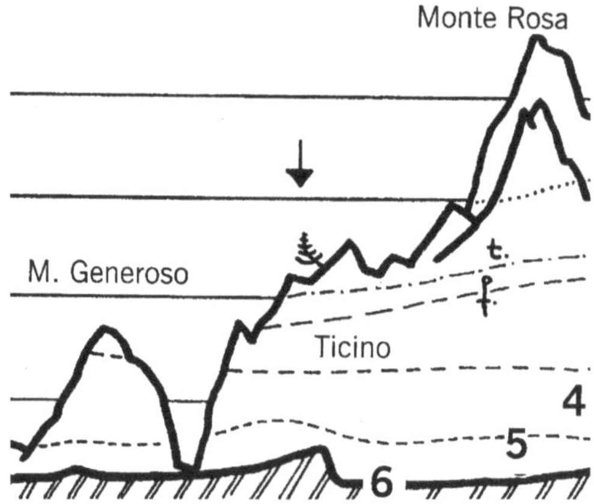

Figure 18.5. Rising tree line in the Southern Swiss Alps. Observation of *Larix Decidua* (arrow) in summer 2005, at 2300 m. Forest line (f) is at 1800 m and tree line (t) at about 2000 m. Height (1 to 4) in km. South is to the left.

The second observation, from September 2005, was made in the Onsernone Valley, not far from Locarno, in the south of Switzerland. In this region the tree line is at a height of about 2000 m a.s.l. One single tree, more then 2 m tall, is growing at 2300 m a.s.l. on a rocky ridge, not only away from snow pressure or avalanche influence but also out of the range of animals.

Forest specialists are not yet sure if the better development of the trees above the actual tree line is just a matter of "returning to more natural conditions" or if there is a warming effect as well (I. Ceschi, pers. com.) (Fig. 18.5).

4. Extreme precipitation events

Single extreme precipitation events cannot prove a climatic change, but they are included in the scenario of climatic change with a warming atmosphere. The moisture content of warmer air as well as the energy potential can increase strongly. Two strong precipitation events occurred in Switzerland and Nepal shortly before the conference of Rome in November 2005. They could be a signal of change in the atmosphere, which will affect mountains too.

4.1. The flood of August 2005 in the Swiss Alps

From August 20 to 23, 2005, the northern part of the Swiss Alps was hit by heavy precipitation, which caused tremendous floods, debris flows and heavy loss of life and properties. The return time of such a meteorological situation was indicated to be 200 to 300 years (MeteoSwiss, 2005). An important part of the floods has been caused by the huge amount of material that was ripped down from the mountains. The rain fell up to an altitude of more than 3000 m a.s.l. Glaciers were covered with only very little snow, left from the previous winter, so the water could not be stored. The runoff water found a lot of debris left behind by quickly retreating glaciers, in the form of morainic material of different sizes. It is probable that slowly melting permafrost at high altitude has contributed to moving material down with the water (MeteoSwiss, 2005).

4.2. The avalanche tragedy of October 20, 2005 in Central Nepal

On October 20, 2005, a huge avalanche hit a big group of mountaineers at Mt Kanguru, North of the Annapurna, in Central Nepal. (Everestnews, 2005). Some 18 people have been killed by the snow mass descending the mountain and destroying the expedition camp, with sherpas, porters and seven French people. Between October 18 and 20, strong precipitation was recorded in Bahirahawa (more than 200 mm) and in Pokhara (about 150 mm) (Manandhar, DHM 2005, pers. com.). Both Nepalese weather stations are located to the southwest of Mt. Kanguru, more or less in the advection zone of the storm. It could be possible that similar amounts of precipitation also hit Mt. Kanguru, which means about 1–2 m of fresh snow. The detailed conditions on that mountain on October 20 are not known, but the fact that professionals were killed – the leader of the French group Daniel Stolzenberg was a mountaineer instructor of Nepali sherpas, with a lot of Himalayan experience – means that the situation must have been quite exceptional.

5. Conclusions

Changes in the mountains are several and of different kind; they concern the soil, flora, fauna, water, ice and the atmosphere in general. They are of quantitative or qualitative aspects. Only a few, mainly personal experiences are presented. Glacier retreat is the most obvious one, in the Arctic, the Alps and in the Himalaya. The reasons for retreat of the glaciers can be several, but one of the main parameters is the rising summer temperature. In all three regions mentioned it is possible to show an increasing warming at the height of the equilibrium line of the glaciers, as shown by several authors and summarised in the IPCC reports (IPCC Climate change, 2001).

Two other examples are considered: the slow changes in the alpine flora and the occurrence of extreme events, in our case two heavy precipitation events that hit the Swiss Alps in August and Central Nepal in October 2005, both with tragic consequences. In Switzerland floods caused loss of life and properties; in Nepal a snow avalanche killed a big climbing group with mountain specialists. A. Roch, a well known avalanche researcher, who worked last century in the Swiss Federal Institute for Snow and Avalanche Research at Davos, said: "The avalanche doesn't know that you are a specialist down there". If we apply this kind of idea to environmental researchers, we should always keep in mind the need to respect Nature first of all.

References

Aellen, M., Kappenberger, G., and Casartelli, G., 1996. Il ghiacciaio del Basodino (Alpi meridionali Svizzere).Geografia Fisica e Dinamica Quaternaria vol. 18.

Blatter, H. and Kappenberger, G., 1988. Mass balance and thermal regime of the Laika Ice Cap, Coburg Island, N.W.T, Canada. Canadian Journal of Glaciology 34 (116), 102–110.

Burga, C., Walther, G., and Beissner, S., 2004. Florenwandel in der alpinen Stufe des Berninagebiets-Ein Klimawandel? Berichte der Reinhold-Tüxen-Gesellschaft, Hannover 16, 57–66.

DHM. 2001. Department of Hydrology and Meteorology, Kathmandu.

DHM. 2005. Department of Hydrology and Meteorology, Kathmandu, http://www.dhm.gov.np

ETHZ, VAW, 2005. http://www.vaw.ethz.ch/research/glaciology/glacier_change/gz_swiss_glacier_monitoring.

Environment Canada, 2005. http://www.climat.meteo.ec.gc.ca/climateData.

EverestNews, 2005. http://www.everestnews.com/stories2005/missing10242005.htm

Google-Earth, Image C 2005. Digital-Globe, http://www.earth.google.com

Hoelzle, M., Haeberli, W., Dischl, M., and Peschke, W., 2003. Secular glacier mass balances derived from cumulative glacier length changes. Global and Planetary Change 36, 295–306.

IPCC: Climate Change, 2001. Contribution of Working Group I to the Third Assessment Report of the Intergovernmental Panel on Climate Change. Cambridge University Press, http://www.ipcc.ch

Kappenberger, G., 1976. Mass balance and movement of the Laika Glacier. Diploma thesis, unpublished 46 pp.

Kappenberger, G., Aellen, M., and Casartelli, G., 1995. Il ghiacciaio del Basodino. Nimbus Nr. 8.

Kappenberger, G. and Kerkmann, J., 1997. Il tempo in montagna. Edizioni Zanichelli, Bologna.

Kappenberger, G., Steinegger, U., Braun, L.B., and Kostka, R., 1993. Recent changes in glacier tongues in the Langtang Khola Basin, Nepal, determined by terrestrial photogrammetry. IAHS Publ. no. 218.

Manandhar, K.H. DHM, 2005. Pers. Com.

MeteoSwiss, 2005. http://www.meteoswiss.ch/web/en/climate.html

Oerlemans, J., 2001. Glaciers and Climate change. Balkema Publishers, Lisse.

Shresta, K.L., Shresta, M.L., Shakya, N.M., et al., 2003. Climate change and water resources of Nepal. Climate Change and Water Resources in South Asia. pp. 259. Asian Agro Dev International, Kathmandu.

University of Wyoming, 2005. http://weather.uwyo.edu/upperair/sounding.html

World Glacier Monitoring Service (WGMS) IUGG (CCS) – UNEP – UNESCO – WMO, 2005. Glacier Mass Balance Bulletin No. 8, Zürich.

WWF Nepal Program, 2005. An overview of Glaciers, Glacier Retreat and subsequent impacts in Nepal, India and China.

Yamada, T., Shiraiwa, T., Kadota, T., et al., 1992. Fluctuation of the glaciers from the 1970s to 1989 in the Khumbu, Shorong and Langtang regions, Nepal Himalayas. Bulletin of Glacier Research 10, 11–19.

SHARE-Asia Scientific Fields: Environmental Indicators of Global Changes

Mountains Witnesses of Global Changes
R. Baudo, G. Tartari, and E. Vuillermoz (Editors)

19

Global scale atmospheric pollution: a regional problem

Ivo Allegrini and Nicola Pirrone

Abstract

Atmospheric pollutants, whether man made or naturally occurring but released in greater concentrations due to man's activities, can now be found at all latitudes and longitudes throughout the boundary layer, and depending on their resistance to photolysis and photo-oxidants, within both the free troposphere and stratosphere. The key to pollutant transport is longevity of the compound concerned and the circulation patterns of the atmosphere. Pollutants that are relatively unreactive and volatile will, with time, be spread across the globe, eventually reaching even the most remote areas. Even semi-volatile pollutants can be transported by successive phases of volatilisation and deposition (multi-hop) until they too are spread across the globe. One of the implications of this world-wide distribution is that pollutants may reach areas where specific climatic, physical or chemical conditions occur that result in increased deposition fluxes to marine and terrestrial surfaces, with the consequence that pollution hot spots may be created at vast distances from the original pollution source. The Polar regions, particularly the Arctic because it is in the same hemisphere as the major industrialised nations, are a good example. Levels of persistent organic pollutants (POPs) and mercury (Hg) in polar wildlife, particularly larger predators, have been known for some time to be higher than the norm at lower latitudes. Mercury is deposited in the Arctic, not just because of its reduced volatility due to low temperatures, but also as a result of specific atmospheric chemical conditions related to the tropospheric ozone depletion events. These events, prompted by the so-called 'Bromine Explosion,' which releases reactive halogen compounds to the troposphere, create conditions in which the atmospheric oxidation of $Hg^0_{(g)}$ occurs with such rapidity that its concentration can fall below detectable limits in a matter of hours. The oxidation products that are much more soluble than elemental Hg are deposited along with particulate matter or dry deposited, and at a time (around polar dawn) when the Arctic ecosystem is entering an active phase with the return of sunlight. There is evidence that Hg released by coal-burning power stations in China can be deposited in rain along the Pacific coast of the U.S. Hg released in the U.S. could potentially be transported with other atmospheric pollutants across the Atlantic to Europe and the Mediterranean. Industrial emissions in northern

ISSN: 0928-2025 DOI: 10.1016/S0928-2025(06)10019-X

Europe due to the general north to south flow of the boundary layer air in summer months have a negative effect on Mediterranean air quality (high O_3 concentrations) and it seems, on atmospheric Hg deposition to the Mediterranean Sea. If the description above seems rather qualitative it is because as yet the tools with which to model (and to a certain extent measure) atmospheric transport, transformation, deposition and re-emission on such a scale are in their infancy. There is a great need for strategically situated monitoring stations, creating a network to sample within the major air mass trajectories, not only at ground level but also at altitude. This would allow models and the emission inventories of pollutant fluxes to be validated in a far more comprehensive manner than isolated measurement campaigns permit. From validated models, it would then be possible to evaluate the potential effects of changing emissions resulting from legislation or changes in, for example, the power generation industry towards renewable energy sources. Finally, it would be possible also to investigate changes in pollution transport and deposition that occur with changing atmospheric transport patterns resulting from climate change.

Mountains Witnesses of Global Changes
R. Baudo, G. Tartari, and E. Vuillermoz (Editors)

20

Platinum group elements and other trace elements in high altitude snow and ice

Giulio Cozzi, Carlo Barbante and Paolo Cescon

Abstract

Atmospheric pollution by heavy metals is an important problem in Europe. Considerable efforts have been made to assess the problem, especially through extensive monitoring programs and emission inventories. A major difficulty with such approaches is that they provide information only for recent decades and do not allow us to go back in time to the pre-industrial period, which is necessary to put recent changes in proper perspective. Information on past changes in atmospheric pollution for heavy metals can only be obtained from atmospheric archives such as peat bogs, lake sediments, or high-altitude alpine snow/ice cores. Here we present time series for various metals (Pb, Pb isotopes, Cr, Mn, Cu, Zn, Co, Ni, Mo, Rh, Pd, Ag, Cd, Sb, Bi, Pt, Au, U), obtained from the analysis of two well-dated snow/ice cores from Colle Gnifetti, which cover ~350-year period from the 1650s to the mid 1990s. These elements were chosen because they are emitted by human activities to the atmosphere and might pose a threat to the environment, especially in populated areas such as Europe. The data are compared with those obtained in areas subject to different anthropic influences, such as the Andes and the Himalaya, where we recently obtained an ice core extending back in time to 400 A.D. Platinum group elements (PGEs) have also been successfully determined in Greenland ice cores, showing how pollution from these elements is a global-scale process.

1. Introduction

Atmospheric pollution by heavy metals is an important problem in Europe and elsewhere in the world. Many efforts have been made to assess such pollution, especially through extensive monitoring programs; but this does not allow assessments back into pre-industrial times so that we can put recent changes into a proper perspective. Only ancient archives such as peat bogs, lake sediments, or high-altitude alpine snow/ice cores can give information on past changes of heavy metals in European atmospheric pollution.

ISSN: 0928-2025 DOI: 10.1016/S0928-2025(06)10020-6

An historical record for many heavy metals, dated from the 17th century to the mid 1990s, has been obtained from the analysis of sections of two ice cores drilled at Monte Rosa on the Swiss–Italian border. The study has focused especially on historical changes of lead because of a lack of glaciochemical data for this element in such a densely populated area as Europe.

1.1. Experimental aspects

The samples analyzed in this study originated from two cores: a 109 m firn/ice core drilled at the Colle Gnifetti in 1982 and a 25 m firn core drilled at the same site in 1995 (4450 m a.s.l., 45°53′33″ N, 7°51′5″ E. The Colle Gnifetti is the uppermost part of the accumulation area of Grenzgletscher, forming a glacier saddle between Zumsteinspitze and Punta Gnifetti at 4400–4550 m a.s.l. in the Monte Rosa massif. Monte Rosa itself, located at the Swiss–Italian border, is the second highest mountain in western Europe (4634 m a.s.l.).

Age dating of the cores was performed by combining several methods; counting of annual layers from continuous concentration profiles of seasonally varying species, such as ammonium, calcium and sulphate; use of stratigraphic markers, such as large Saharan dust events, atmospheric nuclear tests and major volcanic eruptions; ^{210}Pb measurements for the post-1900 period; and a three-dimensional ice flow model. The results show that the upper 83 m of the 109-m snow/ice core analyzed in this work covers a period of 330 years from the 1650s to 1982, whereas the upper 13 m of the 25-m snow core analyzed in this work covers a period of 22 years, from 1972 to 1994.

All pre-analytical operations took place in clean-room conditions. The core sections were first mechanically decontaminated in cold rooms to remove contamination from the core exterior from the drilling operations. The inner cores so obtained were then analyzed using a variety of techniques. Inductively coupled plasma sector field mass spectrometry (ICP–SFMS) was used to determine Cu and Zn in all the sections and Cr, Co, Ni, Mo, Rh, Pd, Ag, Sb, Pt, Au, and U in some of them. Graphite furnace atomic absorption spectrometry (GFAAS) was used to determine Cu, Cd, and Zn in some of the samples. Laser-excited, atomic-fluorescence spectrometry (LEAFS) was used for the determination of Bi in some of the samples. Typical uncertainties ranged from 5 to 20% (RSD). Finally, Pb and Pb isotopes were determined either by ICP–SFMS or thermal ionization mass spectrometry (TIMS).

1.2. Results and discussion

1.2.1. Heavy metals

A striking feature of the data set is that there is a pronounced variability in concentrations. Concentration differences of up to 2–3 orders of magnitude are observed between nearby core sections. This confirms that there are pronounced short-term (inter-annual and intra-annual) variations in heavy metal inputs to Colle Gnifetti. This fact was explained by seasonal changes in the efficiency of vertical transport of boundary layer air to high-alpine sites. Figure 20.1 shows long-term changes in the

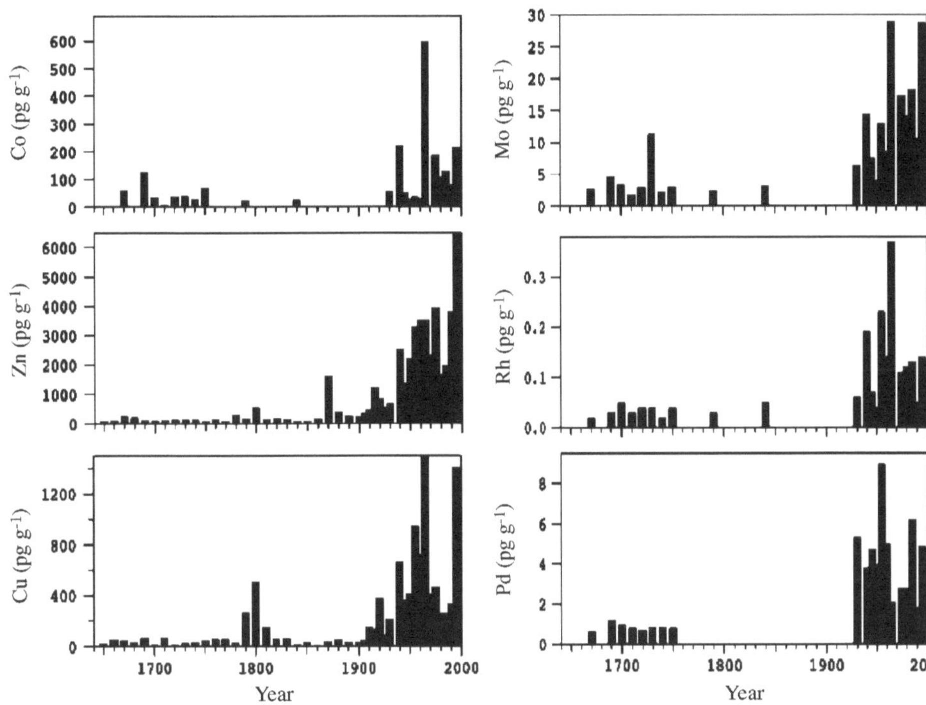

Figure 20.1. Changes in heavy metals concentrations in dated snow/ice from Colle Gniffetti (Monte Rosa massif).

concentration of Cr, Co, Zn, Co, Ni, Mo, Rh, Pd, Ag, Cd, Sb, Pt, Au, and U from the 1650s onward. To isolate the long-term time trends, we have averaged individual data points over periods of several years to compensate for the pronounced short-term variations discussed in the previous section. For most metals, concentrations remained fairly low until the beginning of the 20th century, but then show highly enhanced values especially during the second half of the last century.

Table 20.1 gives the mean factors of increase of concentrations for the various metals from the pre-1700 period to post-1970. The post-1970 period was chosen because it corresponds to the period with the highest concentrations for many metals; conversely, the pre-1700 period was selected because it corresponds to low anthropogenic inputs, before the Industrial Revolution. The highest factor of increase is observed for Cd (~36), followed by Zn (~19), Bi (~16), and Cu and Ni (~10) (Table 20.1). The opposite is observed for Au, with a factor that is close to one. These pronounced increases of concentration for most metals reflect the increasing pollution of the atmosphere by emissions of heavy metals in Europe.

1.2.2. Lead and lead isotopes

Lead concentrations show a large variability ranging from 10 to 19,700 pg g^{-1}, indicating that lead behaves in a manner similar to that of other aerosol constituents in alpine snow and ice. This is caused by the meteorological conditions prevailing in the

Table 20.1. Colle Gnifetti: Average concentrations (in pg g^{-1}) in ice, dated from before 1700, and snow, dated from after 1970; increase factors between these two periods are also reported.

	Average concn before 1700	Average concn after 1970	Increase factor
Cr	144	350	2.4
Cu	39	412	10.6
Zn	171	3176	18.6
Co	70	144	2.1
Ni	24	218	9.1
Mo	3.4	18.0	5.4
Rh	0.030	0.10	3.3
Pd	0.86	3.7	4.2
Ag	0.34	1.16	3.4
Cd	1.6	58	36.3
Sb	21	53	2.5
Bi	0.25[a]	3.9[b]	15.6
Pt	0.15	0.40	2.7
Au	0.16	0.13	0.8
U	2.9	4.7	1.6

[a]No data are available before 1700; the value quoted here is for the 1780s.
[b]No data are available after 1970; the value quoted here is for the 1950s.

Alps, favoring convective transport of the boundary layer air to high-altitude sites such as Colle Gnifetti.

In Fig. 20.2 the lead paleo-record is illustrated, showing an overall increasing trend reaching a maximum concentration in the late 1960s and early 1970s with a factor of increase of 25 between the 1650–1800 value (130 pg g^{-1}) and the 1970s maximum (3160 pg g^{-1}). This is the first historical time series of lead concentrations from a mid-latitude glacier ever published that easily predates the Industrial Revolution. The record mainly reflects changes in fallout of anthropogenic lead from Western Europe, since natural contributions from sources such as rock and soil dust are always minor. A major feature observed in Fig. 20.2 is the large increase in lead concentrations from about the 1930s to the 1970s, and the subsequent decrease to the mid-1990s. It is clearly linked with the rise and fall of the use of lead additives in European countries, which peaked in about 1970, as can be seen in the example of lead emissions from traffic in Switzerland.

Figure 20.3 presents the lead isotopic record $^{206}Pb/^{207}Pb$ for the period 1650–1982. The temporal trend of the isotopic composition is characterized by a constant level of about 1.18 for the period from 1650 to the 1880s, followed by a gradual decrease to a value of about 1.16 in the 1950s. A steeper decrease to a ratio of 1.12 is then observed for the time period 1950–1975. Afterward, a slight recovery to more radiogenic values (higher $^{206}Pb/^{207}Pb$) seems to occur. Ratios of $^{206}Pb/^{207}Pb$ show a so far unexplained higher variability during pre-industrial times (1650–1880). The oldest ice from 1650 to about 1820 shows $^{206}Pb/^{207}Pb$ ratios, which could still be typical of coal combustion or mining and which are just approaching the natural background range, indicating an anthropogenic influence as early as 1650. Between 1820 and 1900 the

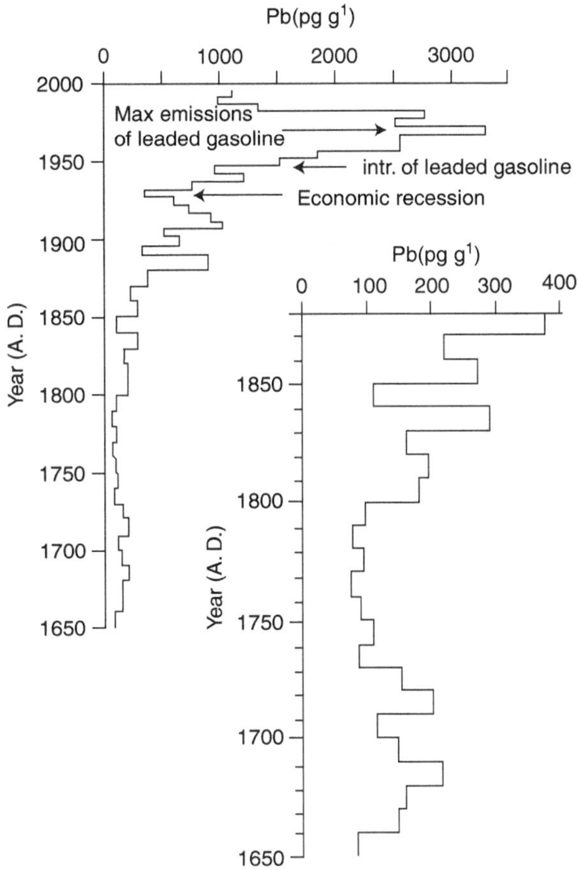

Figure 20.2. Combined lead paleo-concentration record from the Colle Gnifetti cores as 5-year (period 1890–1995) and 10-year averages (period 1650–1890). The period from 1650 to 1880 is additionally shown with a magnified horizontal axis, to take into account the lower lead concentrations.

$^{206}Pb/^{207}Pb$ ratios seem to be composed of anthropogenic lead emitted from coal combustion and mining, and the natural background. In this time period the contribution of crustal material to the isotopic composition was still significant, as indicated by $^{206}Pb/^{207}Pb$ ratios approaching the natural background range during intervals with high concentrations of Al and, thus, high amounts of crustal material (e.g., 1870–1880). From 1930 onward $^{206}Pb/^{207}Pb$ ratios decreased significantly, pointing to additional anthropogenic sources becoming important, e.g., waste incineration and the use of leaded gasoline. In the period 1960–1975 the steepest decline of the overall record was observed, reflecting the increasing amounts of leaded gasoline consumed. From 1973 until the early 1980s $^{206}Pb/^{207}Pb$ ratios were typical for values measured in car exhausts, indicating that this source was dominant. Interestingly, these latter values correspond to the period 1972–1981 of the Italian isotopic lead experiment (IILE), during which lead in gasoline in the Piedmont region of northwest Italy and centered on Turin was replaced with Australian lead, whose

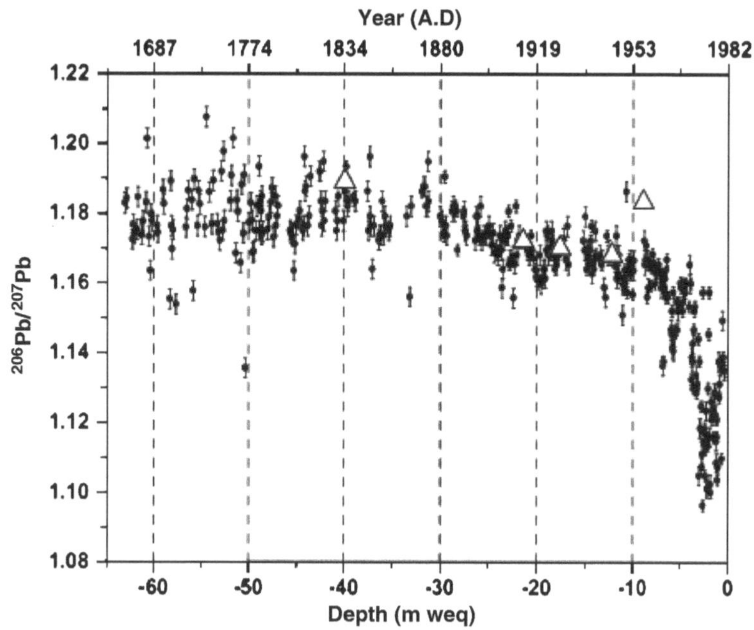

Figure 20.3. Isotopic composition of $^{206}Pb/^{207}Pb$ determined in ice core sections against depth in the glacier. Error bars indicate the relative standard error of the $^{206}Pb/^{207}Pb$ ratio. Open triangles show values obtained by TIMS (uncertainty at 95% confidence interval smaller than triangle size).

$^{206}Pb/^{207}Pb$ ratio ($\sim$1.04) was significantly lower than that for lead used before in that region. The $^{206}Pb/^{207}Pb$ ratio of recent snow, collected between the years 1993 and 1996, shows a reversal of the trend to more radiogenic values as a consequence of the reduced use of leaded gasoline.

2. Conclusions

This study has shown highly enhanced concentrations for most metals in snow/ice dated from the second half of the 20th century, compared with concentrations in ancient ice dated from the 17th and 18th centuries. The highest increase factors from the pre-1700 period to the post-1970 period observed have confirmed the importance of atmospheric pollution by heavy metals in Europe. Furthermore, the first glaciochemical time series with annual resolution has been presented that spans several centuries of lead concentrations and lead isotopic compositions in precipitation in Europe. Lead concentrations in firn dated from the 1970s were 25 times higher than in ice dated from the 17th century, confirming the massive rise in lead pollution in Europe during the last few centuries. A decline of the lead concentration has been observed during the last two decades, i.e., from 1975 to 1994. The lead isotope ratio $^{206}Pb/^{207}Pb$ decreased from about 1.18 in the 17th and 18th centuries to about 1.12 in the 1970s demonstrating that these variations are in good agreement with available information on variations in anthropogenic lead emissions from West European countries, especially from the use of lead additives in gasoline.

Further reading

Barbante, C., Schwikowski, M., Döring, T., et al., 2004. Historical record of European emissions of heavy metals to the atmosphere since the 1650s from Alpine snow/ice drilled near Monte Rosa. Environmental Science and Technology 38, 4085–4090.

Barbante, C., Veysseyre, A., Ferrari, C., et al., 2001. Greenland snow evidence of large scale atmospheric pollution for platinum, palladium and rhodium. Environmental Science and Technology 35, 835–839.

Van de Velde, K., Barbante, C., Cozzi, G., et al., 2000. Changes in the occurrence of silver, gold, platinum, palladium and rhodium in Mont Blanc ice and snow since the 18th century. Atmospheric Environment 34, 3117–3127.

Mountains Witnesses of Global Changes
R. Baudo, G. Tartari, and E. Vuillermoz (Editors)

21

High altitude lakes: limnology and paleolimnology

Andrea Lami, Gabriele A. Tartari, Simona Musazzi, Piero Guilizzoni, Aldo Marchetto, Marina Manca, Angela Boggero, Anna M. Nocentini, Giuseppe Morabito, Gianni Tartari, Licia Guzzella, Roberto Bertoni and Cristiana Callieri

Abstract

The most remote regions of globe represent some of the least disturbed ecosystems, yet they are threatened by air pollution and by climatic change. The Himalaya is one of the most isolated regions in the world and least explored wildernesses outside the Polar Regions; and it is for this reason that the Tibetan Plateau is often referred to as the 'Third Pole'. Limnological survey (including chemistry, biology and sediment core studies) of lakes located between ca. 4500 and 5500 m a.s.l. has been performed from 1992 in the Kumbhu Valley, Nepal. Lake water chemical surveys reveal a constant increase of the ionic content of the lake water probably related to glacier retreat. Modern phytoplankton data compared with previous data point to an increasing trend in lake productivity. Zooplankton, benthos and thechamoebians provide useful biogeographical information. Paleolimnological reconstructions show the potential use of these sites in providing proxy data of past climatic changes in high altitude regions. Data collected of persistent organic pollutants show that the studied sites receive input related to long-range transport pollution. The aims and rationale for the future development of the Ev-K^2-CNR Limnological Information System is discussed.

1. Introduction

The Himalaya and Mt. Everest have fascinated human beings for centuries. This area saw the flourishing of some of the major civilisations, such as the Brahmaputra, the Ganges and the Indus. Since the first ascent by Sir Edmund Hillary and Tenzing Norgay Sherpa in 1953, about 700 people have reached the summit of Mt. Everest. The Himalaya is however important not merely in geographic or mountaineering terms, but also, in common with other mountain regions of the world, they represent a significant global resource. The worldwide importance of mountains in term of

ISSN: 0928-2025 DOI: 10.1016/S0928-2025(06)10021-8

water resources, biodiversity, recreation, tourism and culture of mountain resources is summarised by Messerli (1997).

A large number of international organisations are devoting increasing attention to mountain issues, e.g. the United Nations, the World Bank, the Inter-governmental Panel on Climate Change (IPCC), the Consultative Group on International Agricultural Research (CGIAR) and Diversitas and the International Union of Forestry Research Organisations (IUFRO). A specific point was also included in the Plan for action into the 21st century (Agenda 21, Chapter 13) giving mountain regions a priority in the global environment-development agenda equal to that of other global change topics such as climate change, desertification or deforestation.

Given their three-dimensional nature, mountains encompass the most extensive array of topography, climate, flora and fauna, as well as diversity of human culture, known on earth. Not enough, however, is known about mountain ecosystems. The creation of a worldwide database about mountains is therefore vital for launching programmes contributing to the sustainable development of mountain ecosystems (Agenda 21, Chapter 13). Since Rio de Janeiro 1992 there has been considerable progress in this direction, especially in the Hindu Kush–Himalayan Region (MENRIS–ICIMOD) and in the Alps (Alpine Forum).

The involvement of the Pallanza Institute in mountain lake research dates back to the 1950s (Marchetto, 1998), during the 1980s and 1990s it has taken an active part in several research projects in the Alpine area supported by the European Union, such as the AL:PE (Acidification of Mountain Lakes: Palaeolimnology and Ecology), MOLAR (Mountain Lake Research) programmes and EMERGE (Integrated Project to Evaluate the Impacts of Global Change on European Freshwater Ecosystems). The interdisciplinary joint project 'Ev-K^2-CNR' between Italy and Nepal offered us a unique opportunity to devise and undertake a scientific research programme in the Himalayan region. This research project was made possible through an international agreement between the Italian Foreign Ministry and the Nepal Academy of Science and Technology (NAST).

Limnological research in the Himalayas has been carried out since the beginning of the century (Sars, 1903; Hutchinson, 1937). Up to the 1970s studies were sporadic and oriented towards characterising the biotic communities in lakes and comparing the tropical areas affected by a monsoon climate with temperate zones (Troll, 1959; Hirano, 1963; Ueno, 1966; Löffler, 1969; Zutshi and Vass, 1970). Recent decades have seen a more detailed approach, with greater focus on morphometric, physico–chemical and biological features, primary productivity and trophic status, with particular emphasis on fish production. Results have generally highlighted the very low concentrations of dissolved minerals and nutrients, and the limited plankton assemblages in lakes at altitudes above 4000 m a.s.l. Many of these studies were performed on high altitude lakes in Kashmir and Sikkim in the northwest Himalaya (Khan and Zutshi, 1980; Sharma and Pant, 1985; Vass et al., 1989; Zutshi,, 1991), while there has been comparatively little research in the eastern, Nepalese Himalayas (Löffler, 1969; Aizaki et al., 1987). In Nepal, which is particularly rich in surface water resources, limnological studies have largely involved low altitude lakes (Swar, 1980; Jones et al., 1989; Bhandari, 1993) because of their sensitivity to eutrophication phenomena due to the extensive use made of the water by the local populations.

Interest in high altitude lakes has generally been reserved for large lakes, such as Lake Tilitso on the high altitude lakes (4920 m a.s.l.) in the Himalaya (Aizaki et al., 1987). A study of particular interest for the attention it devotes to small lakes was made by Löffler (1969), and involved 24 lakes at altitudes between 4500 and 5600 m in the Mt. Everest area (Khumbu Valley), providing the first data on morphometry, temperature, chemistry and biology. More recently the multidisciplinary environmental studies, performed within the Ev-K^2-CNR Project, were summarised in Baudo et al. (1998) and Lami and Giussani (1998).

The aim of this paper is therefore to present an overview of the limnological and paleolimnological activities performed up to now, and highlight some of the results obtained, as well as the potential application and future use of an integrated database about the limnology of high altitude freshwater lakes. This is one of the few attempted to produce such a large database on these remote lakes. This database will contribute to filling the gaps in our present knowledge and furthering our understanding of human impact in remote areas.

2. Study area

The area of limnological research referred to in the Limnological Information System (LIS) consists of the watersheds of the Imja Khola and Ngozumpa (Fig. 21.1). These two valleys belong to the southernmost part of the wide watershed of the Dudh Kosi, which drains into the Ganges on the plain near Chatra. The area falls between latitudes N 27°48′ and 28°05′ and longitudes E 86°39′ and 86°59′; it has a surface area of around 650 km^2 and covers 57% of the territory of the Sagarmatha National Park (1148 km^2) in the Khumbu Region of East Nepal. It includes Mt. Everest (8846 m a.s.l.), re-measured in 1992 in the framework of the Ev-K^2-CNR Project (Poretti, 1998), and the Khumbu Valley, which leads to the mountain; this is a sub-basin of the Imja Khola, which drains the south side of Lhotse (8516 m) and the north side of Ama Dablam (6814 m), as well as many other small sub-basins, prominent among which is the one containing the largest lake in the area, Tshola Tsho (Fig. 21.1, Lake 24). The waters of Imja Khola merge to the south (27°50′ N, 86°45′E) with those coming from the long Gokyo Valley. This latter area, which has been included in the research programme only since 1997, contains along the valley floor an interesting group of intermorainic lateral lakes connected by waterfalls, and drains the melt water from the Ngozumpa Glacier on the southern slopes of Cho Oyu Peak (8153 m).

3. Limnological research

3.1. Limnological information system (LIS)

Since 1989 up to nowadays, a large effort has been dedicated to develop the LIS and to perform scientific expeditions with the aim of visiting and sampling the lakes that have been recognised on the map and to populate the lake database (Tartari et al.,

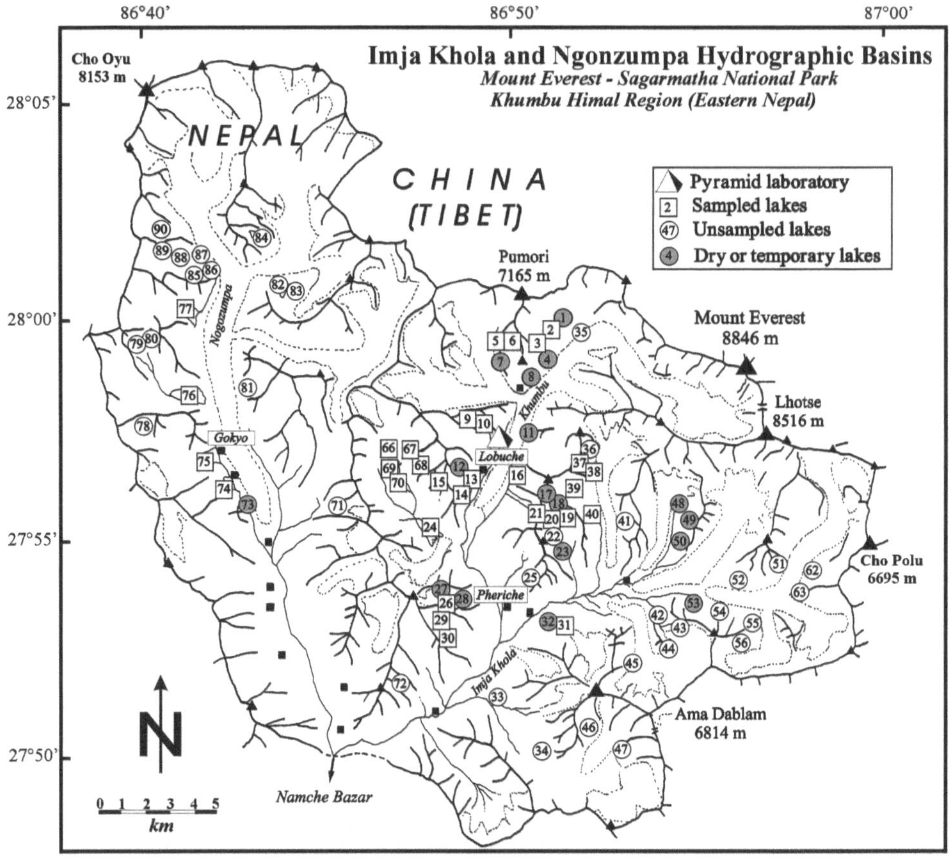

Figure 21.1. Ev-K^2-CNR inventory map of lakes of Imja Khola and Ngonzumpa watershed. The figure shows the location of 83 lakes included on the 1:50,000 'Mount Everest' (National Geographic Society) and 'Khumbu Himal' (Nelles Verlag) maps location of the lakes included in the Ev-K^2-CNR inventory. Modified from Tartari et al. (1998a).

1998a). The present situation is that 48 lakes were visited. Of these lakes 31 were sampled, while 17 turned out to be dry, silted up, frozen etc. Considering that 7 lakes were situated outside the Imja Khola and Ngonzumpa watershed, and that the group comprising lakes 78 to 90 was only recently added to the list to complete the picture of the potential environments included in the watershed, we can conclude that the fieldwork examined most of the lakes (about 70%) of interest in the area already reached by the expeditions. All the lakes are situated between 4460 and 5645 m a.s.l., with an altitude distribution presenting the maximum frequency between 5100 and 5300 m (Fig. 21.2a). The distribution of the lakes sampled also followed exactly the same trend of altitude frequency, confirming their good representativity. As regards size (Fig. 21.2b), most of the lakes in the LIS (2/3) have an area of less than 0.02 km^2; of the others, about half are larger than 0.1 km^2, whereas the rest fall in an intermediate class (0.02–0.1 km^2). The surveys made over the years confirm that from the point of view of the water chemistry, the lakes chosen satisfy the general study aims

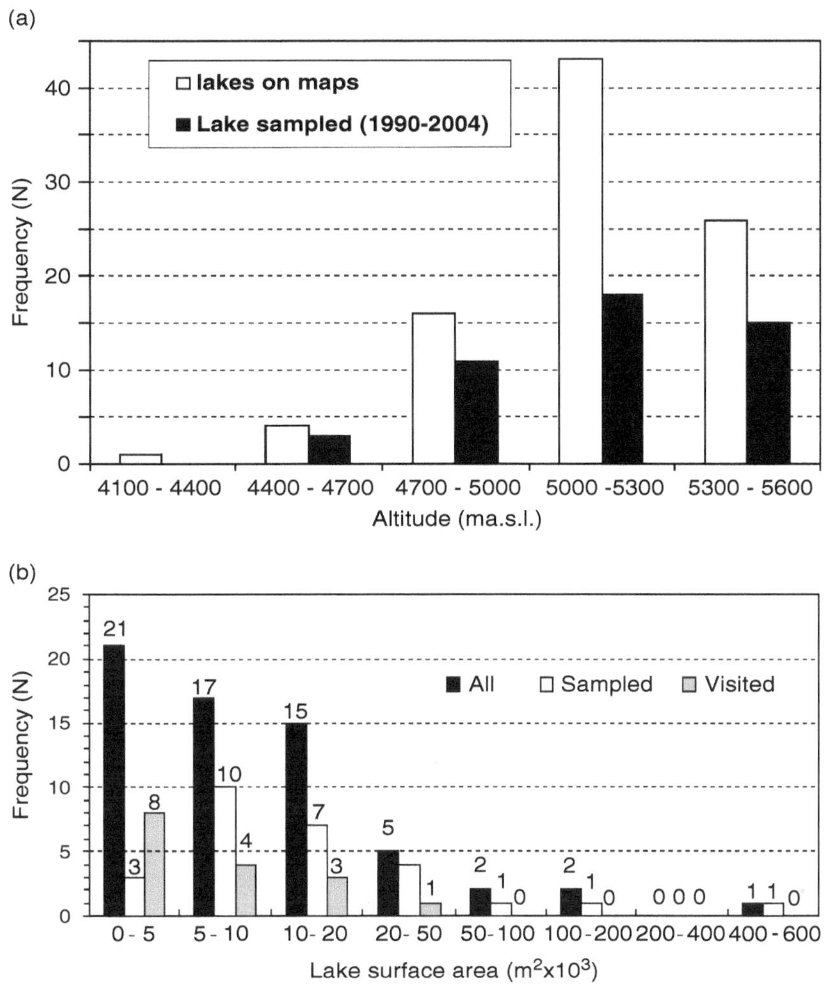

Figure 21.2. Ev-K^2-CNR Limnological Information System distribution with respect to (a) altitude and (b) area. Modified from Tartari et al. (1998a).

envisaged at the outset. The water bodies proved to be mostly clear or with a low silt content (21 lakes), an index of low direct influence from glacial melt water and thus of a fair degree of stability of the lakes. The criterion adopted therefore achieved its objectives and may be regarded as a useful procedure to adopt in other similar cases.

3.2. Geological setting

The study area is located in a complex transition zone between the High Himalaya and Tibet, characterised by different geological units (Bortolami, 1998). The surface area of 27 of the 31 lakes studied occupies a small portion of the watersheds (median value 2%). Watersheds have an extensive ice cover (median 19%). Eight lakes do not have any extensive stretches of glacier in their watersheds, and this is reflected in the presence of glacial silt in their waters observed in the field. A large proportion of ice cover (12–39%) is not matched by obvious turbidity of the water in only four cases. In all 15 lakes where low, intermediate or high turbidity was directly observed, the ice cover is considerable (from 14 to 61%). Altogether, a fair number of the watersheds are covered with detritus and occupied by old (1/5) and recent (1/2) moraines. The presence of moraine deposits, typical of high altitude areas like these, means that underground watercourses are more common than surface drainage. The consequent mineralisation of the waters can affect the chemistry of the lakes, though the widespread incidence of poorly soluble rocks such as granite and gneiss in the Imja Kola and Ngozumpa River valleys suggests that solute enrichment through leaching is not a major factor.

3.3. Limnology

Most of the lakes were sampled from the shore, while some others (LCN 9, 10, 29, 40 and 70) were sampled from an inflatable rubber boat at the three depths on the water column (1 m below the surface, 1 m above the bottom and a third point in between) at the maximum depth site. The chemical characteristics of the lakes show pH values range between 6.2 and 8.2, with 80% of the values between 7.0 and 8.0. The solute content of the waters is generally low, between 130 and 1100 $\mu eq\, l^{-1}$, with a corresponding conductivity interval from 8 to 67 $\mu S\, cm^{-1}$ at 20°C. Conductivity in general is in excellent agreement with the ion concentrations (slope 18.1, R^2 0.959, $P<0.001$), which confirms that no important ion has been neglected. Bicarbonate and calcium are the most important anions and cations in the ion composition of most of the lakes, with a range of variation of 27–422 and 34–450 $\mu eq\, l^{-1}$, respectively. Chloride concentrations are extremely low (1–4 $\mu eq\, l^{-1}$), reflecting the absence of geochemical sources in this area. Despite the great distance of the study area from the sea, the chloride present in the water is derived from the atmospheric transport of sea salt. This is confirmed by the concentrations of chloride in bulk deposition (7 $\mu eq\, l^{-1}$), while as regards sodium (4–32 $\mu eq\, l^{-1}$ and median 19 $\mu eq\, l^{-1}$ in the lake water), it appears to originate from atmospheric transport in only a few cases (10% of the lakes). The contribution of base cations of marine origin can be regarded as negligible for magnesium (5–126 $\mu eq\, l^{-1}$), potassium (3–34 $\mu eq\, l^{-1}$) and also sodium (4–32 $\mu eq\, l^{-1}$). These ions derive mainly from the weathering of rocks in the drainage basins and from glacier erosion.

Inorganic nitrogen is present in very low concentrations in the Khumbu lakes; ammonium is below the detection limit (0.5 $\mu eq\, l^{-1}$), while nitrate shows values equal to or lower than 5 $\mu eq\, l^{-1}$ in 25 of the 31 lakes, with a highest value of 8 $\mu eq\, l^{-1}$. These concentrations of inorganic nitrogen are in good agreement with the median values in bulk depositions (5 $\mu eq\, l^{-1}$). Total nitrogen concentrations, in contrast, are between 120 and 750 $\mu g\, N\, l^{-1}$ (8–54 $\mu eq\, l^{-1}$), with a mean value of 296 $\mu g\, N\, l^{-1}$ (21 $\mu eq\, l^{-1}$), indicating that organic nitrogen is the prevailing form in the water. Nitrate and sulphate have a special relevance in connection to anthropogenic pollution. The very low concentration here measured, in respect to those observed in

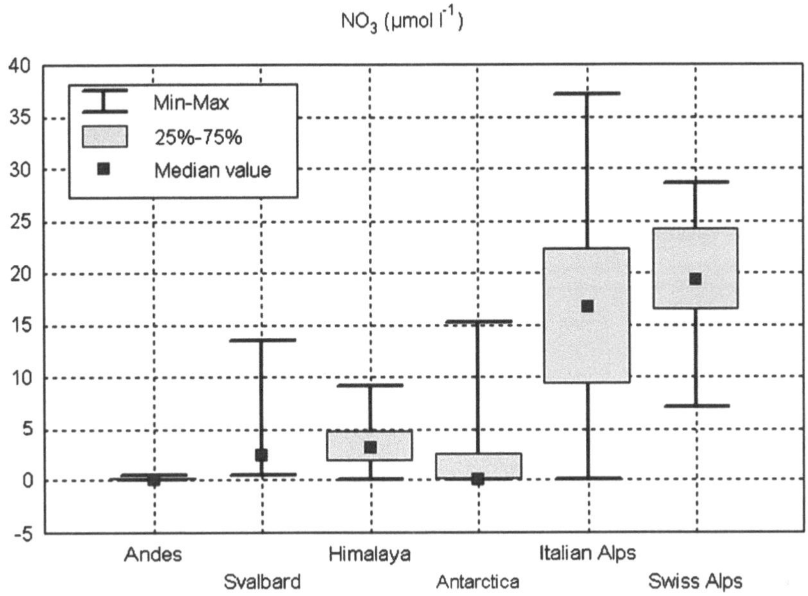

Figure 21.3. Comparison of the nitrate concentrations in different remote areas of the world.

other remote areas of the globe, testifies to the very low anthropogenic impact in these lakes and their potential interest as reference sites to document future human impact (Fig. 21.3).

Phosphorous is present in concentrations lower than 5 $\mu g\,P\,l^{-1}$ in more than 50% of the samples; the highest values measured are probably related to the presence of silt particles released by the ice melt. Silica shows a homogeneous distribution from 0.04 to 1.7 $mg\,Si\,l^{-1}$, with a mean value of 0.63 $mg\,Si\,l^{-1}$.

In addition to the survey of different lakes, a long-term limnological investigation has been carried out on the two lakes close to the Pyramid Laboratory. These data allowed us to provide a good description of the thermal regime of these lakes and the identification of the ice cover duration and lake level fluctuations (Fig. 21.4). Long-term chemical data (Tartari et al., 1998b) allowed us also to document a clear increase in solutes, mainly calcium, magnesium, sulphate and potassium, measured (Fig. 21.5). The consistency in the variations of different chemical variables (e.g. ion concentrations and conductivity) and the quality controls performed on the data exclude the possibility of analytical errors. The close relationship existing between the weathering phenomenon and the presence of glaciers suggests that this increase is partly connected with global atmospheric warming that may have shortened the period of snow cover, thus increasing the period of contact between precipitation and rocks, and may also have caused greater melting of glaciers, with the freeing of glacial silt which, because of its small size, is able to release a greater quantity of ions per unit of weight. These hypotheses need to be verified, both through a closer glaciological study of the area and by sampling other lakes in the same area, for which data from the beginning of the nineties are available, to be sure that the

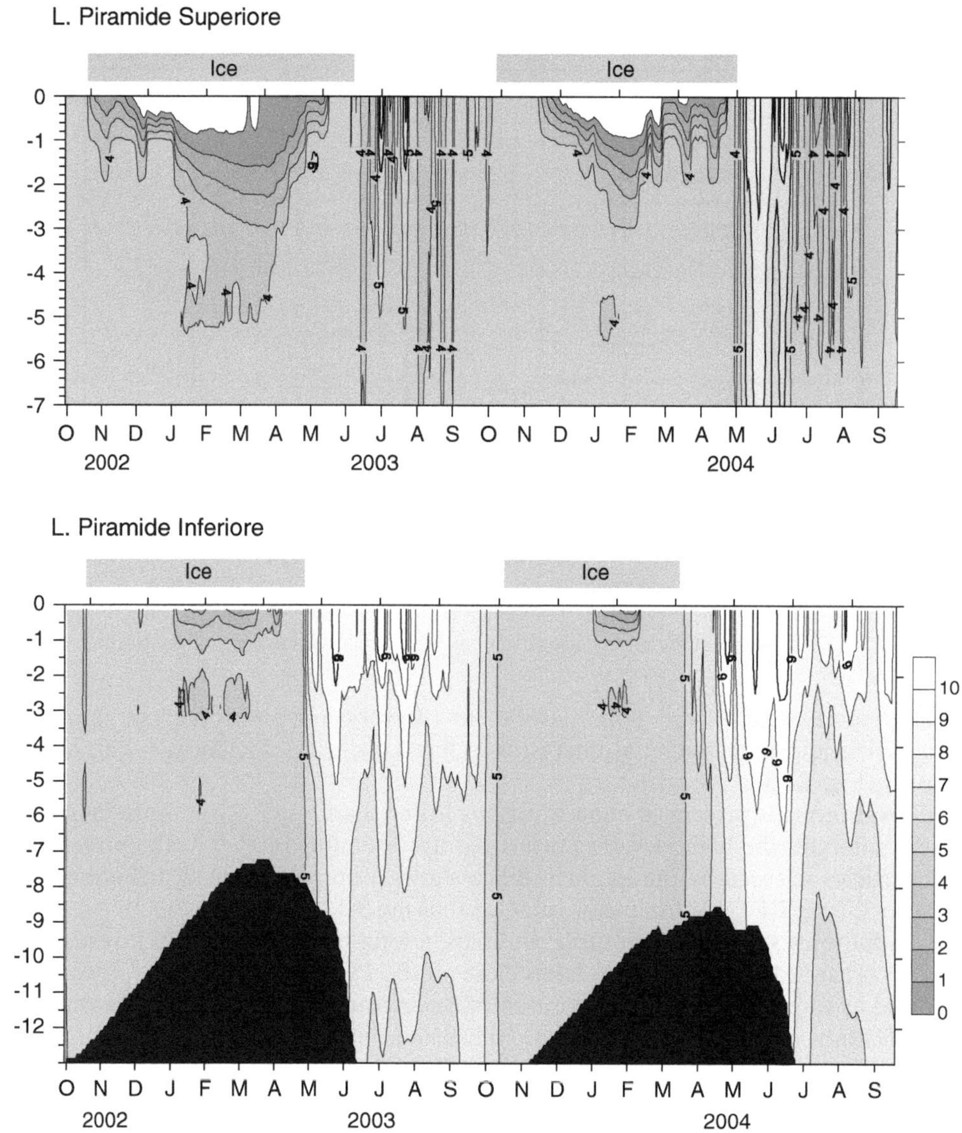

Figure 21.4. The annual lake-water temperature cycle in the two Pyramid lakes.

phenomena do not involve only the two Pyramid lakes. It would also be advisable to complete the studies on atmospheric deposition, with sampling over a whole year, to get a more accurate idea of the role this plays in determining surface water chemistry, with particular reference to inorganic nitrogen and sulphate loads.

Besides chemical analysis, the biotic component (phytoplankton, zooplankton and benthos) was also collected and analysed during the survey performed. Most of the samples were obtained on single occasions from the shore of different water bodies

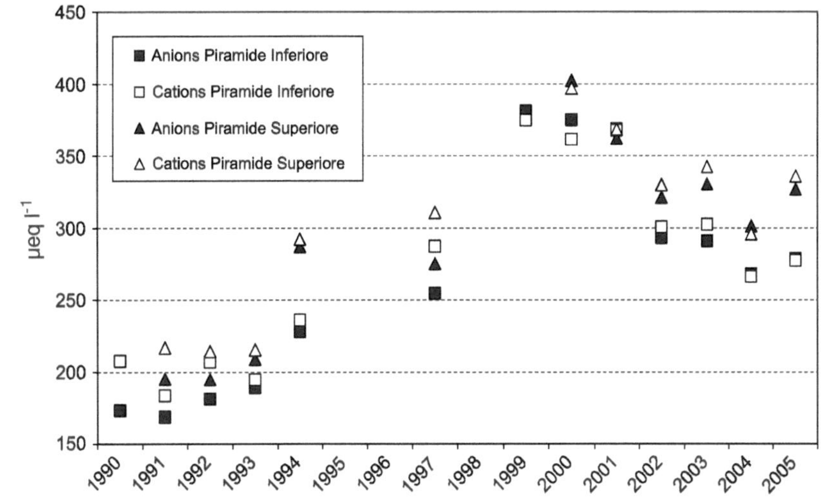

Figure 21.5. Trends of the concentrations of anions and cations of the two Pyramid lakes.

by retrieving bottles or plankton nets thrown by hand on the water surface. For those lakes that were sampled with a rubber boat, samples were collected with a Ruttner bottle and by vertical hauls of a plankton net (126 and 50 μm mesh size).

The phytoplankton population consisted of very few species that were cosmopolitan or difficult to identify. On the other hand, the numbers were high: millions or tens of millions cells per litre, compared to thousands of cells recorded in the same area by earlier research; however, biomass and chlorophyll-*a* were low (e.g. less than $1\,mg\,m^{-3}$ chlorophyll-*a*) because the average cell size was extremely small. Population dynamics resulted in being highly variable; as an example, great changes in phytoplankton assemblages occurred in LPI in 1992 within less than two weeks with a collapse of net phytoplankton and a fivefold increase of ultraplankton.

In Lake Piramide Superiore (LPS) and Lake Piramide Inferiore (LPI), the particulate organic carbon (POC) makes up a considerable part of the total seston or particulate matter (33% and 21% in LPI and LPS), never exceeding a concentration of $200\,\mu g\,C\,l^{-1}$ (Bertoni et al., 1998; Ruggiu et al., 1998). The seston of LPS, closer to the glacier with respect to LPI, has a higher fraction (58%) of inorganic matter. Dissolved organic carbon (DOC) values are also very low, around $0.5\,mg\,l^{-1}$. The chlorophyll-*a* concentration in the 0.2–1 μm size class (picoplankton) is 54% and 33% of the total phytoplankton chlorophyll in LPI and LPS respectively.

Autotrophic picoplankton were present in both Piramide Lakes, though in very low numbers, due to photoinhibition (underwater surface irradiance: $1200–1400\,\mu E\,m^{-2}\,s^{-1}$). The autotrophic cells are *Synechococcus*-type with phycoerythrin as accessory pigment. The heterotrophic cells exceed the autotrophic ones by three orders of magnitude, suggesting a heterotrophy-oriented food web.

Thecamoebians, usually strictly benthic organisms, were present in the samples collected for zooplankton because the net has to be towed close to the bottom; in high mountain lakes, during the day, most zooplankton stay close to the sediment, to minimise the damage by UV radiation and the energetic cost for DNA repair. A detailed description of the thecamoebians specimens found is discussed in Manca et al. (1998); the most abundant taxa are Centropyxidae, considered as the most tolerant and 'pioneer' among the thecamoebians, but also Difflugidae and *Lesquereusia* are present.

Among the Crustacea, the most widely distributed species is an endemic diaptomid of the palearctic genus *Arctodiaptomus* (*Arctodiaptomus jurisovitchi*). Apparently, it lacks only from one lake (LCN69). In one case (LCN26) it is the only zooplankter found. All developmental stages were represented in the various samples analysed, with last stages of copepodites and adults generally most abundant. Daphniidae are represented by the dark and large *Daphnia tibetana* and *Simocephalus vetulus*, as well as by a pale *Daphnia* of the *longispina* group. The latter is present in seven lakes, four of which were among those with the highest silica concentration and pH within 7.1–7.7 units. Its occurrence has been associated with low transparency (Manca et al., 1998) or to inhabit lakes with milky waters (e.g. LCN2) or waters rich in suspended solids (LCN75, LCN76 and LCN77, although lacking from other organic rich lakes), as well as clear-water lakes, where refuge on the bottom is allowed by a dense bed of mosses (LCN40). In one case it also occurs together with the dark *tibetana* (LCN66). As suggested by Hutchinson (1937), the former probably inhabits the deepest waters, whereas the latter swims in the littoral zone. One of the most interesting traits of these types of environments is the co-occurrence of a copepod and a *Daphnia*.

The macrobenthic fauna recorded in the lakes studied mainly consisted of Insecta belonging to Diptera Chironomidae, followed by Oligochaeta. Other Insecta groups, such as Plecoptera and Trichoptera, or other taxonomic entities such as Acari Hydracarina and Turbellaria, appear to be relatively uncommon. Chironomids are mostly composed of Diamesinae and Chironominae, followed by Orthocladiinae. The Diamesinae belongs mainly to the genus *Pseudodiamesa*, probably represented by the species *Pseudodiamesa nepalensis* Reiss and P. *branickii* (Nowicki), reported by Reiss (1968) and Löffler (1969) in some places in the Nepal Himalayas above an altitude of 5000 m. The Chironominae are represented by the genus *Micropsectra*, which is the dominant genus of Tanytarsini tribe in Nepal, especially above 2000 m (Roback and Coffman 1987). *Micropsectra* larvae were found in all the lakes with the exception of LCN13, where the chironomids are made up of Orthocladiinae, at least in the littoral.

During our expeditions we found a remarkable number of species, most of which typical of extreme environments and, hence, interesting *per se*. Noteworthy is also the finding of so-called cosmopolitan species, whose occurrence is not explained by means of common dispersal mechanisms. The latter is particularly important for the phytoplankton community, which is more typified by a rarefaction of species, than by the occurrence of peculiar ones. The phytoplankton numbers, far greater than in earlier investigations, indicate a possible nutrient enrichment, which would be of great interest considering the scopes of our activity in the region. The zooplankton samples were rich in all the developmental stages of species, which in some cases were never found before. On a whole, they are invaluable to taxonomy and biogeography, as well as to the study of the typology of lakes. The types of Crustacea assemblages, with large daphnias coexisting with copepods and Anostraca in fish-less lakes are among the most interesting in the world, for the study of food interactions on which there is an increasing interest (Gliwicz and Stibor, 1993). The presence and the diversity of the thecamoebians encourage further studies in these lakes to provide additional information on their ecological niches. A low number of systematic entities characterise the macrobenthic fauna of these lakes. Chironomids and Oligochaets dominate, with cold sthenothermic species that can tolerate the extreme physical and chemical conditions of these high altitude environments. Low temperatures have a strong effect on the life cycles of the different taxa. As for chironomids, they influence the phenology of the emergences and the reproduction of adults, which have to be completed during the short period of free waters. Other factors involved in the low diversity of the macrobenthos are the low level of nutrients in the water and of organic matter in the sediments, which are essentially formed of very fine elements. To fill the lack of information on the taxonomy and ecology of the macrobenthic fauna of these high altitude lakes, further research is needed, also from other sites in the same region.

3.4. Palaeolimnology

Lake sediments are natural archives of climatic and environment-related proxies (e.g. photosynthetic pigments, pollen, diatoms and organic geochemistry) on the response of a lake and its catchment to anthropogenic and climatic changes. In particular, the wide variety of environments along the Pole–Equator–Pole (PEP II, IGBP–PAGES) transect makes the lake sediments from the Himalayan region very suitable for palaeoclimatic reconstruction and modelling (Wake and Mayewski, 1996). During several expeditions from 1992 to the Pyramid Laboratory in Nepal (5050 m a.s.l.), a number of small lakes were sampled for the study of geochemical and biological fossil remains (Guilizzoni et al., 1998; Lami et al., 1998). The lakes in this remote area are particularly suitable because, for example, the climate signals are maximised due to the limited importance of human impact (Smol et al., 1991). A tentative comparison between the Holocene fluctuations of glaciers in the Himalaya and the Karakoram (Pakistan, India, Nepal; Röthlisberger and Geyh, 1985) in the last 3000 years based on the paper from Smiraglia (1997) and some of two proxies studied in LPI (Musazzi, 2005) is shown in Fig. 21.6. Our dated cores also suggest a number of fluctuations that can be related to warmer or colder periods. Wide variations in chemical and biological parameters are common in cores from most lakes, and may be a consequence of climatic forcing. The algal response to climatic change will depend in these systems on a number of factors, among which the duration and depth of ice-cover and water-level fluctuations in the lake are probably the most important ones. Evidence of changes in lake levels in the highland lakes of Tibet and in the Pokhara Basin were reported by De Terra and Hutchinson (1934) and Inouchi et al. (1995). These authors considered several physical properties and chemical parameters and inferred the water-level fluctuations during the last 1000 years, and consequently the flood events and the variations in the amount of melting water.

Past environmental conditions were also inferred by *Daphnia* body size and abundance estimates, in addition to an analysis of changes in the Cladocera assemblage

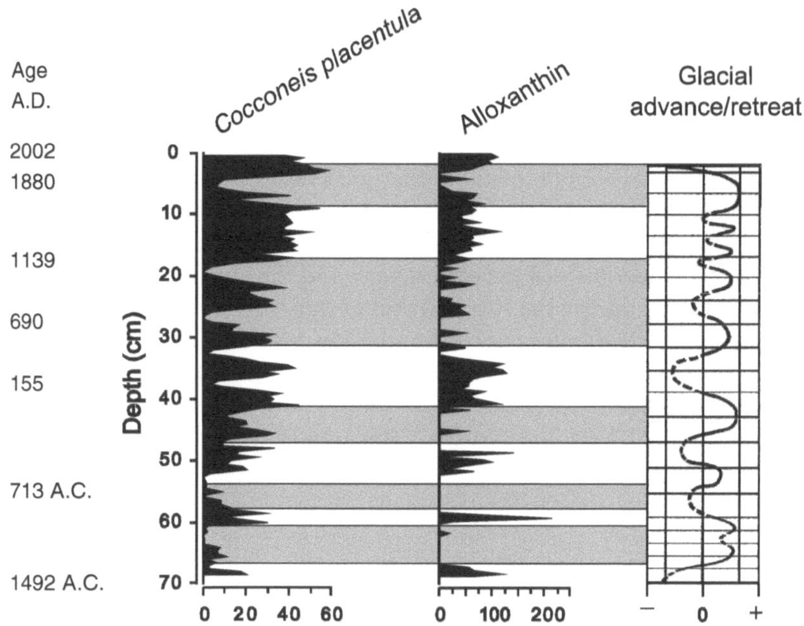

Figure 21.6. Schematic reconstruction (inferred from a diatom and an algal carotenoid typical of planktonic species) of different climatic phase based on core PIR INF 02-3 and their comparison with glacier dynamics in the Himalayan region.

(Manca and Comoli, 2004). A combined analysis of modern zooplankton and fossil Cladocera assemblages from a Himalayan lake, Lake 40, revealed that the endemic *Daphnia tibetana* disappeared in the late-1980s, after persisting as the only *Daphnia* species for almost 3000 years. The substitution of the original species appears to be related to an increase in the production of mucilaginous aggregates from filamentous green algae, and is contrary to the general tendency for non-pigmented species to be lost, probably as an effect of increased UV radiation due to climate change.

Therefore, the observed changes in all the parameters in the study of lakes have to be associated directly or indirectly with the climate changes occurring in the area and shown by the observed advances and retreats of the major glaciers. A cooling event affects the duration of seasonal ice-cover, which in turn affects the characteristics of these small lakes. The suggestion emerging from our results is that these lakes have considerable potential for providing proxy data of past climatic changes in high altitude regions as well as exciting research opportunities for paleolimnologists.

We have also analysed the presence of persistent organic pollutants (POPs) in the sediment of some of the survey lakes (Teti et al., 2005). In Fig. 21.7 a comparison of the superficial concentration of PCB in some remote areas of the world is shown. The distribution of the different congeners among different remote areas is similar and it is related to the altitudinal gradient; this supports the hypothesis of the cold trap effect (Calamari et al., 1991). Himalayan sites have a lower concentration compared to other European sites that are much nearer to the pollutants sources. The sediment core profile of DDT revealed that only the metabolite pp'DDE was found, whereas

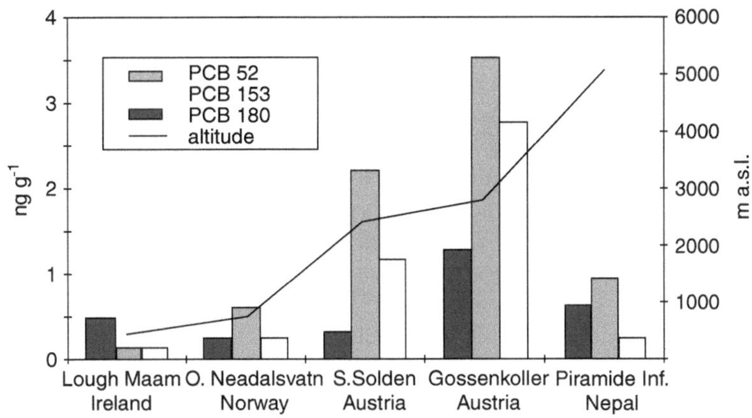

Figure 21.7. Comparison of PCB distribution in surface sediment in different lakes located in remote areas and at different altitude. Modified from Teti et al. (2005).

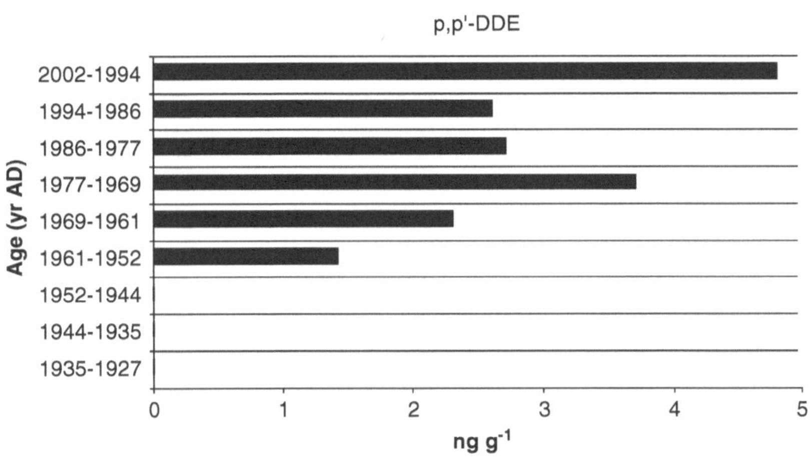

Figure 21.8. Concentration of pp'DDE along the sediment core (PIRINF 02/1) from Lake Piramide Inferiore. Modified from Teti et al. (2005).

the parental compound, pp'DDT, was not detected (Fig. 21.8). This consideration allowed us to conclude that these sites do not receive pollution from local sources, but they receive pollutants from long-range transport; the profile along the sediment core reflect quite well the well known historical trends in the use of DDT.

4. Conclusion

Despite the recommendation reported in the Chapter 13 (Sustainable Mountain Development) of the Agenda adopted at the United Nations Conference on Environment and Development (UNCED) in Rio de Janeiro 1992, which highlighted the lack of specific knowledge of mountain ecosystems, mountains still need a dedicated research approach as noted in a recent publication (Huber et al., 2005) in order to assure a sustainable use of the resources and a sufficient protection of these natural environments.

Our research effort is to integrate lakes and wetland regions at a catchment scale and to focus on the key drivers of aquatic system change (nutrients, acid deposition, toxic substances) and their interaction with global drivers such as climate using time-series analysis, paleolimnology, experiments and process modelling at different time scales (seasons/years and decades/millennia).

A central activity is the development of an innovative toolkit for integrated catchment analysis and modelling to simulate hydrological, hydrochemical and ecological processes at the catchment scale for use in assessing the potential impact of global change under different climate and socio–economic scenarios. A unified system of ecological indicators for monitoring freshwater ecosystem health, and new methods for defining reference conditions and restoration strategies will also be developed. This will fully involve users and stakeholders and will be demonstrated at study catchments.

At present the Ev-K^2-CNR LIS will focus on reaching an ecological understanding of ecosystems at multiple spatial and temporal scales, by integration in a GIS database of the persistent lake bodies in the Sagarmatha National Park, all the information (geology, meteorology, hydrochemistry, hydrobiology, glaciology etc.) collected over a decade of investigations. We intend to develop well-designed, well-documented databases that are accessible to the broader scientific community.

On a longer perspective we intend our effort to be directly involved in:

- creating a network of sites to gain general ecological knowledge through the synthesis of information obtained from long-term research and development of theory;
- creating a legacy of well designed and well documented long-term observations, experiments and archives of samples and specimens;
- providing knowledge to the broader ecological community, general public, resource managers and policy makers to address complex environmental challenges;
- developing studies on the specific mechanisms that influence the pollutants as a support to a modelling description of long-range transport via the atmosphere;
- following the long-term evolution (decades to millennia) of some of these lakes to investigate the natural (climatic) or anthropogenic (pollution) impact;
- developing a cadre of scientists who are equipped to conduct long-term, collaborative research to address complex ecological problems.

Acknowledgements

Special thanks goes to all the Nepalese people that made possible the field work in such a remote environment. We are very grateful to the Ev-K^2-CNR Committee and RONAST for having supported part of the reported studies.

References

Aizaki, M., Terashima, A., Nakahara, H., et al., 1987. Trophic status of Tilitso, a high altitude Himalayan lake. Hydrobiologia 153, 217–224.

Baudo, R., Tartari, G., and Munawar, M., 1998. Top of the World Environmental Research: Mount Everest-Himalayan Ecosystem. Backhuys Publishers, Leiden, pp. 290.

Bertoni, R., Callieri, C., and Contesini, M., 1998. Organic carbon and microorganisms in two Nepalese lakes. In: Lami, A. and Giussani, G. (Eds), Limnology of high altitude lakes in the Mt. Everest Region (Nepal), Memorie dell' Istituto Italiano di Idrobiologia 57. CNR – Istituto per lo Studio degli Ecosistemi, Verbania, Italia, pp. 99–106.

Bhandari, B. 1993. The current status of wetland in Nepal. In: K. ILEC Foundation, Japan. (Ed.), Towards wise use of Asian wetlands. Asian wetland Symposium, pp. 103–111.

Bortolami, G., 1998. Geology of the Khumbu Region, Mt. Everest, Nepal. In: Lami, A. and Giussani, G. (Eds), Limnology of high altitude lakes in the Mt. Everest Region (Nepal), Memorie dell' Istituto Italiano di Idrobiologia 57. CNR – Istituto per lo Studio degli Ecosistemi, Verbania, Italia, pp. 41–49.

Calamari, D., Bacci, E., Focardi, S., et al., 1991. Role of plant biomass in the global environmental partitioning of chlorinated hydrocarbons. Environmental Science and Technology 25, 1489–1495.

De Terra, H. and Hutchinson, G.E., 1934. Evidence of recent climatic changes shown by Tibetan highland lakes. Geographical Journal 84, 311–320.

Gliwicz, Z.M. and Stibor, H., 1993. Egg predation by copepods in Daphnia brood cavities. Oecologia 95, 295–298.

Guilizzoni, P., Lami, A., Smith, J.D., et al., 1998. Palaeolimnological analysis of four Himalayan lakes (Khumbu Valley, Nepal). In: Tartari, G., Baudo, R., and Munawar, M. (Eds), Top of the Word, Mount Everest-Himalaya Ecosystem, Ecovision Word Monographic series. pp. 189–217.

Hirano, M., 1963. Freshwater algae from the Nepal Himalaya, collected by a member of the Japanese climbing expedition. Contributions from the biological laboratory Kyoto University, p. 16.

Huber, U.M., Bugmann, H.K.M., and Reasoner, M.A., 2005. Global Change and Mountain Regions. An Overview of Current Knowledge. Springer, Berlin, Heidelberg, New York, pp. 250.

Hutchinson, G.E., 1937. Limnological studies in Indian Tibet. Internationale Revue gesamten Hydrobiologie 35, 134–177.

Inouchi, Y., Tokuoka, T., Nakamura, T., and Prakash Das, U., 1995. Climatic changes during the last ca. 1000 years recorded in the sediment of the lakes around Pokhara, central Nepal. Internat. Himalayan/Tibetan Plateau Palaeoclimate Workshop, Kathmandu, April 2–7.

Jones, J.R., Knowlton, M.F., and Swar, D.B., 1989. Limnological reconnaissance of waterbodies in central and southern Nepal. Hydrobiologia 184, 171–189.

Khan, M.A. and Zutshi, D.P., 1980. Contribution to high altitude limnology of the Himalayan system. I. Limnology and primary productivity of the plankton community of Nilang Lake, Kashmir. Hydrobiologia 75, 102–112.

Lami, A. and Giussani, G., 1998. Limnology of high altitude lakes in the Mt. Everest Region (Himalayas, Nepal). Memorie dell' Istituto Italiano di Idrobiologia 57, 244.

Lami, A., Guilizzoni, P., Marchetto, A., et al., 1998. Palaeolimnological evidence of environmental changes in some high altitude Himalayan lakes (Nepal). In: Lami, A. and Giussani, G. (Eds), Limnology of high altitude lakes in the Mt. Everest Region (Himalayas, Nepal), Memorie dell' Istituto Italiano di Idrobiologia 57. CNR – Istituto per lo Studio degli Ecosistemi, Verbania, Italia, pp. 107–130.

Löffler, H., 1969. High altitude lakes in Mt. Everest region. Verhandlung Internationale Vereinigung für theteoretische und angewande Limnologie 17, 373–385.

Manca, M. and Comoli, P., 2004. Reconstructing long-term changes in Daphnia's body size from subfossil remains in sediments of a small lake in the Himalayas. Journal of Paleolimnology 32, 95–107.

Manca, M., Ruggiu, D., Panzani, P., et al., 1998. Report on a collection of aquatic organisms from high mountain lakes in the Khumbu Valley (Nepalese Himalayas). In: Lami, A. and Giussani, G. (Eds), Limnology of high altitude lakes in the Mt. Everest Region (Nepal), Memorie dell' Istituto Italiano di Idrobiologia 57. CNR – Istituto per lo Studio degli Ecosistemi, Verbania, Italia, pp. 77–98.

Marchetto, A., 1998. The study of high mountain lakes in the activity of the Istituto Italiano di Idrobiologia. In: Lami, A. and Giussani, G. (Eds), Limnology of high altitude lakes in the Mt. Everest

Region (Nepal), Memorie dell' Istituto Italiano di Idrobiologia 57. CNR – Istituto per lo Studio degli Ecosistemi, Verbania, Italia, pp. 1–10.
Messerli, B., 1997. The global mountain problematique (Abstracts). European Conference on Environmental and Societal Change, pp. 2–3.
Musazzi, S., 2005. Evoluzione del paleoambiente e del paleoclima nel Tardo Olocene in due aree remote (Svalbard e Himalaya) attraverso l'analisi dei sedimenti lacustri. Dottorato di Ricerca in Ecologia – XVII Ciclo. Università degli Studi di Parma, Parma, I, p. 168.
Poretti, G., 1998. Geophysical, geological and geographical features of the Himalayas. In: Baudo, R., Tartari, G., and Munawar, M. (Eds), Top of the World Environmental Research: Mount Everest – Himalayan Ecosystem. Ecovision World Monograph Series, Backhuys Publ., Leiden, The Netherlands, pp. 19–34.
Reiss, F., 1968. Neue Chironomiden Arten (Diptera) aus Nepal. Khumbu Himal 3, 55–77.
Roback, S.S. and Coffman, W.P., 1987. Results of the Nepal alpine zone. Research project, Chironomidae (Diptera). Proceedings of the Academy of Natural Sciences of Philadelphia 139, 87–158.
Röthlisberger, R. and Geyh, M.A., 1985. Glacier variations in Himalayas and Karakorum. Zeitshrift für Gletscherkunde und Glazialgeologie. 21, 237–249.
Ruggiu, D., Bertoni, R., Callieri, C., et al., 1998. Assessment of biota in lakes from the Khumbu Valley, High Himalayas. In: Baudo, R., Tartari, G., and Munawar, M. (Eds), Top of the World Environmental Research: Mount Everest-Himalayan Ecosystem. Ecovision World Monograph Series. Backhuys Publishers, Leiden, pp. 219–233.
Sars, G., 1903. On the crustacean fauna of central Asia. Pt. II Cladocera. Annales Musei Zoologie de l'Accademic Imperial Scientifique de St. Petersbourg 8, 157–194.
Sharma, P.C. and Pant, M.C., 1985. Species composition of zooplancton in two Kaumann Himalayan lakes (U.P. India). Archives of Hydobiology 102, 387–403.
Smiraglia, C., 1997. Glaciers and glaciology of Himalaya. In: Baudo, R., Tartari, G., and Munawar, M. (Eds), Top of the World Environmental Research: Mount Everest-Himalayan Ecosystem, Ecovision World Monograph Series. Backhuys Publishers, Leiden, pp. 65–100.
Smol, J.P., Walker, I.R., and Leavitt, P.R., 1991. Paleolimnology and hindcasting climatic trends. Verh. int. Ver. Limnol. 24, 1240–1246.
Swar, D.B., 1980. Present status of limnological studies and research in Nepal. In: Mori, S. and Ikusima, I. (Eds), Proc. First Workshop on "Promotion of Limnology in Developing Countries". XXI SIL Congress, Kyoto, Japan, pp. 43–47.
Tartari, G., Previtali, L., and Tartari, G.A., 1998a. Genesis of the lake cadastre of Khumbu Himal Region (Sagarmatha National Park, East Nepal). In: Lami, A. and Giussani, G. (Eds), Limnology of high altitude lakes in the Mt. Everest Region (Nepal), Memorie dell' Istituto Italiano di Idrobiologia 57. CNR – Istituto per lo Studio degli Ecosistemi, Verbania, Italia, pp. 139–149.
Tartari, G.A., Tartari, G., and Mosello, R., 1998b. Water chemistry of high altitude lakes in the Khumbu and Imja Kola valleys (Nepalese Himalayas). In: Lami, A. and Giussani, G. (Eds), Limnology of high altitude lakes in the Mt. Everest Region (Nepal), Memorie dell' Istituto Italiano di Idrobiologia 57. CNR – Istituto per lo Studio degli Ecosistemi, Verbania, Italia, pp. 51–76.
Teti, P., Guzzella, L., Roscioli, C., and De Paolis, A., 2005. I composti organoclorurati nei sedimenti di laghi remoti e del Lago Maggiore. Acqua Aria 4, 24–29.
Troll, C., 1959. Die tropischen Gebie, 93 pp.
Ueno, M., 1966. Cladocera and Copepoda from Nepal. Japanese Journal of Zoology 15, 95–100.
Vass, K.K., Wanganeo, A., Raina, H.S., et al., 1989. Summer limnology and fisheries of high mountain lakes of Kashmir Himalayas. Archives of Hydrobiology 114, 603–620.
Wake, C.P. and Mayewski, P.A., 1996. Himalayan interdisciplinary Paleoclimate Project. PAGES Workshop report, series 96–1, 96 pp.
Zutshi, D.P., 1991. Limnology of high altitude lakes of Himalayan region. Verh. int. Ver. Limnol. 24, 1077–1080.
Zutshi, D.P. and Vass, K.K., 1970. High altitude lakes of Kashmir. Ichthiologica 10, 12–15.

Mountains Witnesses of Global Changes
R. Baudo, G. Tartari, and E. Vuillermoz (Editors)

22

Elemental characterization of Himalayan airborne particulate matter collected at 5100 m a.s.l

Enrico Rizzio, Giuseppe Giaveri, Luigi Bergamaschi, Antonella Profumo, Gianni Tartari and Mario Gallorini

Abstract

The elemental composition of the airborne particulate matter collected at CNR Ev-K^2-Laboratory-Observatory at 5100 m a.s.l. in the Himalaya–Khumbu Valley, Nepal, has been evaluated. Samples of total suspended particles (TSP), as well as of the particle size fractions PM 10 and PM 2.5, were collected by pump aspiration onto filters and analyzed by INAA and GF–AAS for the determination of 30 elements. Enrichment factors have been calculated as well, by the analysis of the local soils. Samplings were carried out in the pre-monsoon period of March–May 2002 and the post-monsoon period of October 2003 in order to observe seasonal variations of the amount and the composition of the air particulates.

1. Introduction

In 1990, following an agreement with the Royal Nepal Academy of Science and Technology (RONAST), the Italian National Research Council (CNR) installed the scientific laboratory (Pyramid Laboratory-Observatory) at 5050 m a.s.l. in the Himalayan region. The laboratory is located in one of the highest inhabited sites in the world, the Khumbu valley in Sagarmatha National Park, at the foot of Mt. Everest. Since that moment, several multidisciplinary research activities have been initiated in the framework of the CNR scientific program of Ev-K^2-CNR. Among other environmental-related researches, the task project RATEAP (remote areas trace elements atmospheric pollution) was started in 2001 and aims at obtaining information about the elemental composition of airborne particulate matter of the high altitude remote areas such as those of the Himalayan region, free of any industrial or anthropogenic sources. Furthermore, this research can provide additional information to the studies of the so-called Asian Brown Cloud (ABC), which is the haze hovering the Indo–Asian–Pacific region, as documented (Ramanathan and Crutzen, 1999) by the INDOEX–Indian Ocean Experiment.

ISSN: 0928-2025 DOI: 10.1016/S0928-2025(06)10022-X

In this work the characterization and the evaluation of trace-element (TE) concentrations in the air particulate fractions, total suspended particles (TSP), PM 10 and PM 2.5, have been obtained. The results may be considered as background concentration ranges of the TE present in the airborne particulates of unpolluted high altitude remote areas. In addition, the analysis of the finest fraction PM 2.5 can give important information about long-distance transportation phenomena responsible for the distribution of the pollutants far from the emission sources (Maenhaut et al., 1999; Querol et al., 2002; Artiñano et al., 2003; Yli-Tuomi et al., 2003).

Samplings were carried out during two missions at the Ev-K^2-CNR laboratory, from March to May 2002 (pre-monsoon period) and in October 2003 (post-monsoon period), allowing a comparison between the two seasons. Samples of local soils, surrounding the sampling area, have been collected as well for the analysis necessary to calculate the enrichment factors (EF).

2. Materials and methods

The filters and the soil samples have been submitted to instrumental neutron activation analysis (INAA) for the determination of 30 elements. All lead and cadmium determinations and, in some cases, nickel and copper, have been carried out by graphite furnace atomic absorption spectroscopy (GF–AAS). Data quality control on the analytical procedures was accomplished by means of certified reference materials: SRM 2783 Air Particulate on Filter Media, SRM 1632c Trace Element in Coal from NIST (USA). During the sampling periods, temperature (°C), pressure (hPa), relative humidity (%), wind velocity ($m\,s^{-1}$) and precipitation (mm) were recorded hourly by the automatic weather station (LSI-Lastem, Milan, Italy) located near the Pyramid Laboratory-Observatory.

2.1. Airborne particulate sampling

Three different air particle size fractions were sampled i.e.: TSP, particles with an equivalent aerodynamic diameter $<10\,\mu m$ (PM 10) and particles $<2.5\,\mu m$ (PM 2.5). The air particulate matter was collected onto 0.5 μm pore size high purity Teflon® filters with a 47 mm diameter (Zefluor TM, PALL–USA) using an aspiration sampling system expressly developed to operate at low atmospheric pressure (Zambelli s.r.l. – Bareggio–Milan, Italy). Samples were collected in the pre-monsoon period (March–May 2002) and in the post-monsoon period (October 2003). Minimum sampling time of 120 h (5 days) was adopted in order to collect an amount of airborne particulate suitable for the analysis, corresponding to a minimum of 120 m^3 (about 67.1 Nm^3) of air for each sample. All samplings were carried out in similar weather conditions (excluding rain and snow periods), while the flow rates and pump functioning were continuously checked. All filter-handling procedures were carried out inside of the Pyramid laboratory in clean conditions (gloves and plastic tweezers) to avoid possible contaminations. At the end of each sampling, the filters were sealed in individual numbered plastic containers. Five independent field-blanks consisting of filters handled in the same manner of the samples (including 10 s of air sampling, loading and dismantling from the filter-holders) were collected as well in each sampling campaign and stored (Rizzio et al., 2000).

2.2. *Soils and lichen sampling*

Three different samples of superficial soil (about 200 g each) were collected at about 3–5 cm depth from five selected points (4 cardinal points surrounding the Pyramid laboratory and in the site of the air sampling). All sampling procedures were carried out using plastic tools to avoid possible metals contamination. The three soil samples, collected at each point, were pooled obtaining five final samples that had been stored in pre-washed (HNO_3/Milli–Q water) polyethylene containers.

In proximity to the area of the Pyramid (between 4700 and 5100 m a.s.l.), lichen samples were also collected to obtain preliminary information about the possibility of their use as biomonitors of atmospheric pollution by metals. The lichen species, collected by following the sampling protocol used in previous studies (Bergamaschi et al., 2002), were: *Hypogymnia* sp, *Umbilicaria* sp and *Alectoria* sp, this last collected nearby the sampling site of the air particulates. The quality control on the analyses of the soils and lichens was carried out by analyzing, in the same conditions, the certified reference materials SRM8704 Buffalo River Sediment, SRM2709 San Joaquin Soil, SRM1547 Peach Leaves from NIST (USA) and CRM Lichen 482 from BCR (EU).

2.3. *Analyses*

INAA was used for the determination of most of the trace elements, while GF–AAS was employed in the analysis of Pb, Cd, Cu and Ni. The blanks arising from the filters have been deeply evaluated and measured. In these types of studies the filter blank contribution constitutes a key point since many TE may be present in the filters at levels close to those in the collected mass of the air particulate. In collecting airborne particulate matter in very low polluted areas, the ratio of air dust mass/filter blank could be very critical (Dams, 1992; Heller-Zeisler et al., 1999; Rizzio et al., 2000). All the analytical procedures adopted for the analysis of air particulate samples, soils, blanks and standards are given elsewhere (Rizzio et al., 1999).

3. Results and discussion

The elemental concentration values in ng/Nm^3 of 30 elements determined in samples of TSP, PM 10 and PM 2.5 collected at the Ev-K^2-CNR Pyramid in the pre-monsoon period (March–May 2002) and in the post-monsoon period (October 2003) are listed in Table 22.1. In general, and in order to have an idea about the concentrations present in this area of the Himalaya, the values can be compared with the ranges of those obtained in previous sampling campaigns in rural areas (Ispra)

Table 22.1. Trace elements content in the different particle-size fractions of air particulate matter collected at 5100 m a.s.l. in the Himalaya (Khumbu valley Ev-K^2-CNR Pyramid) in the pre- and post-monsoon periods. Concentration in ng/Nm^3.

Element	Pre-monsoon period March–May 2002			Post-monsoon period October 2003		
	TSP Mean	PM 10 Mean	PM 2.5 Mean	TSP Mean	PM 10 Mean	PM 2.5 Mean
As	0.14±0.04	0.13±0.05	0.04±0.02	0.06±0.02	0.04±0.01	0.007±0.001
Br	2.1±0.8	1.9±0.7	1.9±0.7	2.5±0.7	1.6±0.7	1.1±0.7
Ca	483±121	380±77	65±20	67±29	45±23	28±11
Cd	0.04±0.03	0.03±0.01	0.015±0.009	0.03±0.02	0.03±0.02	0.024±0.001
Ce	1.1±0.6	0.8±0.2	0.11±0.07	0.14±0.06	0.10±0.04	0.02±0.01
Cl	175±110	40±22	22±8	52±38	18±10	4±2
Co	0.4±0.1	0.4±0.1	0.20±0.09	0.10±00.4	0.07±0.02	0.022±0.006
Cr	3.65±1.75	3.13±2.0	1.76±0.9	0.38±0.12	0.32±0.14	0.14±0.08
Cs	0.22±0.07	0.18±0.04	0.016±0.006	0.17±0.04	0.16±0.03	0.017±0.006
Cu	3.7±1.4	2.9±1.1	1.9±0.8	0.6±0.3	0.6±0.3	0.4±0.2
Fe	500±185	435±108	63±30	126±46	87±32	17±2
Hf	0.036±0.006	0.024±0.007	0.008±0.003	0.012±0.008	0.011±0.008	0.008±0.002
Hg	<0.01	<0.01	<0.01	<0.01	<0.01	<0.01
I	1.13±0.44	0.83±0.62	0.44±0.21	2.03±0.6	1.9±0.2	0.9±0.2
La	0.4±0.1	0.40±0.08	0.10±0.03	0.10±0.03	0.05±0.004	0.012±0.004

Mg	490 ± 220	452 ± 130	44 ± 24	152 ± 56	132 ± 50	30 ± 16
Mn	11.4 ± 4.3	8.3 ± 1.2	1.1 ± 0.3	2.3 ± 0.9	1.60 ± 0.05	0.3 ± 0.1
Mo	< 0.2	< 0.2	< 0.2	< 0.2	< 0.2	< 0.2
Ni	1.5 ± 0.7	1.2 ± 0.7	0.8 ± 0.4	0.5 ± 0.5	0.3 ± 0.1	< 0.1
Pb	4.4 ± 2.4	3.7 ± 2.6	2.6 ± 2.0	2.9 ± 1.4	1.9 ± 0.9	1.7 ± 0.8
Rb	1.9 ± 0.6	1.7 ± 0.5	0.2 ± 0.1	1.2 ± 0.5	0.8 ± 0.3	0.15 ± 0.06
Sb	0.4 ± 0.2	0.4 ± 0.2	0.2 ± 0.1	0.10 ± 0.04	0.08 ± 0.02	0.07 ± 0.03
Sc	0.16 ± 0.05	0.12 ± 0.03	0.022 ± 0.009	0.048 ± 0.018	0.035 ± 0.010	0.007 ± 0.001
Se	0.17 ± 0.05	0.09 ± 0.06	0.03 ± 0.01	0.06 ± 0.03	0.04 ± 0.017	0.03 ± 0.02
Ta	0.05 ± 0.03	0.02 ± 0.01	< 0.01	< 0.01	< 0.01	< 0.01
Th	0.21 ± 0.05	0.18 ± 0.02	0.02 ± 0.01	0.06 ± 0.02	0.04 ± 0.02	0.005 ± 0.003
Ti	73.7 ± 39.5	53.6 ± 13.7	7.45 ± 2.96	8.35 ± 3.06	4.25 ± 2.15	0.83 ± 0.32
U	0.026 ± 0.02	0.018 ± 0.01	< 0.01	< 0.01	< 0.01	< 0.01
V	1.3 ± 0.2	1.1 ± 0.2	0.25 ± 0.05	0.3 ± 0.1	0.25 ± 0.08	0.02 ± 0.01
Zn	11.2 ± 5.4	6.5 ± 3.0	5.8 ± 2.1	4.1 ± 1.7	2.2 ± 0.9	1.9 ± 1.1

Note: Values obtained from eight individual samplings.

and in an urban area (Milan downtown) of North Italy reported in Table 22.2 (Gallorini et al., 1999; Rizzio et al., 1999; Rizzio et al., 2001; Bem et al., 2003). Obviously, the concentrations found at the Pyramid are sensibly lower (orders of magnitude in comparison to those of Milan downtown). Considering the TSP, the Himalayan values are, to some extent, comparable only with those of Ispra (rural) and elements such as As, Br, Cd, Ni, Pb, Sb, Se and Zn, some of which may be considered associated to atmospheric pollution processes, are significantly lower. Only few elements such as Ce, Rb, Sc and Th show values that are slightly similar or higher, reflecting a possible soil-dust influence coming from a different soil background composition.

Table 22.2. Ranges of elemental content in the TSP and in the PM10 fractions of air particulate matter found in Ispra (rural areas) and Milan (urban areas), Italy. Winters 1996–1999. Concentration in ng/Nm^3.

Element	TSP		PM 10	
	Ispra (rural areas)	Milan downtown	Ispra (rural areas)	Milan downtown
As	0.5–1.6	0.5–12.5	0.8–1.8	6.0–11
Br	7.5–16.9	43–908	8–14	280–758
Cd	0.4–0.9	1.7–6.3	0.3–0.7	4.5–6.8
Ce	0.2–0.6	0.3–3.1	0.3–0.4	1.0–1.9
Cl	185–421	150–2600	270–364	1700–2400
Co	0.4–1.2	0.8–19	0.5–1.0	3.5–15
Cr	4.1–10.3	2–298	5.5–9.1	52–250
Cs	0.1–0.3	0.1–3.8	0.2–0.3	0.5–1.9
Cu	6.5–15.8	6–153	7.7–12.2	52–120
Fe	312–834	300–4067	510–638	1600–2870
Hf	0.02–0.08	0.05–0.56	n.d.	0.2–0.4
Hg	0.03–0.15	0.1–1.8	0.03–0.07	0.9–1.6
I	1.6–4.3	5–18	2.1–4.1	11–17
K	270–578	690–1650	312–422	750–1188
La	0.2–0.7	0.4–2.4	0.3–0.6	0.8–1.3
Mg	120–730	1900–3870	290–449	500–1900
Mn	9–28	16–330	11–19	45–210
Mo	0.2–0.7	7.5–14	0.3–0.6	6.1–11.5
Ni	6.7–15	3.9–167	6.8–12.5	30–110
Pb	61–154	75–3200	68–112	512–1900
Rb	0.3–1.1	3.2–8.9	0.4–0.6	1.6–2.6
Sb	3.2–6.9	14–156	4.0–6.1	55–125
Se	0.8–1.6	0.3–1.9	0.5–1.3	0.9–1.3
Sc	0.03–0.07	0.5–3.7	0.03–0.05	0.7–1.8
Ta	n.d	n.d.	n.d.	n.d.
Ti	12–35	40–132	18–28	55–89
Th	0.02–0.1	0.02–1.1	0.03–0.04	0.02–0.45
V	2.3–6.3	9.1–75	3.6–5.6	28–61
Zn	125–198	31–329	110–170	209–250

Note: Values obtained from literature (Gallorini et al., 1999; Rizzio et al.,1999).

The comparison between the Himalayan concentration values occurring in the two different periods (pre- and post-monsoon) shows a significant seasonal variation. Excluding bromine and iodine, all the other elements present lower concentrations in the post-monsoon period. This is particularly evident for Ca, Ce, Cr, Cu, Mn, Ti (more then 5 times) and for Cl, Co, Fe, Hf, La, Mg, Ni, Sc, Se, V and Zn (more than 3 times). The decreasing trend of the concentrations is similar in all the three fractions TSP, PM 10 and PM 2.5; nevertheless, for the major part of the elements, the distribution between the particle size fractions is changed, showing, in the post-monsoon period, an enrichment in the PM 2.5 fraction as reported in Table 22.3. Thus, in the post-monsoon period, the elemental concentration of the airborne particulates is lower and a consistent part of the elements are associated with the finest particle-size fraction.

These findings may be explained considering the climate conditions (rain) and the air mass trajectories associated with the two periods. The post-monsoon airborne samplings were carried out in October after a prolonged period of summer rain precipitation (Fig. 22.1). This washing-away effect has contributed to the decrease of the total airborne particulates, especially the coarse fraction. On the other hand, no decrease of iodine and bromine concentrations has been noticed in the post-monsoon period. In particular, the iodine concentration is sensibly higher. This should be correlated to the predominant air mass trajectories present in the two sampling periods. In fact, the summer monsoon (period before the October sampling) has a component coming from the Indian Ocean (Arabian Sea) that is absent in the pre-monsoon period. The iodine content in the airborne particulates can be associated to that part of the air mass passing over the sea surface (Vogt et al., 1999) and submitted to long distance transport in the high layers of the Himalayan troposphere. According to this point of view, iodine determinations may be used as a marker of this phenomenon.

3.1. Enrichment factors

The EF has been calculated according to the following equation: EF = Cx/Cn (air particulate): Cx/Cn (background); where Cx is the concentration of the X element whose EF is to be determined and Cn is the concentration of the normalizing element assumed to be uniquely characteristic of the background (Hernandez et al., 2003). In this work, the ambient consists of samples of air particulate matter whereas the background consists of samples of the surrounding soils. As a normalizing element, scandium has been used. The EF data have been calculated using the mean values of the concentrations obtained from the analyses.

Table 22.4 presents the EF values of 23 elements calculated in the Himalayan airborne particulate samples, as well as in the lichen samples. In the airborne particulates the EFs of elements such as Br, Cd, Sb, Cu, Zn and, to some extent, Pb and Ni, are significantly higher than 1 and may be considered incompletely originating from the local soils. These elements may be related to long-range transport processes. Most of them come from far sources and are transported with the finest particulates in the high troposphere layers. From this point of view, Cd and Sb have the highest EF values that could be considered marker elements of these processes.

Table 22.3. Variation (%) of the elemental concentrations (ng/Nm^3) and variation (%) of the elemental repartitions in PM 10 and PM 2.5 fractions, of the Himalayan airborne particulate matter collected in pre-monsoon (March–May 2002) and in post-monsoon (October 2003) periods.

Element	Variation of the concentrations between pre- and post-monsoon			Variation of the elemental repartition in PM 10 and in PM 2.5					
	Post-monsoon			% in the PM 10 vs. TSP			% in the PM 2.5 vs. PM 10		
	TSP variation %	PM 10 variation %	PM 2.5 variation %	Pre-mon. %	Post-mon. %	Variation %	Pre-mon. %	Post-mon. %	Variation %
As	−54	−69	−82	93	63	−32	31	18	−42
Br	+18	−14	−2.5	92	65	−29	60	70	+17
Ca	−86	−88	−57	79	67	−15	17	62	+264
Cd	−25	n.v.	n.v.	75	93	+24	50	80	+60
Ce	−87	−87	−81	68	71	+4	14	20	+43
Cl	−70	−55	−82	23	35	+52	55	22	−60
Co	−70	−82	−89	93	70	−25	50	31	−38
Cr	−89	−89	−92	86	84	−2	56	44	−21
Cs	−23	−11	n.v.	82	94	+15	9	11	+22
Cu	−82	−79	−78	77	89	−16	68	72	+6
Fe	−75	−80	−73	87	69	−21	14	20	+43
Hf	−75	−54	n.v.	67	92	−37	33	73	+121
I	+84	+137	+125	73	94	+29	53	47	−11

La	−77	−87	n.v.	93	50	−46	25	24	−4
Mg	−69	−70	−54	92	87	−5	10	23	+130
Mn	−80	−81	−70	73	70	−4	13	21	+62
Ni	−64	−73	−86	81	59	−27	66	34	−48
Pb	−34	−48	−34	84	67	−20	70	88	+26
Rb	−40	−51	−25	89	72	−19	12	18	+50
Sb	−76	−78	−68	86	80	−7	61	88	+44
Sc	−70	−70	−68	75	73	−3	18	20	+11
Se	−65	−56	−7	53	67	+26	33	70	+112
Th	−71	−79	−75	86	62	−28	11	14	+27
Ti	−89	−92	−89	73	51	−30	14	20	+43
U	−46	−44	n.v.	69	71	+3	n.v.	n.v.	n.v.
V	−75	−77	−92	86	78	−9	22	7	−68
Zn	−63	−66	−67	58	53	−9	89	86	−4

Note: From data reported in Table 22.1; n.v. not valuable.

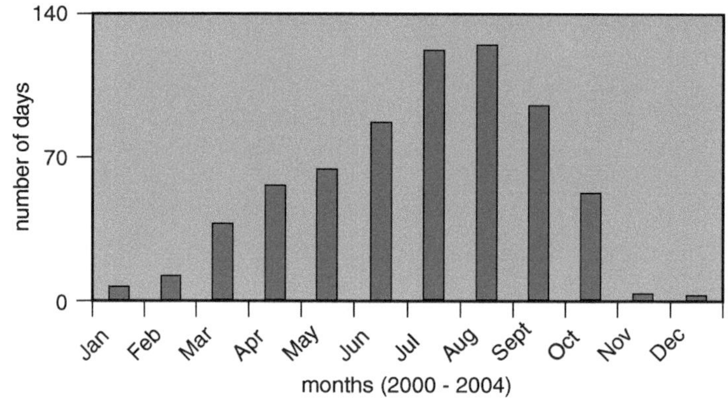

Figure 22.1. Seasonal frequency of precipitation at Ev-K^2-CNR Pyramid in years 2000–2004.

Table 22.4. Comparison between the trace elements EF calculated in airborne particulate matter and in lichens collected at high altitude in the Himalaya (Khumbu valley – Ev-K^2-CNR Pyramid, March–May 2002).

Element	Enrichment Factors			
	Airborne particulates (Ev-K^2-CNR Pyramid)	Himalayan lichens (4700–5100 m. a.s. l.)		
	TSP	*Hypogymnia* sp	*Umbilicaria* sp	*Alectoria* sp
As	2.9	1.9	8.0	6.4
Br	43	151	55	166
Cd	59	32	37	42
Ce	0.4	2.1	1.5	0.2
Co	3.2	2.5	1.5	0.2
Cr	7.0	1.1	0.9	0.4
Cs	0.5	0.5	0.6	0.4
Cu	21	2.3	8.1	8.4
Fe	1.1	1.0	0.9	0.5
Hf	0.3	1.0	0.6	0.3
K	0.8	1.3	1.7	1.6
La	0.4	1.8	1.3	0.4
Mg	0.7	0.7	0.5	0.3
Mn	1.7	1.6	0.6	0.8
Ni	7.2	n.d.	n.d.	n.d.
Pb	9.0	3.7	5.3	1.9
Rb	0.5	1.0	0.8	0.3
Sb	320	120	70	170
Sc	1.0	1.0	1.0	1.0
Se	4.3	1.3	0.1	0.6
Th	0.3	1.9	1.5	0.3
V	1.6	0.5	0.6	n.d.
Zn	14	9.6	22	8.9

Note: Values obtained from the elemental analysis of local soils and normalized versus Sc concentrations.

3.2. Lichens as biomonitors

The EF values calculated in the airborne samples have been compared to those of lichens collected (May 1999) in the same areas for a previous study on Himalayan lichens. All details about lichens sampling and analysis are given elsewhere (Bergamaschi et al., 2002; Bergamaschi et al., 2004). The lichen species considered in this work are; *Hypogymnia* sp, *Umbilicaria* sp and *Alectoria* sp, this last collected at 5000 m a.s.l. in proximity to the EV-K^2-CNR Pyramid. The results are reported in Table 22.4 where it can be observed that the elements with highest EF values are the same in the air particulates and in the lichens. This agreement may suggest that lichens could be used as permanent biomonitors for elements such as Cd, Br, Sb, Zn and Pb.

4. Conclusions

The first and preliminary elemental characterization of the airborne particulate matter collected in the Himalayan region at 5100 m a.s.l., during the pre-monsoon and post-monsoon periods (2002/2003) has been accomplished. For both periods, the concentrations in ng/Nm^3 of 30 elements have been measured in the different particle size fractions (total suspended particles: TSP, PM 10 and PM 2.5). Most of the elements were determined by INAA that has demonstrated its great capability in performing multi-elemental analyses on fractions of milligrams of air dust collected onto filters (Biegalski and Landsberger, 1995; Landsberger and Biegalski, 1995; Landsberger et al., 1997; Zeisler et al., 1997; Gallorini, 2000). All the concentration values found in this work are very low and can be taken as baseline values of unpolluted high altitude remote areas. Nevertheless, some elements present in the finest particles show significant high EF values and should be considered not completely originating from the local soils and submitted to long-distance transportation phenomena.

According to these findings and with the air mass trajectories, some elements such as Br, Sb, Cd and Zn have to be considered mainly coming from western natural and/or anthropic sources far from the Ev-K^2-CNR Pyramid. A sensible seasonal variation between pre- and post-monsoon periods has been noticed for all elements present in the airborne particulates, with their concentrations much lower in the post-monsoon period. On the contrary, iodine and to some extent bromine show an enrichment of their concentrations that may be explained with the arrival of a southern air mass component of the summer monsoon coming from the Arabian Sea. Finally, the quite good correspondence of EF values of many elements found in lichens and in the airborne particulates may also suggest the use of lichens as permanent trace-metal biomonitors in this part of the world (Smodiš and Bleise, 2002).

Acknowledgments

This work was carried out within the framework of the Ev-K^2-CNR "Scientific and Technological Research in Himalaya and Karakorum" Project in collaboration with RONAST as foreseen by the Memorandum of Understanding between the

Government of the Kingdom of Nepal and the Government of the Republic of Italy. The research conducted was also made possible thanks to contributions from the Italian National Research Council and the Italian Ministry of Foreign Affairs. Special acknowledgments are given to Agostino Da Polenza, Beth Shommer and the staff of the Mountain Equipe s.r.l. (Bergamo-Italy).

Special thanks must be given to Dr. Bonasoni (CNR-ISAC) and Dr. Bertolani (EPSON-Meteo) for evaluating and supplying the climate and air mass trajectories data.

References

Artiñano, B., Salvador, P., Alonso, D.G., et al., 2003. Anthropogenic and natural influence on the PM 10 and PM2.5 aerosol in Madrid (Spain). Analysis of high concentration episodes. Environmental Pollution 125, 453–465.

Bem, H., Gallorini, M., Rizzio, E., and Krzemiska, M., 2003. Comparative studies on the concentrations of some elements in the urban air particulate matter in Lodz city of Poland and in Milan, Italy. Environment International 29 (4), 423–428.

Bergamaschi, L., Rizzio, E., Giaveri, G., et al., 2004. Determination of baseline element composition of lichens using samples from high elevations. Chemosphere 55, 933–939.

Bergamaschi, L., Rizzio, E., Valcuvia, M.G., et al., 2002. Determination of trace elements and evaluation of their enrichment factors in Himalayan lichens. Environmental Pollution 120 (1), 137–144.

Biegalski, S. and Landsberger, S., 1995. Improved detection limits for trace elements on aerosol filters using Compton suppression epithermal irradiation techniques. Journal of Radioanalytical and Nuclear Chemistry 19, 195–204.

Dams, R., 1992. Nuclear activation techniques for the determination of trace elements in atmospheric aerosols particulates and sludges samples. Pure Applied Chemistry 64 (7), 991–1003.

Gallorini, M., 2000. Trace elements in atmospheric pollution processes: the contribution of the neutron activation analysis. In: Spurny, K.R. (Ed.), Aerosol Chemical Processes in the Environment. Lewis-CRC Press, Boca Raton, Florida, pp. 431–455.

Gallorini, M., Rizzio, E., Birattari, C., et al., 1999. Content of trace elements in the respirable fractions of the air particulate of urban and rural areas monitored by neutron activation analysis. Biological Trace Element Research 71 (2), 209–222.

Heller-Zeisler, S.F., Ondov, J.M., and Zeisler, R., 1999. Collection and characterization of a bulk PM 2.5 air particulate matter material for use in reference materials. Biological Trace Element Research 71–72, 195–202.

Hernandez, L., Probst, A., Probst, J.L., and Ulrich, E., 2003. Heavy metals distribution in some French forest soils: evidence for the atmospheric contamination. Science of the Total Environment 312, 195–219.

Landsberger, S. and Biegalski, S., 1995. Analysis of inorganic particulate pollutants by nuclear methods. In: Kouimtzic, T. and Samara, C. (Eds), Handbook of Environmental Chemistry, Vol.: Airborne Particulate Matter 4 part D. Springer-Verlag, N.Y., pp. 175–200.

Landsberger, S., Zhang, P., Wu, D., and Chatt, A., 1997. Analysis of the arctic aerosols for a ten year period using various neutron activation analysis methods. Journal of Radioanalytical and Nuclear Chemistry 217, 11–15.

Maenhaut, W., Rajta, I., François, F., et al., 1999. Long-term atmospheric aerosol study in the Finnish arctic: chemical composition, sources types and source regions. Journal of Aerosol Science Suppl. 1, 87–89.

Querol, X., Alastuey, A.A., Jesús de la Rosa, J., et al., 2002. Source apportionment analysis of atmospheric particulate in an industrialised urban site in southwestern Spain. Atmospheric Environment 36, 3113–3125.

Ramanathan R. and Crutzen, P.J., 1999. Project Asian Brown Cloud. Air Pollution in the Indo-Asia-Pacific Region: Impact on Climate and the Environment – INDOEX; http://www-indoex.ucsd.edu

Rizzio, E., Bergamaschi, G., Profumo, A., and Gallorini, M., 2001. The use of the neutron activation analysis for particle size fractionation and chemical characterization of trace elements in urban air particulate matter. Journal of Radioanalytical and Nuclear Chemistry 248 (1), 21–28.

Rizzio, E., Giaveri, G., Arginelli, D., et al., 1999. Trace elements total content and particle sizes distribution in the air particulate matter of a rural-residential area in North Italy investigated by instrumental neutron activation analysis. Science of the Total Environment 226 (1), 47–56.

Rizzio, E., Giaveri, G., and Gallorini, M., 2000. Some analytical problems encountered for trace elements determination in the airborne particulate matter of urban and rural areas. Science of the Total Environment 256 (1), 11–22.

Smodiš, B. and Bleise, A., 2002. Internationally harmonized approach to biomonitoring trace element atmospheric deposition. Environmental Pollution 120, 3–10.

Vogt, R., Sander, R., Von Glasow, R., and Crutzen, P.J., 1999. Iodine chemistry and its role in halogen activation and ozone loss in the marine boundary layer: a model study. Journal of Atmospheric Chemistry 32, 375–395.

Yli-Tuomi, T., Venditte, L., Philip, P.K., et al., 2003. Composition of the Finnish arctic aerosol: collection and analysis of historic filter samples. Atmospheric Environment 37, 2355–2364.

Zeisler, R., Haselberger, N., Makarewicz, M., et al., 1997. Nuclear techniques applied to air particulate matter studies. Journal of Radioanalytical and Nuclear Chemistry 217, 5–12.

Mountains Witnesses of Global Changes
R. Baudo, G. Tartari, and E. Vuillermoz (Editors)

23

Interactions between solar ultraviolet radiation and climatic warming in alpine lakes

Ruben Sommaruga

Abstract

Solar ultraviolet (UV) radiation has been a crucial environmental factor since the origin of alpine lakes (i.e., lakes above treeline) ca. 10,000 years ago, because those ecosystems receive high solar UV fluxes due to a thinner ozone column and usually lower aerosol scattering. Another important change that takes place with increasing elevation is the reduction of in-lake colored dissolved organic carbon (cDOC) concentrations, which reflect the dominance of small and sparsely vegetated watersheds at high altitude. Concomitantly with the reduction in cDOC concentrations, there is a significant decrease in the UV absorption capacity of this carbon pool (i.e., reduced content in humic substances). The consequence of these qualitative and quantitative changes of cDOC is that most high-mountain lakes (with the exception of those fed by glacier streams) rank among the most UV-transparent aquatic systems, with UVB (290–320 nm wavelength) penetration depths of up to ca. 30 m. Climatic warming has been particularly pronounced in mountain regions. For example, in the northern Alps, the mean air temperature has increased by 1°C since 1985. Several environmental changes associated with climatic warming, such as glacier retreat and the timing, extent and duration of ice and snow cover, are already taking place in several mountainous regions. Alpine lakes are particularly sensitive to climatic variability because several crucial ecosystem processes are directly affected by those changes. Based on observational analysis, possible changes in alpine lakes are presented. On a short timescale, the decrease in the snow- and ice-cover duration will lead to an increase in UV stress on the ecosystem. However, on a longer timescale, the increase in mean air temperature could favor the development of terrestrial vegetation in the catchment, and result in an increase of cDOC export to alpine lakes, particularly to those located near the treeline. As a consequence, underwater UV transparency and UV stress on these ecosystems will be reduced. This scenario is supported by the finding that the alpine-nival flora in some mountainous regions, such as in the western Austrian and eastern Swiss Alps has moved upward in recent years. On the other hand, glacier retreat may cause turbid lakes to become more UV transparent.

ISSN: 0928-2025 DOI: 10.1016/S0928-2025(06)10023-1

The magnitude of these changes will be more important for underwater UVB exposure levels than those expected from ozone reduction at mid-latitudes.

24

Global land ice monitoring from space (GLIMS) project regional center for Southwest Asia (Afghanistan and Pakistan)

John F. Shroder Jr., Michael P. Bishop, Henry N.N. Bulley, Umesh K. Haritashya and Jeffrey A. Olsenholler

Abstract

Concerns over world-wide loss of ice have resulted in the Global Land Ice Measurements from Space (GLIMS) Project wherein glaciers are being mapped and monitored from space with the ASTER satellite sensor. A combined American and Japanese satellite system launched in 1999 on the Terra rocket allowed all countries with glaciers to receive free, large-scale (15 m resolution) satellite imagery. The Governments of Afghanistan and Pakistan expressed little initial interest to the U.S. Geological Survey (USGS) or to the U.S. Aeronautics and Space Administration (NASA) who were funding GLIMS. The University of Nebraska at Omaha (UNOmaha), with its three decades of research association in both countries, became the Southwest Asia (Afghanistan and Pakistan) GLIMS Regional Center for the Hindu Kush and Western Himalaya. This paper is a brief discussion and presentation of a few glaciers we have been studying.

In the Hindu Kush and Pamir ranges of Afghanistan we selected a transect of glaciers from west to east, that include (1) Foladi Glacier in Koh-i-Baba Range; (2) Mir Samir glaciers in central Hindu Kush; (3) Sakhi Glacier in Koh-i-Bandakha range in north-central Hindu Kush; (4) Keshnikhan Glacier at the mouth of Wakhan Corridor; and (5) Little Pamir glaciers in the Wakhan Corridor. All glaciers were mapped by geoscientists in the past half century and are now being reassessed for change detection. In general smaller, lower-altitude glaciers were already below the climatic equilibrium line 40 years ago when first mapped, but were protected in shadowed cirques; many are now wasting away, although deconvoluting original cartographic error from real change is problematic. Nonetheless, evidence of serious glacial retreat has major implications for downstream melt-water irrigation in this chronically drought-torn region. In Pakistan focus is primarily upon five main areas in a west-to-east transect, starting with (1) glaciers of Tirich Mir in the northwest; (2) Gorshai Glacier in Swat; (3) Batura Glacier and others in Hunza; (4) glaciers of the Nanga Parbat Himalaya; and (5) Biafo and Baltoro Glacier in the northeast. Because

ISSN: 0928-2025 DOI: 10.1016/S0928-2025(06)10024-3

these glaciers are all from higher altitude areas than most of those in Afghanistan, they receive more nourishment but are still undergoing significant downwasting, and some termini are backwasting. Most significant related events, however, are the debuttressing of valley walls that can cause massive landslides, and glacier lake outburst floods (GLOF), which threaten much of the Himalaya. Overall between Afghanistan and Pakistan we can say that the loss of significant glacier ice in coming decades may be becoming progressively more serious unless global warming ultimately generates greater marine evaporation that augments precipitation.

The GLIMS Project will continue monitoring glaciers but the task must be passed on to newly trained specialists from Afghanistan and Pakistan. New satellite systems, perhaps even the new European cryosat system, must also be deployed for this purpose. World-wide glacier monitoring underway for decades is at last capable of achieving significant results with high resolution, stereographic satellite imagery in the GLIMS Project.

1. Introduction

The Global Land Ice Measurements from Space (GLIMS) Project (www.glims.org/) was established by the U.S. Geological Survey (USGS) and the U.S. Aeronautics and Space Administration (NASA) so that glaciers world-wide could be mapped and monitored from space with the ASTER satellite sensor system that was launched on the Terra rocket (Kieffer et al., 2000; Bishop et al., 2004). The Governments of Afghanistan and Pakistan were not originally interested in the project so the University of Nebraska at Omaha (UNOmaha), with its three decades of geoscience research association in both countries, became the Southwest Asia (Afghanistan and Pakistan) GLIMS Regional Center for the Hindu Kush and Western Himalaya. Once sufficient Southwest Asian scientists are trained in GLIMS methodology, the project will be continued more directly by scientists in Afghanistan and Pakistan.

In the Hindu Kush and Pamir ranges of Afghanistan (Fig. 24.1) we selected from a general total of >3000 glaciers a sample transect of ice masses from west to east, that include (1) Foladi Glacier in Koh-i-Baba Range; (2) the Mir Samir glaciers in the central Hindu Kush; (3) Sakhi Glacier in the Koh-i-Bandakha range in north-central Hindu Kush; (4) Keshnikhan Glacier at the mouth of Wakhan Corridor; and (5) the glaciers of the Little Pamir Mountains in the northeast Wakhan Corridor (Shroder and Bishop, 2006a). All these glaciers have been mapped by others in the past half century and are now being reassessed by GLIMS Project scientists for change detection. In general the smaller, lower-altitude glaciers were already below the climatic equilibrium line 40 years ago when first mapped, but were protected in shadowed cirques; many small ones are now wasting away, although deconvoluting the original cartographic variance from real change caused by climatic warming is problematic. Nonetheless, evidence of serious glacial retreat has major implications for downstream melt-water irrigation in this chronically drought-torn region (Cyranoski, 2005).

In Pakistan with >5218 glaciers (Mool et al., 2003, 2005) the focus of our work is primarily upon six areas in a west-to-east transect (Fig. 24.1), starting with (1) the glaciers of Tirich Mir in the northwest; (2) Gorshai Glacier in Swat; (3) Batura

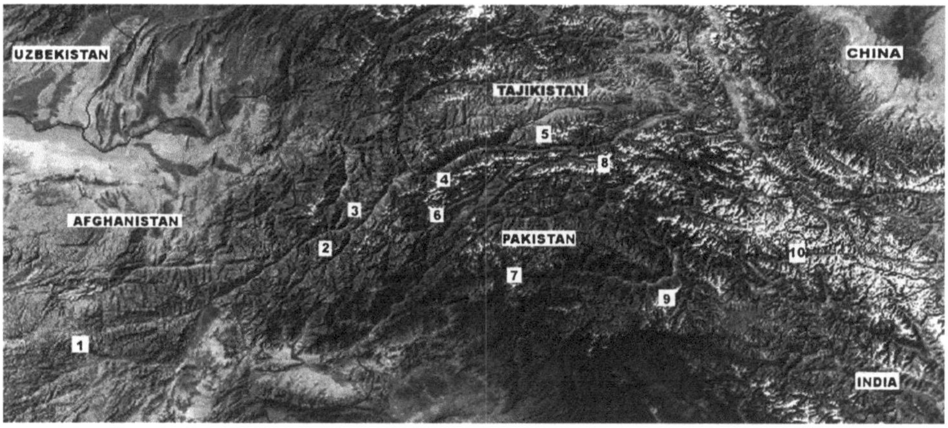

Figure 24.1. Location map of mountain ranges and glaciers of Afghanistan and Pakistan. Glaciers enumerated in Afghanistan from west to east are (1) Foladi Glacier and neighbors in Koh-i-Baba range; (2) Mir Samir Glacier and neighbors in north-central Hindu Kush; (3) Sakhi Glacier and neighbors in the Koh-i-Bandaka range of the north-central Hindu Kush; (4) Keshnikhan Glacier and neighbors in the Wakhan Hindu Kush; (5) Issik and Zemestan Glaciers of the Wakhan Pamir ranges. Glaciers in Pakistan from west to east are (6) Tirich Glacier on Tirich Mir mountain; (7) Gorshai Glacier in Swat; (8) Batura Glacier in the Hunza Karakoram; (9) glaciers of Nanga Parbat; (10) Biafo Glacier and Baltoro Glacier in the K2 Karakoram Himalaya.

Glacier and others in the Hunza Karakoram; (4) glaciers of the Nanga Parbat Himalaya; and (5) Biafo and Baltoro Glacier in the Karakoram to the northeast of Pakistan (Shroder and Bishop, 2006b). Because these glaciers are from higher altitude areas than most of those in Afghanistan, they receive more nourishment but are still undergoing significant downwasting, and some termini are backwasting. Significant related events are the debuttressing of valley walls that can cause massive landslides, and glacier lake outburst floods (GLOF) that threaten much of the Himalaya. Overall the loss of significant glacier ice in coming decades in Afghanistan and Pakistan is likely to become much more serious in many ways. On the other hand, some limited observations recently (Hewitt, 2005; Fowler and Archer, 2006; Lau and Kim, 2007) indicate that climate change may also decrease melting, increase precipitation, and cause renewed glacier advance.

World-wide glacier monitoring underway piecemeal for decades is capable of achieving significant synoptic results with high resolution, stereographic satellite imagery in the GLIMS Project. In addition the ASTER satellite imagery is also available for mapping in emergencies, such as the disastrous earthquake of 10/08/2005 in which >73,000 people died as buildings collapsed and massive landslides occurred throughout the foothill region of the Western Himalaya near Muzaffarabad, Pakistan (Fig. 24.2).

This paper is a brief survey of extensive work by the authors and many others on the glaciers of Afghanistan and Pakistan. Because of the needs for only a summary overview herein, many important details and the prior work of others have been left out. Nevertheless, this paper can serve as an overview of some general trends that will allow other interested people to focus in on the more detailed work expressed elsewhere.

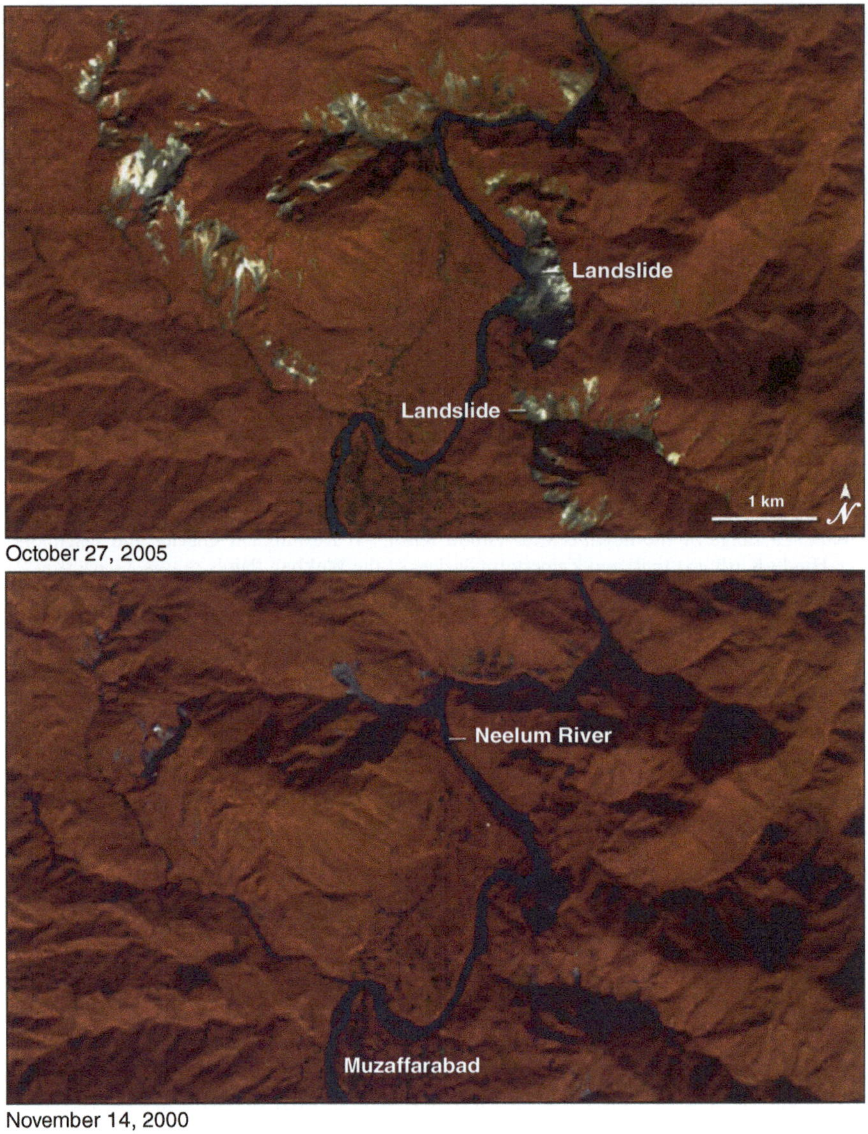

Figure 24.2. Before (bottom, 14/11/2000) and after (top, 27/10/2005) ASTER satellite imagery of huge landslides around Muzaffarabad, Pakistan, following the massive (magnitude 7.3) and destructive (>73,000 fatalities) earthquake of 8/10/2005. Such landslides dammed nearby rivers to produce two large lakes 15–24 m deep that as they fill up and overflow could breach catastrophically and wreak havoc downstream.

2. Glaciers of Afghanistan

The Hindu Kush and Pamir Mountains of Afghanistan rise in altitude from <500 m a.s.l. on the Iranian border on the west to >7000 m a.s.l. on the Pakistan border on the east. The glaciers of Afghanistan occur in the rugged alpine topography of the central and northeast mountains and their melt waters drain out radially, southeast into the Indus drainage basin, southwest into Afghan–Iran Plateau endorheic basin, and north into the Turkestan endorheic basin (Shroder, 1980; Shroder, 1989b). These melt waters are vital to the arid country of Afghanistan and assessing the snow and ice sources in the mountains are viewed as essential to the future of irrigation in the country.

The snowline contours for Afghanistan range from 4600 to 5200 m a.s.l., with the highest values at both the northeast Wakhan Pamir and central southwest ends of the Hindu Kush where precipitation is less (Wissman, 1960; Shroder and Bishop, 2006a). Glaciers in Afghanistan are generally small in the central Hindu Kush, but larger toward the higher northeast. The climatic snowline is above many peaks and the glaciers exist there only where ablation is retarded in the shadows of the peaks (Grötzbach and Rathjens, 1969). The result is a stronger north-facing glacierization in much of the central Hindu Kush but this aspect tendency is less strong into the higher peaks of the Wakhan. Breckle and Frey (1976a,b) observed a strong glacierization facing towards the east and southeast close to the Pakistan border where the summer monsoon produces protective cloud shading and more precipitation. This part of Afghanistan commonly has part of the intertropical convergence zone (ITCZ) across it in summer and is regularly influenced by monsoon precipitation (Sivall, 1977; Shroder, 1989a).

Most glaciers in Afghanistan are mountain or valley types that occur in cirques, simple basins, and short valley segments. The common avalanche nourishment sources also produce much debris-covered ice and rock glaciers. The longest and largest valley glaciers in Afghanistan occur on the south side of the entrance to the Wakhan Corridor. The mountains on the border with Pakistan rise to well over 6000 m a.s.l., and at least 15 of the north-flowing glaciers of the area are over 10 km long. The largest are some 75–100 km^2 in surface area. For example, Qadzi Deh glacier flows ~14 km down to an altitude of ~3580 m a.s.l. and is covered extensively with black slate and argillite fragments derived from the Wakhan Formation (Buchroithner, 1978; Desio, 1975; Shroder and Bishop, 2006a). This debris produces anomalous black glacier surfaces that are prominent on the satellite imagery. Downwasting and backwasting are thought to be a dominant characteristic throughout much of the Himalaya (Mayewski and Jeschke, 1979) and the Hindu Kush (Shroder, 1980).

3. Mapping in Afghanistan

Topographic maps are useful development tools, so in the late 1950s the southern two-third of the country was flown for stereographic photographs by the U.S.A. and the northern one-third by the U.S.S.R. (Glicken, 1960; Reiner, 1966). In a Cold-War competition complete map series were made by both governments for the Afghan Cartographic Institute in Kabul who republished the maps. Subsequent Cold-War machinations on all sides precluded ease of availability of maps until the dissolution of the Soviet empire in the early 1990s, at which time economic collapse necessitated sale of maps of their areas of interest. Cyrillic-alphabet topographic maps at scales of

1:50,000, 1:100,000, and 1:200,000 then became available, as well as US Department of Defense (DOD) maps at 1:100,000, from which reasonably accurate location and size assessments of glaciers could be made. The Soviet maps, however, had a haphazard and inconsistent approach so that care must be exercised for definitive delineations of the ice as it was in the late 1950s when the stereo photos were first obtained. The DOD cartographers noted clean ice as white with dashed blue outline and contours, and debris-covered ice as brown dots, but in comparison, Soviet mapmakers entirely ignored some small glaciers and commonly showed both debris-covered and clean ice with undifferentiated blue contours (Fig. 24.3).

Only one large-scale glaciological sketch map and two definitive glacier maps were ever produced in Afghanistan, and included the Mir Samir area (Gilbert et al., 1969) Keshnikhan, and the Wakhan Pamir ranges (Exploration Pamir 75, 1978a,b). These maps have detailed moraine configurations, transient snowlines, accurate topography, and important glacier landforms. The great detail of these maps at 1:25,000 scales makes them excellent sources for comparative analysis of the satellite imagery.

4. Selected glaciers of Afghanistan

Five glacierized areas can provide some assessment of glacial conditions across the country. The westernmost glaciers of Afghanistan occur above the Yawkawlang graben valley ~100 km northwest of Bamiyan. The small glaciers of the centrally located Koh-i-Baba range are also among the westernmost in Afghanistan, whereas Mir Samir and Koh-i-Bandaka in the north-central part of the country have glaciers of small and intermediate size. Glaciers near the mouth of the Wakhan Corridor, such as Keshnikhan, have extensive debris covers in their lower reaches. In the high Wakhan Pamir the glaciers are among the largest in the country in spite of having low precipitation.

4.1. Koh-i-Baba

About 18 small glaciers occur in the Koh-i-Baba range south of Bamiyan where the famous huge carved stone Buddhas were destroyed by the Taliban. All glaciers have north exposures and average only ~0.5 km^2 in area (Shroder and Giardino, 1978; Shroder and Bishop, 2006a). The lowermost elevations of exposed-ice termini range 4075–4657 m a.s.l. and average 4365 m a.s.l. Most glaciers terminate in what are classified as debris-covered glacier fronts, ice-cored moraines, or ice-cored rock glaciers, depending on subtle geomorphologic differences purported to be the result of different

Figure 24.3. Comparison of US DOD (3a, left) and Soviet (3b, right) topographic maps of Mir Samir glacierized area showing varying quality and quantity of mapping of glaciers. The DOD map shows glacier ice as white ground with dashed blue outline and contours, as well as debris-covered ice and moraine. The Soviet map has several areas that are treated as rock but that are actually ice, as well as areas of debris that are treated as clean ice. For example, note the debris-covered ice on the upper northwest side of the DOD map that is shown as clean ice on the Soviet map (horizontal arrows). The main Mir Samir Glacier studied by British glaciologists in indicated by vertical arrow.

(a)

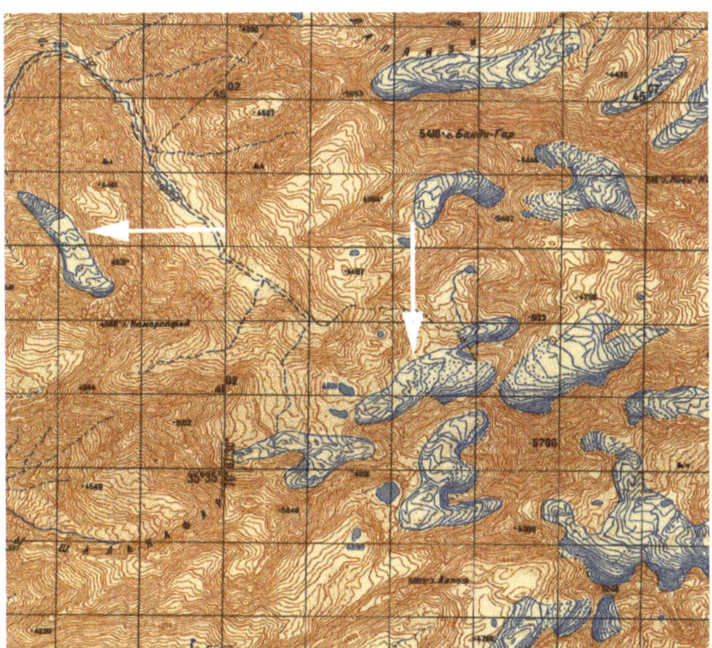

(b)

Figure 24.4. Foladi Glacier in July 1978 (photo by JS). Note the strongly concave surface that is indicative of downwasting. The shadow of the pointed peak on the upper right can be seen in Figure 25.5b.

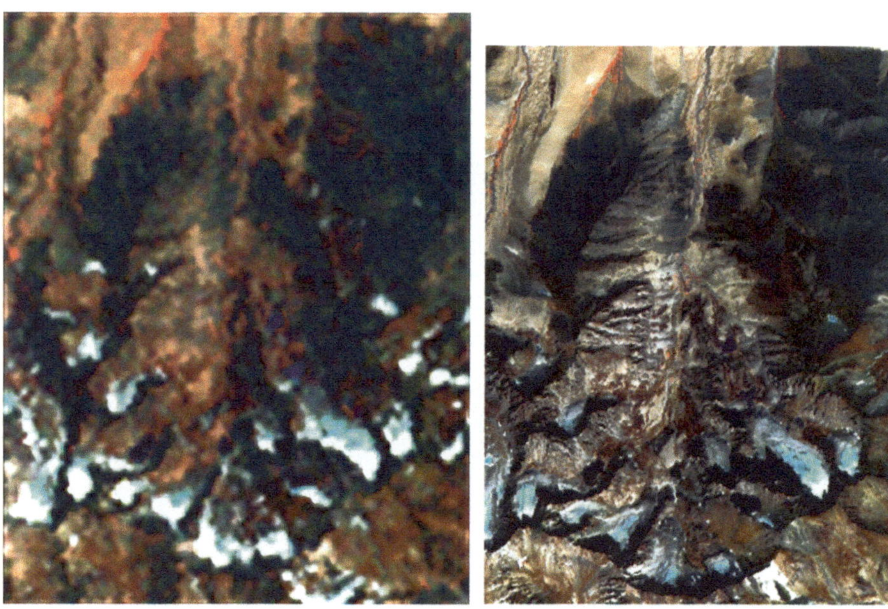

p165r36_1m19730819_vnir
(a)

AST_L1B_00310022003061744_10162003100406_vnir
(b)

Figure 24.5. Satellite images of Foladi Glacier on 08/19/1973 (a), and on 10/04/2003 (b). Even though the 2003 scene on the right (b) was taken in October close to the end of the melt season and the scene from three decades earlier on the left (a) was taken in late August earlier in the melt season, the decline in overall glacier mass over time can be seen clearly.

mechanics of emplacement. These debris-laden termini range 3850–4600 m a.s.l. and average ~4230 m a.s.l.

Field reconnaissance of Foladi Glacier in July 1978 revealed that the glacier surface was strongly concave up and entrenched behind its end moraine (Fig. 24.3). Mean annual precipitation is only ~600 mm and the snowline is above the tops of most of the surrounding peaks (Grötzbach and Rathjens, 1969) so the glacier survives because it is protected by the shadows of the surrounding peaks. Comparison of satellite imagery from 8/19/1973 (Landsat) and 10/04/2003 (ASTER), while 2 months apart at the end of the melt season, show considerable change of all the nearby glaciers over the ensuing three decades, with several glaciers having no remaining firn fields and considerable downwasting and backwasting evident (Figs. 24.4 and 24.5).

4.2. Mir Samir

The area of Mir Samir peak (5809 m a.s.l.) was visited by a British glaciological team in 1965 to map moraines with lichenometry and to collect basic mass balance data (Gilbert et al., 1969). At that time the Yakhchaal-i-Gharb (West) Glacier in the cirque just a km northwest of the main peak was at ~4800–4950 m a.s.l. It was an oval shape in plan view, with a half-km long narrow strip of glacial ice connected to it along the southwest margin of the cirque headwall cliff. Both the DOD and Soviet topographic maps made

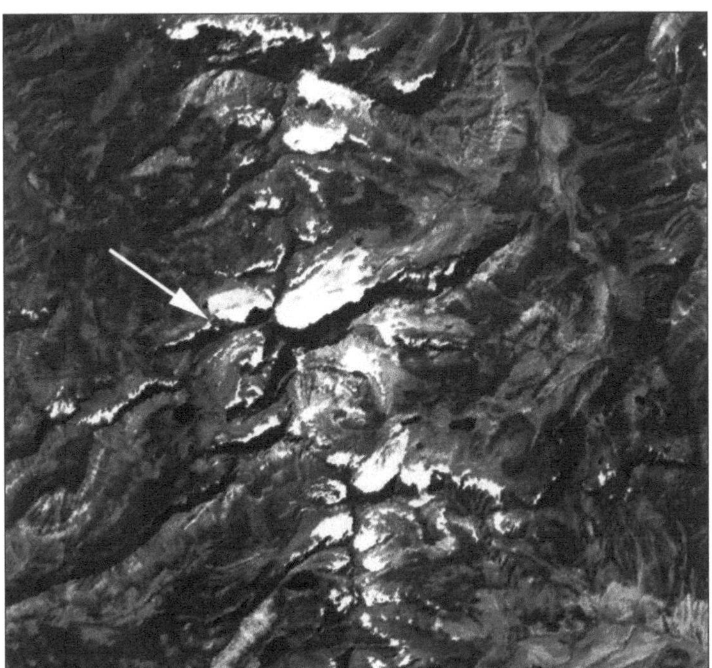

Figure 24.6. ASTER image of 26/8/2004 showing glaciers of Fig. 24.3. Note (arrow) rock rib that now divides Mir Samir Glacier into two. Compare with Fig. 24.3 for most probable (within limits of cartographic error) mid-twentieth century condition of ice masses.

from aerial photographs taken some 5 years earlier also show the glacier to be a long NE–SW trending oval shape with the narrow strip being even wider (Fig. 24.3). By the time of the ASTER satellite imagery of 26/8/2004 the narrow strip of glacier ice was detached from the main glacier by a newly exposed rock rib between them (Fig. 24.6). Other evidence of nearby glacier downwasting and backwasting abounds in comparison to the prior maps of some 40-odd years before (Figs. 24.3 and 24.6).

4.3. Koh-i-Bandaka

The Bandaka massif in the north-central Hindu Kush has six major peaks >6000 m in height, with a number of others almost as high. The Sakhi Glacier begins in a northwest-facing cirque below Sakhi peak (6414 m) before turning through a sharp bend to the south down to its terminus. Sakhi Glacier is relatively debris free except

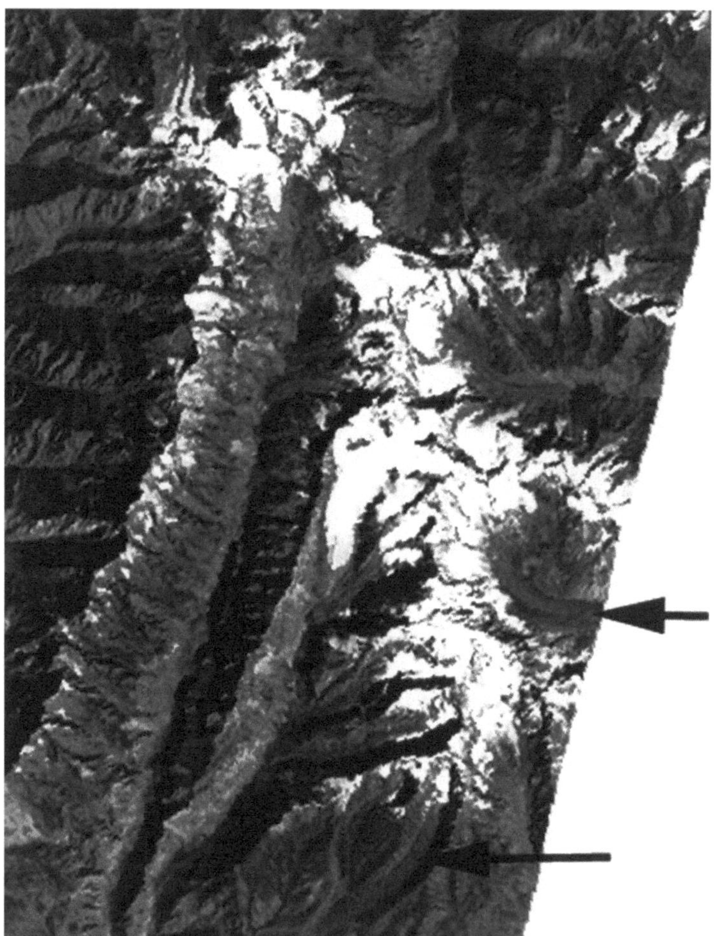

Figure 24.7. Landsat MSS satellite image (10/9/1972) of Koh-i-Bandaka massif in north-central Hindu Kush. Note strongly debris-covered glaciers (arrows).

at its terminus but many of the other glaciers around the Bandaka massif are extensively debris covered. Inasmuch as this area is also the center of the main Hindu Kush zone of highest magnitude and frequency seismicity, the extensive debris covers are most likely from earthquake-triggered avalanches that mobilize and transport copious rocks to glacier surfaces (Fig. 24.7).

4.4. Koh-i-Keshnikhan

Keshnikhan Glacier at the entrance to the Wakhan Corridor has been studied extensively (Braslau, 1972; Braslau and Bussom, 1978a,b; Shroder and Giardino, 1978). It is a relatively steep, small mass, 4 km long and ~15 km^2 in area. The equilibrium line in 1970 was judged to be ~5100–5200 m a.s.l. because the transient snowline was ~5000 m a.s.l. (Braslau, 1972). The actual terminus occurred at ~4400 m a.s.l. and an extensive moraine extended from there down to an altitude of 2600 m a.s.l. A major break in slope at ~3600 m a.s.l. apparently convinced the Soviet cartographers that the glacier-ice terminus was at that altitude, so that is where they incorrectly mapped it (Fig. 24.8).

4.5. Wakhan Pamir

The part of the Little Pamir mountains in the middle of the Wakhan Corridor has nine peaks >6000 m high. In spite of a mean annual precipitation of <100 mm (Lalande et al., 1974), the glaciers in this area are the most areally extensive in Afghanistan, apparently because of the general high altitude areas of snow catchment. The core of the range is light-colored granitic rock of late Cretaceous–Tertiary age, with peripheral rocks of the Wakhan Formation described by Buchroithner (1978), with resulting light-colored granitic moraines. The whole region was mapped and studied by an Austrian expedition in 1975 (Exploration Pamir 75, 1978a,b; Patzelt, 1978).

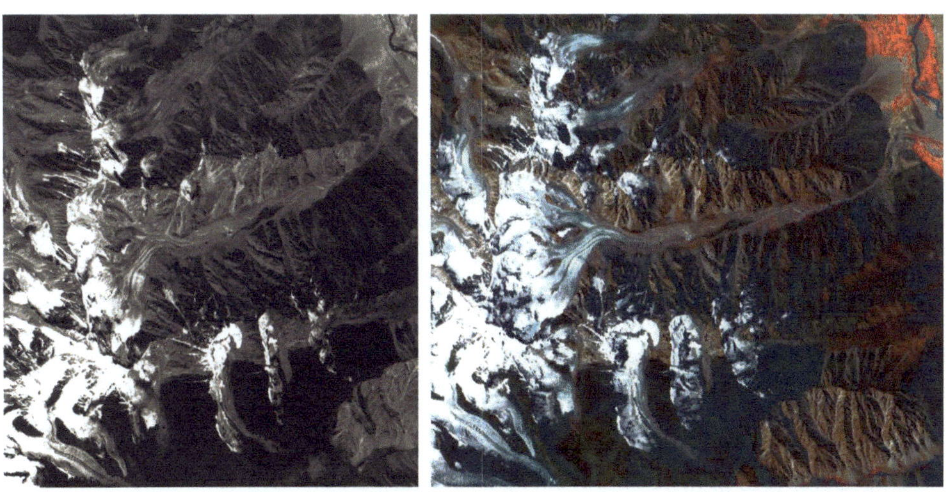

ASTER Stereogram, AST_L1B_003_08312000062611_09182003202055

Figure 24.8. Stereopair of ASTER images (8/31/2000) of Keshnikhan Glacier area.

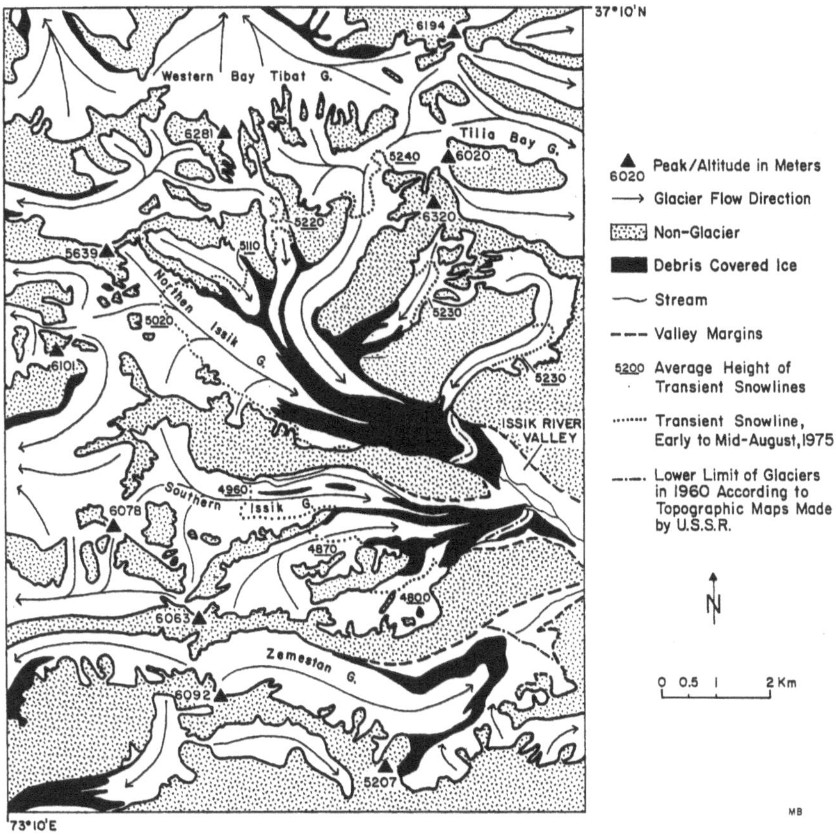

Figure 24.9. Map (9a) of glaciers of central Wakhan Pamir that were mapped separately by Austrian and by Soviet cartographers and (9b) a Return Beam Vidicon (RBV) camera scene of the same area in the 1970s.

The northern and southern Issik Glaciers, and Zemestan Glacier had transient snowlines ranging from ~5240–4800 m a.s.l. in 1975, with debris covered termini at ~4700–4500 m a.s.l. Soviet cartographers had also mapped the termini at various altitudes hundreds of m above and below the actual locations (Fig. 24.9). Comparison of multiple generations of satellite imagery shows a few downwasting and backwasting variations, particularly the contact of white ice with end moraine on the northern Issik Glacier (Fig. 24.10).

5. Selected glaciers of Pakistan

Glaciers of northern Pakistan are some of the largest and longest mid-latitude glaciers on Earth. Field- and space-based glacier studies in this region are necessary to elucidate their role in providing melt-water for irrigation, hazard potential, in erosion and geodynamics, and their sensitivity to climate forcing. Repeated field surveys of numerous glaciers have provided reference data and verification for information

RBV3 30534-05053, subscene D
P162R34

Figure 24.9. (Continued)

extraction from satellite imagery (Shroder and Bishop, 2006b; Bishop et al., 2000, 2004). Five areas were selected for case-study assessment in this paper in east–west and north–south transects to give an idea of conditions.

6. Mapping in Pakistan

Because of the relatively greater accessibility and popularity of climbing in the Karakoram and western Himalaya, many of the glacierized areas of Pakistan have been mapped and analyzed far more than Afghanistan. Early topographic maps made by cartographers from the Indian Survey of the British Empire included various topographic series at 1:63,360 (Inch to the Mile Series) and 1:253,440 (Quarter Inch Series). Some glacier and topographic mapping by British and foreign climbing teams was undertaken, as for example, that by E. E. Shipton's Karakoram Expedition of 1937 (Shipton, 1938) the work of the Royal Geographical Society (1939), and the many German teams on Nanga Parbat in the 1930s (Finsterwalder, 1936). After partition in 1947, the now separate governments of India and Pakistan continued to improve and update the old British mapping they had inherited but the exact delineation of glacierization was apparently not regarded as a very high priority so mapping was deficient. Furthermore, because of the long continued hostilities between India and Pakistan, such topographic maps were

L1A_003_08172003060405_09082003191808

Figure 24.10. Glaciers in Wakhan Pamir in ASTER scene of 17/08/03. Note some diminution of northern white ice stream of Northern Issik Glacier compared with RBV scene of Fig. 24.9b.

treated as state secrets and continue to be most difficult to obtain. Foreign topographic mapping (Soviet 1:100,000) and foreign climbing teams have continued to dominate the highest quality available mapping, with Austrians and Germans at Hasanabad Glacier in Hunza (Anonymous, 1995), Americans and Poles at K2 (Molnaar, 1977; Wala, 1994); British in Hunza and elsewhere (Miller, 1984), Chinese at Batura Glacier in Hunza (Academica Sinica, 1978) and K2 Mountain (Academy of Sciences, n.d.), Italians also at K2 Mountain (Anonymous, 1969), and so forth. Now with the advent of space-based satellite-image making (U.S.G.S., 1997a,b; Geosystems Polka, 2003, 2004), the unknown aspects of the topography are far less easy to conceal by the military and so information on the glaciers is able to be obtained reasonably easily.

7. Selected glaciers of Pakistan

7.1. Tirich Mir

The glaciers of Tirich Mir on the Afghanistan–Pakistan border have received little direct scientific study, although apparent recent negative mass balances have

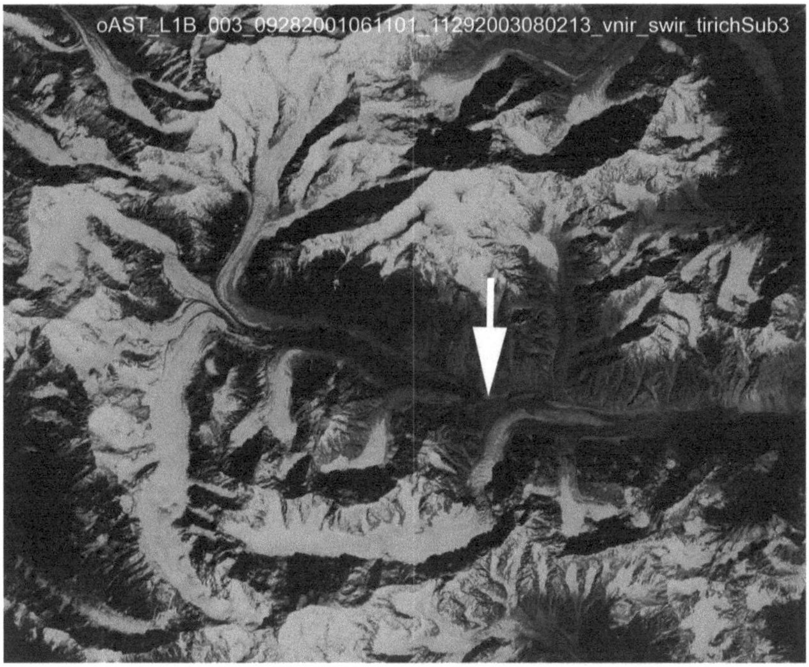

Figure 24.11. ASTER scene of 28/9/2001 of Upper and Lower Tirich Glaciers that are now disconnected from each other (arrow).

caused a disconnection between Upper and Lower Tirich glaciers (Fig. 24.11). In general we can say that the glaciers of Tirich Mir could provide more useful information, especially because of the strong contrasts available between debris covers of light-colored granitic lithologies versus the black slates and argillites (Fig. 24.11).

7.2. *Batura Glacier*

Batura Glacier, easily accessible off the Karakoram Highway (KKH) in Hunza, has been visited many times for > 120 years and its terminus position recorded (Goudie, et al., 1984; Bishop et al., 1995, 1998, 1999). Recently its terminus has largely stagnated and no longer threatens the KKH as it once did. The changing frontal positions of white-ice streams surrounded by debris covers on the glacier are viewed as sensitive measures of change in the otherwise obscuring increase in debris through extensive downwasting (Fig. 24.12).

7.3. *Gorshai Glacier*

Gorshair Glacier in Swat (Fig. 24.13) was visited by the senior author in 1984 (Shroder and Bishop, 2006b). The small size of the glacier, its lack of debris cover,

Figure 24.12. Landsat Thematic Mapper (TM) scene of Batura Glacier in 15/7/1979. Batura Glacier is the largest glacier at the top of the scene.

Figure 24.13. ASTER scene of 13 October 2003 showing Gorshai Glacier in the Swat Himalaya at about latitude 35° 28′ 30″; longitude 72° 43′. It is the white glacier in the right-hand middle of the image.

and its relative ease of access on good roads through the town of Matilton might indicate its potential usefulness as a first benchmark glacier (Kaser and Fountain, 2001) in Pakistan, unless in the intervening years downwasting or other changes have made it less accessible.

Figure 24.14. Satellite image of Nanga Parbat showing chief glaciers clockwise surrounding the ninth highest (8126 m) mountain in the world (1: Raikot; 2: Buldar; 3: Siachen; 4: Chungpar: Tarshing; 5: Bahzin; 6: Tap; 7: Shagiri; 8: Rupal; 9: Diamir; 10: Patro.

7.4. *Nanga Parbat Glaciers*

Nanga Parbat, the ninth highest mountain in the world (8,125 m a.s.l.), is accessible because of its close proximity to the KKH (Fig. 24.14). Detailed mapping and assessment of the ice on its flanks has been undertaken since the 1930s. The glaciers of Raikot, Sachen, Chungpar, Bazhin, Tap, Shagiri, Rupal, and Diamir that surround the massif have been investigated by a number of people (Gardner, 1986; Gardner and Jones, 1993; Bishop et al., 1999, 2000; Shroder et al., 2000). Outburst floods have been occurring in some places for many decades which tend to remove terminal moraines (Shroder et al., 1998), but otherwise the great height of the mountain and its consequent considerable avalanche flux maintains not only significant debris loads but also reasonably constant mass balances so that terminus fluctuations are generally minimal.

7.5. *Biafo Glacier*

This glacier on the way to K2 Mountain has been visited by many people for more than a century and a half and its frontal position has been monitored extensively

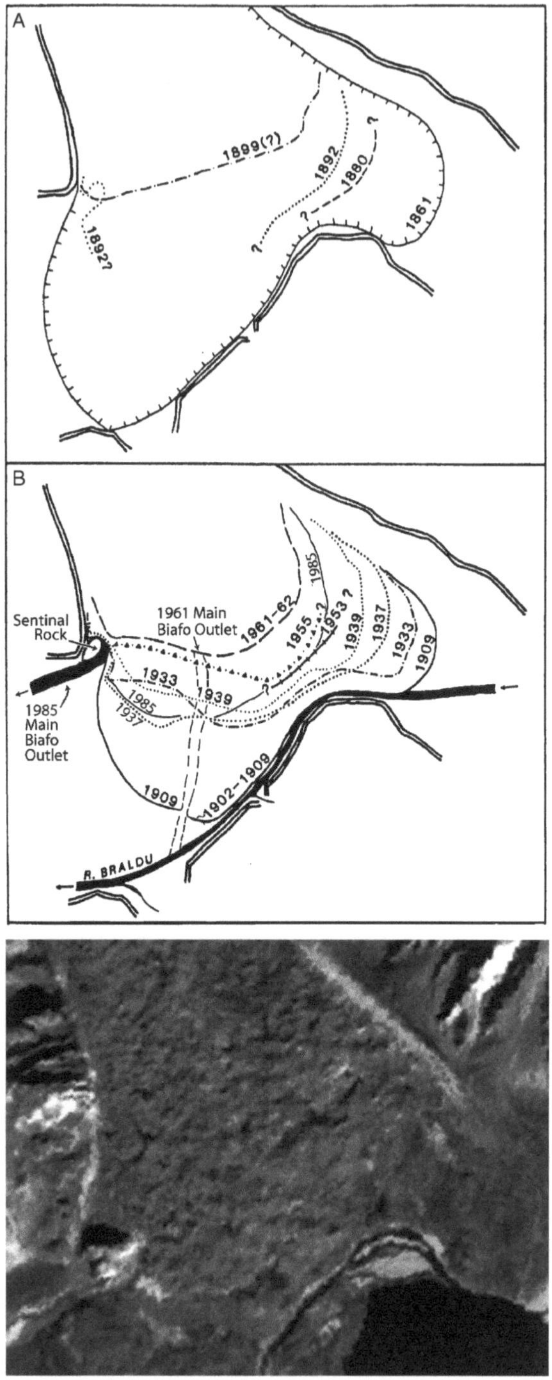

Figure 24.15. Terminus changes (15a) of Biafo Glacier that have been mapped and described over the years by many authors (Shroder and Bishop, 2007a, b, in press). (15b) ASTER image of the glacier terminus for comparison. Velocities measured in summer 2005 on the right side of the glacier (left in this picture) were 5 m/yr, and on the left (right in this picture) were 1 m/yr.

Figure 24.16. Baltoro Glacier scene showing profuse supraglacial debris from presumed recently increased downwasting and concentration of rock fragments.

(Fig. 24.15) (Hewitt and Young, 1990; Shroder and Bishop, 2006b). We have noticed in recent years that the ice velocities near the terminus are quite slow (~1 m/yr) on the northeast (left) side, whereas they are faster (~5 m/yr) on the southwest (right) side from whence the main melt-water stream now issues since its diversion from the middle of the terminus to the right (west) after 1961 (Fig. 24.15).

7.6. Baltoro Glacier

This huge glacier, at ~58 km in length is similar in size to the Biafo and Batura glaciers, yet the many >8000 m peaks in its catchment lend an aura of even larger size (Fig. 24.16). The terminus position of this glacier has hardly changed more than a few hundred m over the past century and a half of observation time (Mayer et al., 2006), yet in only the past several years of ASTER satellite image acquisition, both significant downwasting and debris cover seem to have increased (Fig. 24.16). Changes in elevation at Concordia seem to be a few tens of meters at most, however (Mayer et al., 2006).

8. Conclusion

The many glaciers of Afghanistan and Pakistan constitute a vital natural resource whose melt waters are the irrigation lifeblood of many millions of people. Mapping of these resources through traditional air photographs and ground surveys has been a fundamental exploration and assessment tool that is now superseded by new generations of satellite images. Future monitoring of the vital snow and ice resource base must proceed with synoptic coverage of these remote and difficult areas of the Hindu Kush and Himalaya. Issues of global change and ongoing natural hazards associated with changing mountain ice and snow mass will be critical in coming years (Barnett et al., 2005; Milly et al., 2005). Satellite-image assessment of these regions will be an essential element for using these mountains as witnesses for global change.

References

Academica Sinica, 1978. The Map of Batura Glacier, 1:60,000. Institute of Glaciology, Cryopedology and Desert Research, Lanzhou, China.

Academy of Sciences, (n.d., late 20th Century). K2 (Mount Qogri), 1:100,000. Lanzhou Institute of Glaciology and Geocryology, China.

Anonymous, 1969. Ghiacciaio Baltoro, 1:100,000. Italian Expeditions, 1929, 1954.

Anonymous, 1995. Hunza–Karakorum, 1:100,000. Deutsch—Österreischen Himalaya—Karakorum Expedition 1954 and Deutsche Karakorum Expedition 1959, München, Germany.

Barnett, T.P., Adam, J.C., and Lettenmaier, D.P., 2005. Potential impacts of a warming climate on water availability in snow-dominated regions. Nature 438, 303–309.

Bishop, M.P., Shroder, J.F., Jr., and Ward, J.L., 1995. SPOT multispectral analysis for producing supraglacial debris-load estimates for Batura Glacier. Geocarto International 10, 81–90.

Bishop, M.P., Shroder, J.F., Jr., Hickman, B.L., and Copland, L., 1998. Scale-dependant analysis of satellite imagery for characterization of glacier surfaces in the Karakoram Himalaya. In: Walsh, S. and Butler, D. (Eds) Special Volume on "Remote Sensing in Geomorphology," Geomorphology, Vol. 21, pp. 217–232.

Bishop, M.P., Shroder, J.F., Jr., and Hickman, B.L., 1999. SPOT panchromatic imagery and neural networks for information extraction in a complex mountain environment. Geocarto International 14, 17–26.

Bishop, M.P., Kargel, J.S., Kieffer, H.H., MacKinnon, D.J., Raup, B.H., and Shroder, J.F. Jr., 2000. Remote-sensing science and technology for studying glacier processes in High Asia. Ann. Glaciol. 31, 164–170.

Bishop, M.P., Barry, R.G., Bush, A.B.G., et al., 2004. Global land-ice measurements from space (GLIMS): remote sensing and GIS investigations of the earth's cryosphere. Geocarto Int. 19, 57–84.

Braslau, D., 1972. The glaciers of Keshnikhan. In: Gratzl, K. (Ed.), Hindukusch-Osterreichische Forschungsexpedition in den Wakhan 1970. Akademische Druck- u. Verlagsanstalt, Graz, Austria, pp. 112–116.

Braslau, D. and Bussom, D.E., 1978a. A glacier inventory method using Landsat MSS CCT. In: Rundquist, D.C. (Ed.), The Use of Landsat Digital Information for Assessing Glacier Inventory Parameters: An Evaluation. Final Report, Project Sponsored by Temporary Technical Secretariat for World Glacier Inventory, International Commission of Snow and Ice, and United Nations Educational, Scientific and Cultural Organization, pp. 23–40.

Braslau, D. and Bussom, D.E., 1978b. Landsat sensing of glaciers with application to mass-balance and runoff. In: Colbeck, S.C., and Ray, M. (Eds), Proceedings of a Meeting on Modeling of Snow Cover Runoff 26–28 September 1978, Hanover New Hampshire. U.S. Army Cold Regions Research and Engineering Laboratory, CRREL SR 79–36, pp. 77–82.

Breckle, S.W. and Frey, W., 1976a. Die hochsten Berge im Zentralen Hindukusch. Afghanistan Journal 3 (3), 91–94.

Breckle, S.W. and Frey, W., 1976b. Beobachtungen zur heutigen Vergletscherung der Hauptkette des Zentralen Hindukusch. Afghanistan J. 3 (3), 95–100.

Buchroithner, M.F., 1978. Zur Geologie des Afghanischen Pamir. In: Senarclens de Grancy, R. and Kostka, R. (Eds), Grosser Pamir. Akademische Druck- u. Verlagsanstalt, Graz, Austria, pp. 85–118.

Cyranoski, D., 2005. The long-range forecast. Nature 438, 275–276.

Desio, A., 1975. Notes on the pleistocene of central Badakhshan. In: Desio, A. (Ed.), Geology of Central Badakhshan (North-East Afghanistan) and Surrounding Countries. Italian Expedition to the Karakorum (K2) and Hindu Kush, Scientific Reports, v. 3, Geology-petrology. E.J., Brill, Leiden, pp. 339–409.

Exploration Pamir 75, 1978a, Darrah-e Issik-e Bala Glacier Map, 1:25,000 scale. Freytag-Berndt und Artaria, Vienna (also In: Senarclens de Grancy, R. and Kostka, R. (Eds), Grosser Pamir. Akademische Druck- u. Verlagsanstalt, Graz, Austria).

Exploration Pamir 75, 1978b, Koh-e Pamir Topographic Map, 1:50,000 scale. Freytag-Berndt und Artaria, Vienna (also In: Senarclens de Grancy, R. and Kostka, R. (Eds), Grosser Pamir. Akademische Druck- u. Verlagsanstalt, Graz, Austria).

Finsterwalder, R., 1936. Karte der Nanga Parbat Gruppe, 1:50,000, Deutsche Himalaya Expedition 1934, Berlin, Germany.

Fowler, H.J. and Archer, D.R., 2006. Conflicting signals of climate change in the upper Indus Basin. Journal of Climate 19 (17), 4276–4294.

Gardner, J.S., 1986. Recent fluctuations of Rakhiot Glacier, Nanga Parbat, Punjab Himalaya, Pakistan. J. Glaciol. 32 (112), 527–529.

Gardner, J.S. and Jones, N.K., 1993. Sediment transport and yield at the Raikot Glacier, Nanga Parbat, Punjab Himalaya. In: Shroder, J.F. (Ed.), Himalaya to the Sea: Geology, Geomorphology and the Quaternary. Routledge Press, London, UK, pp. 184–197.

Geosystems Polka, 2003, 2004. K2 and Baltoro Glacier in the Karakorum, Satellite Image Map 1:80,000. Krakow, Poland.

Gilbert, O., Jamieson, D., Lister, H., and Pendlington, A., 1969. Regime of an Afghan glacier. Journal of Glaciology 8 (52), 51–65.

Glicken, M., 1960. Making a map of Afghanistan. Photogramm. Eng. 26 (5), 743–745.

Goudie, A.S., Jones, D.K.C., and Brunsden, D., 1984. Recent fluctuations in some glaciers of the Western Karakoram Mountains, Hunza, Pakistan. In: Miller, K.J. (Ed.) The International Karakoram Project, Vol. 2, pp. 411–455.

Grötzbach, E. and Rathjens, C., 1969. Die heutige und die jungpleistozäne Vergletscherung des Afghanischen Hindukusch. Zeitschrift für Geomorphologie, Supplementband 8, 58–75.

Hewitt, K., 2005. The Karakoram Anomaly? Glacier expansion and the 'elevation effect,' Karakoram Himalaya. Mountain Research and Development 25 (4), 332–340.

Hewitt, K. and Young, G.J., 1990. Snow and ice hydrology project: Upper Indus Basin. Canadian Centre (Technical Report).

Kaser, G. and Fountain, A., 2001. A Manual for Monitoring the Mass Balance of Mountain Glaciers: With Particular Attention to Low Latitude Characteristics. ICSI/UNESCO/HKH—Friend program 12.01.

Kieffer, H., Kargel, J.S., Barry, R., et al., 2000. Satellite Measurements of Glaciers and Ice Sheets. EOS Transactions, American Geophysical Union. 81 (24), 265–271.

Lalande, P., Herman, N.M., and Zillhardt, J., 1974, Cartes climatiques de l'Afghanistan. Kaboul, L'Institut de Meteorologie, Publication no. 4, v. 1, p. 47 and v. 2, maps.

Lau, K.M. and Kim, K.M., 2007, Does aerosol weaken or strengthen the Asian monsoon water cycle? Mountains, Witnesses of Global Change, this volume.

Mayer, C., Lambrecht, A., Belò, M., et al., 2006. Glaciological characteristics of the ablation zone of Baltoro Glacier, Karakoram, Pakistan. Annals of Glaciology 43 (1), 123–131.

Mayewski, P.A. and Jeschke, P.A., 1979. Himalayan and trans-Himalayan glacier fluctuations since AD 1812. Arctic Alpine Res. 11 (3), 267–287.

Miller, K.J., 1984. The International Karakoram Project. Cambridge University Press, Chichester, UK, Vol. 1 and 2.

Milly, P.C., Dunne, K.A., and Vecchia, A.V., 2005. Global pattern of trends in streamflow and water availability in a changing climate. Nature 438, 347–350.

Molnaar, D., 1977. The Baltoro Region Karakoram Himalaya; Kashmir Pakistan and Sinkiang China, 1:250,000. 1975 American Expedition.

Mool, P.K., Bajracharya, S.R., Roohi, R., Ashraf, A., 2003. Astor Basin, Pakistan Himalaya: Inventory of Glaciers and Glacial Lakes and the Identification of Potential Glacial Lake Outburst Floods (GLOFs) Affected by Global Warming in the Mountains of Himalaya Region. Pakistan Agricultural Research

Council, Asia-Pacific Network for Global Change Research, Global Change System for Analysis, Research and Training, International Centre for Integrated Mountain Development, United Nations Environmental Programme/Regional Resource Centre for Asia and the Pacific. Document published on CD-ROM.

Mool, P.K., Bajracharya, S.R., Shrestha, B., et al., 2005. Indus Basin: Inventory of Glaciers and Glacial Lakes and the Identification of Potential Glacial Lake Outburst Floods (GLOFs) Affected by Global Warming in the Mountains of Himalaya Region. Pakistan Agricultural Research Council (PARC), International Centre for Integrated Mountain Development (ICIMOD), Asia-Pacific Network for Global Change Research (APN), Global Change System for Analysis, Research and Training (START), United Nations Environmental Programme (UNEP). Document published on CD-ROM.

Patzelt, G., 1978. Gletscherkundliche Untersuchlungen im 'Grossen Pamir'. In: Senarclens de Grancy, R. and Kostka, R. (Eds), Grosser Pamir. Akademische Druck- u. Verlagsanstalt, Graz, Austria, pp. 131–149.

Reiner, E., 1966. Die Kartographie in Afghanistan. Kartographische Nachrichten 16 (4), 137–145.

Royal Geographic Society, 1939. The Karakoram, 1:750,000.

Shipton, E.E., 1938. Blank on the Map. Hodder & Stoughton, London, U.K.

Shroder, J.F., Jr. and Giardino, J.R., 1978. Progress on Rock Glacier Research. Transactions Nebraska Academy of Sciences 6, 51–54.

Shroder, J.F., Jr., 1980. Special Problems of Glacial Inventory in Afghanistan, World Glacier Inventory Proceedings Reideralp Workshop, September 1978 (IAHS–AISH) Publication No. 126, Hydrological Sciences Bulletin, 142–147.

Shroder, J.F., Jr., 1989a. Slope failure: extent and economic significance in Afghanistan and Pakistan. In: Brabb, E.E., Harrod, B.L., and Balkema, A.A. (Eds), Landslides: Extent and Economic Significance in the World. Balkema, Rotterdam, Netherlands, pp. 325–341.

Shroder, J.F., Jr., 1989b. Glacierized areas of Afghanistan. In: Haeberli, W., Bosch, H., Scherler, K., Ostrem, G., and Wallen, C.C. (Eds), World Glacier Inventory, Status 1988, C39–C40 and C346–C353, IAHS (ICSI)-UNEP-UNESCO. Teufen, Switzerland.

Shroder, J.F., Jr., Bishop, M.P., and Scheppy, R., 1998. Catastrophic Flood Flushing of Sediment, Western Himalaya, Pakistan. In: Kalvoda, J. and Rosenfeld, C.L. (Eds), Geomorphological Hazards in High Mountains Areas. Kluwer Academic Publishers, Norwell, pp. 27–48.

Shroder, J.F., Jr., Bishop, M.P., Sloan, V., and Copland, L., 2000. Debris-Covered Glaciers and Rock Glaciers in the Nanga Parbat Himalaya, Pakistan. Geografiska Annaler 82A, 17–31.

Shroder, J.F., Jr. and Bishop, M.P., 2006a. Satellite glacier inventory of Afghanistan. In: Williams, R.S., Jr. and Ferrigno, J.G. (Eds), Satellite Image Atlas of Glaciers. U.S. Geological Survey Professional Paper 1386-F.

Shroder, J.F., Jr. and Bishop, M.P., 2006b. Satellite glacier analysis of glaciers of northern Pakistan. In: Williams, R.S., Jr. and Ferrigno, J.G. (Eds), Satellite Image Atlas of Glaciers. U.S. Geological Survey Professional Paper 1386-F.

Sivall, T.R., 1977. Synoptic-Climatological Study of the Asian Summer Monsoon in Afghanistan. Geografiska Analer 59A (1–2), 67–87.

U.S. Geological Survey, 1997a. Satellite Image Map I—2587-B, N.W. (Landsat mosaics), Frontier Pakistan.

U.S. Geological Survey, 1997b. Satellite Image Map I—2587-C, (Landsat mosaics), Northern Areas Pakistan.

Wala, J., 1994. The Eight-Thousand-Metre Peaks of the Karakoram, 1:50,000. Orographical Sketch Map, Krakow, Poland.

Wissman, H. v., 1960. Die heutige Vergletscherung und Schneegrenze in Hochasien. Abhandlungen der Mathematisch-Naturwissenschaftlichen Klasse 1959 (14), 1,101–1,407.

Mountains Witnesses of Global Changes
R. Baudo, G. Tartari, and E. Vuillermoz (Editors)

25

Remote sensing and GIS for alpine glacier change detection in the Himalaya

Michael P. Bishop, John F. Shroder Jr., Umesh K. Haritashya and Henry N.N. Bulley

Abstract

Concerns over greenhouse-gas forcing and warmer temperatures have initiated research into understanding climate forcing and associated Earth-system responses. Alpine glacier fluctuations are directly and indirectly related to climate change. Consequently, it is essential to be able to assess glacier fluctuations from space and determine the causal mechanisms responsible for change. Although satellite imagery and topographic information can be used for alpine glacier mapping, interpreting causal mechanisms for changing glacial boundary conditions and climate is difficult, as there is a significant disconnect between information on boundary conditions and process mechanics. Therefore, information integration and computer-assisted approaches to glacier mapping, parameter estimation, and numerical modeling are required to produce reliable results that go beyond traditional image interpretation and mapping. Only in this way can a multitude of forcing factors and interrelated processes be evaluated in an objective way to quantitatively ascertain the role of climate on glacier fluctuations in the Himalaya.

1. Introduction

Concerns over greenhouse-gas forcing and warmer temperatures have initiated research into understanding climate forcing and associated Earth-system responses. There is considerable scientific debate regarding climate forcing and landscape response, as complex geodynamics regulate feedback mechanisms that couple climatic, tectonic and surface processes (Molnar and England, 1990; Ruddiman, 1997; Bush, 2000; Zeitler et al., 2001b; Bishop et al., 2002).

A significant component in the coupling of Earth's systems involves the cryosphere, as glacier-related feedback mechanisms govern atmospheric, hydrospheric and lithospheric response (Bush, 2000; Shroder and Bishop, 2000; Meier and Wahr, 2002). Specifically, snow and ice mass distributions partially regulate atmospheric

ISSN: 0928-2025 DOI: 10.1016/S0928-2025(06)10025-5

properties (Henderson-Sellers and Pitman, 1992; Kaser, 2001), sea level variations (Meier, 1984; Haeberli et al., 1998; Lambeck and Chappell, 2001; Meier and Wahr, 2002), surface and regional hydrology (Schaper et al., 1999; Mattson, 2000), erosion (Harbor and Warburton, 1992, 1993; Hallet et al., 1996), and topographic evolution (Molnar and England, 1990; Brozovik et al., 1997; Bishop et al., 2002). Consequently, scientists have recognized the significance of understanding glacier fluctuations and their use as direct and indirect indicators of climate change (Kotlyakov et al., 1991; Seltzer, 1993; Haeberli and Beniston, 1998; Maisch, 2000). In addition, the international scientific community now recognizes the need to assess glacier fluctuations at a global scale, to elucidate the complex scale-dependent interactions involving climate forcing and glacier response (Haeberli et al., 1998; Meier and Dyurgerov, 2002). Furthermore, it is essential that we identify and characterize those regions that are changing most rapidly and having the most significant impact on sea level, water resources, economics and geopolitics (Haeberli, 1998).

Mountain environments are known for their complexity and sensitivity to climate change (Beniston, 1994; Meier and Dyurgerov, 2002). Numerous mountain systems have been identified as "critical regions" and include Alaska, Patagonia and the Himalaya (Haeberli, 1998; Meier and Dyurgerov, 2002). Within the Himalaya, alpine glaciers are thought to be very sensitive to climate forcing due to the altitude range and the variability in debris cover (Nakawo et al., 1997). Furthermore, such high-altitude geodynamic systems are thought to be the direct result of climate forcing (Molnar and England, 1990; Bishop et al., 2002), although climate versus tectonic causation is still being debated (e.g., Raymo et al., 1988; Raymo and Ruddiman, 1992).

Central to these and other glaciological arguments is obtaining a fundamental understanding of the feedbacks between climate forcing and glacier response (Dyurgerov and Meier, 2000). McClung and Armstrong (1993) have indicated that detailed studies of a few well-monitored glaciers do not permit characterization of regional mass-balance trend, the advance/retreat behavior of glaciers, or regional and global extrapolation. Therefore, approaches other than isolated field studies are required to generate information about glacier distribution and ice volumes, mass-balance gradients, regional mass-balance trend, and landscape factors that significantly control ablation and ice-flow dynamics. An integrated approach to studying alpine glaciers must be accomplished using new developments in remote sensing and geographic information science (GIScience) (Haeberli et al., 1998; Bishop et al., 2004; Bishop and Shroder, 2004).

Although such an approach has been recognized as essential in order to make progress, there are numerous science and geographic information technology (GIT) issues (Bishop and Shroder, 2004). Similarly, there are inherent limitations that must be recognized and circumvented in order to generate critical thematic and quantitative information necessary to drive numerical models and improve our understanding of climate-glacier systems. While static spatial snapshots of topography and glaciers are useful, they are of limited value in explaining complex mountain landscape systems (Bishop and Shroder, 2004). Consequently, the objective of this paper is to address the nature of alpine glacier dynamics and discuss the issues associated with information extraction from remote sensing and geographic information systems (GISs). Specifically, we place the complexity of such systems into perspective and show the disconnect between complex system responses and information generated from traditional remote sensing and GIS-based approaches. We then highlight the capabilities and limitations of utilizing satellite imagery and GISs to produce information required for assessing alpine glaciers and climate change research. A discussion on the integration and contribution of GIScience to numerical modeling follows. We finish with a discussion of application issues and recommendations for future research.

2. Background

It is essential to place climate-glacier interactions into the context of mountain geodynamics. Globally significant interactions between climate, surface processes, and tectonics have recently been proposed to explain climate change and mountain building (Raymo et al., 1988; Molnar and England, 1990; Koons, 1995; Avouac and Burov, 1996). Much research has focused on climate versus tectonic forcing (Raymo et al., 1988; Molnar and England, 1990) and the dominant surface processes responsible for relief production and topographic evolution (Montgomery, 1994; Burbank et al., 1996; Brozovik et al., 1997; Whipple and Tucker, 1999).

Research indicates that climate-driven surface processes are capable of reducing lithospheric mass, thereby inducing isostatic uplift or accelerating tectonic uplift, collectively resulting in relief production and complex topography (Montgomery, 1994; Shroder and Bishop, 2000) which can alter regional and micro-climatic conditions. The spatio-temporal scale-dependent linkages are complex, and global climate-topography relationships are controversial. Furthermore, the operational scale-dependencies of surface processes that govern the topography and feedback mechanisms are not well understood (Bishop et al., 1998a,b; Shroder and Bishop, 1998; Bishop and Shroder, 2000).

Alpine glaciers play a significant role in geodynamics, as they directly and indirectly respond to climatic and lithospheric processes. In addition, they partially control climatic and lithospheric processes through numerous feedback mechanisms (Fig. 25.1). Radiative and atmospheric forcings control accumulation and ablation, while glacier erosion modifies the topography, which also governs ice-flow dynamics and ablation. The glacier mass balance is regulated by the spatio-temporal variations in ablation and accumulation with altitude. Glacier mass balance is an important parameter because it controls ice-flow dynamics and the magnitude of glacier erosion. Alpine glaciers are known as being effective erosion agents (Shuster et al., 2005).

Over time, a reduction in lithospheric mass produces isostatic adjustment, and in special cases, the advection of lithospheric mass causes local tectonic uplift (Zeitler et al., 2001a,b). More relief potentially generates more snow accumulation that increases erosion. More erosion generates more relief, more landsliding, and more supraglacial debris, which reduces ablation. Consequently, glacier dynamics are regulated by climate, surface processes and tectonics in such a way that glacier changes can be caused by external climate forcing, glacial dynamics, surface process coupling,

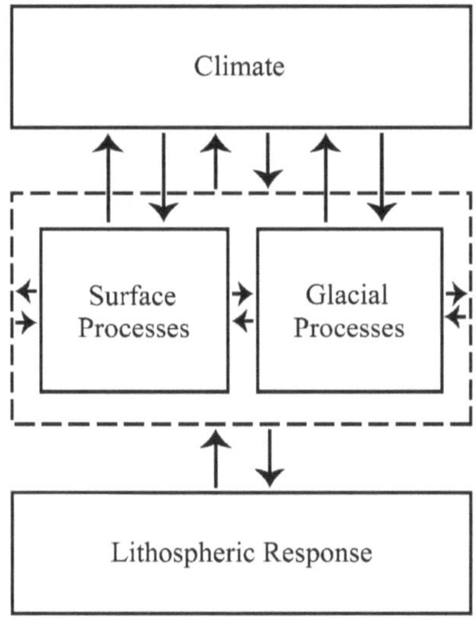

Figure 25.1. Conceptual diagram of the interrelationships between climate, surface processes and tectonics. The arrows depict the directional influences on processes, while the dashed box represents topography. Notice the critical role that topography plays in governing a variety of processes, as it integrates the results of numerous processes and governs process in a variety of complex feedback mechanisms.

external lithospheric forcing, and local topographic boundary conditions in a complex scale-dependent hierarchy of feedback mechanisms.

From a climate-change perspective it is necessary to map and assess alpine glacial fluctuations (Bishop et al., 2004; Kargel et al., 2005). There is strong evidence that many glaciers are retreating and downwasting world wide (Kargel et al., 2005). The response time of glaciers varies, as small glaciers respond to relatively short-term climate fluctuations, whereas large glaciers respond to climatic conditions of the past, with response times potentially ranging from 10 to 1000 years. Observations also indicate that many glaciers are advancing, either due to positive mass balance or surging activity. In some instances, retreating termini positions are offset by accumulation, such that changing terminus geometry may not be indicative of negative mass balance. Similarly, many glaciers may have relatively stable terminus positions because of very heavy debris loads, although significant downwasting may be occurring (Shroder and Bishop, in press).

Remote sensing- and GIS-based studies are ideal for detecting areal variations in alpine glaciers. High-resolution multi-temporal satellite imagery can be effectively utilized to map the differences in the spatial extent of white-ice glacier surfaces. This is possible because of spectral differentiation between snow/ice and vegetation and rock/sediment. Mapping debris-covered glaciers is notoriously difficult, although mapping can be facilitated by using topographic information (Bishop et al., 2001, 2004; Kargel et al., 2005). Furthermore, digital elevation models (DEMs) can be generated from stereo-pairs of satellite imagery that enable analysis of multi-date DEMs. Collectively, this permits change detection and the generation of downwasting and ice-volume estimates. Interpreting causal mechanisms for changing glacial boundary conditions and climate is difficult, as there is a significant disconnect between glacier boundary changes and process mechanics.

Given that space-based information is usually extracted empirically to provide insights into climate change, careful attention must be given to data preprocessing, algorithms and analysis procedures. Unfortunately, remote sensing of mountain environments is extremely difficult due to complex radiation transfer processes, and information extraction methodologies that can be strongly influenced by the nature and limitations of algorithms and approaches (Bishop et al., 2004). Consequently, scientists who utilize satellite imagery and satellite-derived DEMs for assessing glaciers must be aware of the issues and limitations associated with change detection of alpine glaciers. Furthermore, it is critical to begin integrated formulation of GIScience and numerical modeling efforts in order to provide valid interpretations and insights into causal mechanisms of change (Bishop and Shroder, 2004).

3. Remote sensing and GIScience

Obtaining spatial and temporal data in mountain environments is exceptionally difficult, although geodetic, hydrological, climate and other field-based data are frequently collected by scientists and resource managers. The remoteness, logistical difficulties, and frequent governmental restrictions commonly prohibit data collection campaigns in the Himalaya. Scientists at international conferences routinely discuss the issue of data collection stations in order to produce time series of key parameters. As important as time-series data can be, there is also a need to account for the spatial variability in key parameters such as albedo, topography, surface temperature, and land-cover variations. Satellite-based sensors have routinely been relied upon to produce thematic information and quantitative estimates of surface biophysical parameters (Bishop et al., 2004). Key variables for glacier change-detection studies include topography, landcover, albedo, ice-velocity fields and surface temperature variation.

3.1. Topography

Increasingly, scientists are depending upon satellite-derived topographic information. Multi-spectral (off-nadir viewing) and radar sensors can be used to generate DEMs (Fig. 25.2). Accurate DEMs are essential in change-detection studies of glaciers because multi-spectral imagery needs to be ortho-rectified for human- and computer-assisted analysis and interpretation. Furthermore, DEMs are required to address numerous radiation transfer issues such as atmospheric correction and anisotropic-reflectance correction. Similarly, geomorphometric analysis of glaciers is becoming more routine, and debris-covered glacier mapping and numerical-modeling efforts depend upon geomorphometric parameters derived from a high-quality DEM.

Unfortunately, research indicates that systematic biases are usually present when comparing multi-temporal DEMs generated from off-nadir viewing sensors (Rivera

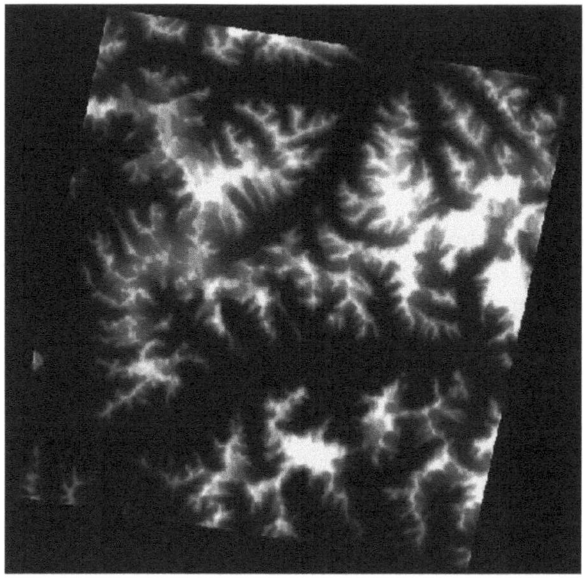

Figure 25.2. ASTER-derived DEM of the Karakoram Himalaya near K2 Mountain in northern Pakistan. Such high-resolution (30 m) DEMs can be used for radiation transfer modeling, change-detection studies, geomorphometry and landform mapping, glacier parameter estimation and numerical modeling.

et al., 2005; Berthier et al., in press). This includes linear and nonlinear biases that are the result of changes in viewing geometry and/or spacecraft parameters (Fig. 25.3). Consequently, systematic biases must be characterized off-glacier in order to model and remove bias over glacier surfaces. Furthermore, error estimates are required to determine if computed differences are the result of error, or real changes in glacier surface altitude.

3.2. Geometric corrections

It is necessary to reduce the spatial distortions that are inherent in satellite imagery as a result of sensor and orbital geometry and the topography of the Earth. Much progress has been made in geometric calibration, which is primarily concerned with repositioning pixel locations in the image array, to a known reference grid. In general, geometric correction involves the following three steps:

- Selection of a suitable mathematical distortion model.
- Coordinate transformation. Pixel locations are altered by image rotation and pixel array coordinates are transformed to a coordinate system.
- Spatial interpolation. Interpolated pixel values are generated because of image rotation. Standard interpolation algorithms include nearest neighbor, bilinear, and cubic-convolution.

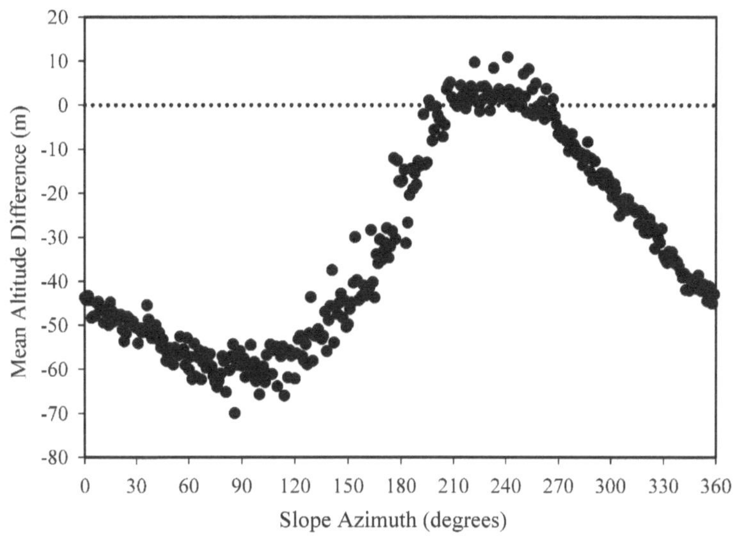

Figure 25.3. Scatterplot of mean altitude difference generated from a SRTM DEM (2000) and ASTER-derived DEM (2004). Mean values were computed off-glacier over the Nanga Parbat Massif in Pakistan. Variations in viewing geometry and/or spacecraft orbital parameters result in a non-linear bias that significantly deviates from zero.

For alpine glacier applications, it is important to note that the topography can introduce significant spatial distortions in the imagery and ortho-rectification is required so that topographic distortion is removed on a pixel by pixel basis (Bishop et al., 1998b, 2004). Ortho-rectification requires the use of a DEM so that each pixel location can be corrected for relief displacement. The procedure generally requires the selection of ground control points (GCPs), although numerous software packages permit this to be a semi-automated procedure.

3.3. Radiometric calibration

Satellite imagery should be accurately radiometrically calibrated so that multi-spectral and multi-temporal data can be used to assess glacier surfaces. Unfortunately, unlike geometric calibration, accurate radiometric calibration of satellite imagery is problematic due to complex surface-atmosphere interactions (Kimes and Kirchner, 1981; Chavez, 1989; Thorne et al., 1998). In general, radiometric calibration involves:

- A conversion of image digital numbers (DN) to at-satellite radiance values (L_λ^0).
- Atmospheric correction to remove the influence of atmospheric attenuation and additive path-radiance (L_λ^p).
- Anisotropic-reflectance correction to modify radiance values to account for irradiance variations and bi-directional reflectance caused by topography and land cover.
- A conversion from surface radiance (L_λ) to surface reflectance (ρ_λ).

A DN-to-radiance conversion represents a linear transformation that is applied to every pixel. The transformation makes use of pre-launch or post-launch calibration coefficients, and removes the gain and offset effects introduced by the sensor. This initial calibration procedure is necessary in quantitative studies as spectral characteristics can be significantly distorted by system gain and offset conditions (Chavez, 1996).

Atmospheric correction of satellite imagery can be a complicated task. It involves accounting for the influence of the upward atmospheric attenuation and additive path radiance which influences the at-satellite radiance. It is not always necessary to perform atmospheric correction depending upon the scientific objectives. In general, quantitative biophysical remote sensing studies require atmospheric correction, whereas some glacier mapping applications can be successfully accomplished without this correction.

In high-mountain environments it is important to perform anisotropic-reflectance correction to reduce spectral variations caused by the topography (Bishop et al., 2004). It is commonly difficult to delineate glacier boundaries due to cast shadows and the surrounding topography, as mesoscale relief and topographic shielding influence the direct and diffuse-skylight irradiance. Although most Earth-scientists attempt to address this issue using image transformations (e.g., ratios and principal component analysis), the influence of topography on spectral response is a complicated problem that has yet to be effectively resolved (Bishop and Colby, 2002; Bishop et al., 2004). It can potentially be addressed using radiation-transfer models that account for multi-scale topographic effects. For more detailed information on anisotropic-reflectance correction see Bishop et al. (2003, 2004).

For multi-temporal change-detection studies, at-satellite radiance values can be normalized to surface reflectance. This is needed to account for changes in at-satellite radiance caused by the atmosphere and acquisition time that incorporates the Earth–sun distance (*d*) and solar geometry, such that

$$L_{\lambda} = \frac{(L_{\lambda}^{0} - L_{\lambda}^{p})}{\tau_{\lambda}^{\uparrow}(\theta_{\upsilon})}, \tag{1}$$

where $\tau_{\lambda}^{\uparrow}$ is the beam transmittance of the atmosphere in the upward direction, and θ_{υ} is the view angle of the sensor. The surface radiance can then be converted to reflectance:

$$\rho_{\lambda} = \frac{\pi L_{\lambda} d^{2}}{E_{\lambda} \cos \theta_{s}}, \tag{2}$$

where E_{λ} is the surface spectral irradiance and θ_{s} is the solar zenith angle. Spectral irradiance values can be obtained from radiation-transfer models that account for topographic variation. This is especially important if surface albedo estimates are to be derived from satellite imagery.

3.4. Glacier mapping

Glacier mapping can be accomplished using human- and computer-assisted analysis of satellite imagery. Many researchers are actively involved in producing inventories of alpine glaciers throughout the world (Bishop et al., 2004; Kargel et al., 2005). Hand-digitization of alpine glaciers from false-color composites and/or enhanced imagery represents a common approach in delineating glaciers and detecting changes in spatial geometry. This approach, however, is fraught with difficulties because it is subjective and does not adequately account for radiation-transfer processes and analyst a priori experience.

To demonstrate this, laboratory analysts digitized two alpine glaciers in the Nanga Parbat Himalaya (Fig. 25.4). The results clearly reveal misinterpretation caused by spectral variations and by differences in the geographic and domain expertise of the analysts. Furthermore, it is very difficult, if not impossible to generate repeatable results for a glacier. Part of the problem is that Himalayan glaciers are heavily debris covered and it is difficult to determine the terminus boundaries and accumulation areas due to snow cover and spectral saturation at high altitude.

Computer-assisted analysis procedures can be used to assist in the classification process, although it is very difficult to utilize spectral information alone. A common technique is to utilize image ratios, as this reduces the influence of topography and the thresholding of ratio values can be used to classify snow and ice. This technique has been used by the Canadian and Swiss Global Land Ice Measurement from Space (GLIMS) regional centers.

Figure 25.4. ASTER false-color composite of the eastern side of Nanga Parbat Mountain in Pakistan. ASTER VNIR data were utilized to digitize the boundaries of the Lotang Glacier to the north and the Sachen Glacier directly south of Lotang. The results of five independent interpretations by different lab analysts for each glacier are displayed and depict significant variation in the areal extent of these glaciers. Problems areas include the terminus and the accumulation areas.

Another more advanced technique is to utilize artificial neural networks (ANN). Neural networks have numerous advantages over traditional statistical classification techniques, although numerous issues regarding training and ANN structure must be empirically determined (Bishop et al., 1999). Nevertheless, ANN technology can produce superior results and map entire scenes of snow and ice (Fig. 25.5). This capability is very important for change-detection studies and for assessing changes in seasonal and yearly snow-line variations.

These spectral-classification techniques, however, do not address the issue of mapping debris-covered glaciers (Bishop et al., 2001, 2004; Kargel et al., 2005). Consequently, topographic information has to be integrated into mapping and change-detection studies.

3.5. Geomorphometry

Basic gemorphometric parameters represent the first- and second-order derivatives of the elevation field. These parameters include slope, slope aspect, and various curvature metrics, which can depict the boundaries of debris-covered alpine glaciers (Fig. 25.6). Combinations of geomorphometric parameters and satellite imagery provide a more robust approach towards accurate digitization of alpine glaciers.

More sophisticated approaches to mapping and delineating glaciers make use of object-oriented analysis (Bishop et al., 2001), where objects represent the partitioning or segmentation of the landscape into discrete spatial entities based upon geomorphometric criteria. For example, utilizing slope azimuth (ϕ), it is possible to map the spatial-orientation structure of the topography. This can be accomplished by utilizing an unsupervised clustering algorithm on the geomorphometric parameters $\sin\phi$ and $\cos\phi$ (Fig. 25.7). By accounting for changes in the slope angle with altitude, constrained by the spatial distribution of a slope-aspect object, topo-sequence information can be generated for each object and compared. Figure 25.7

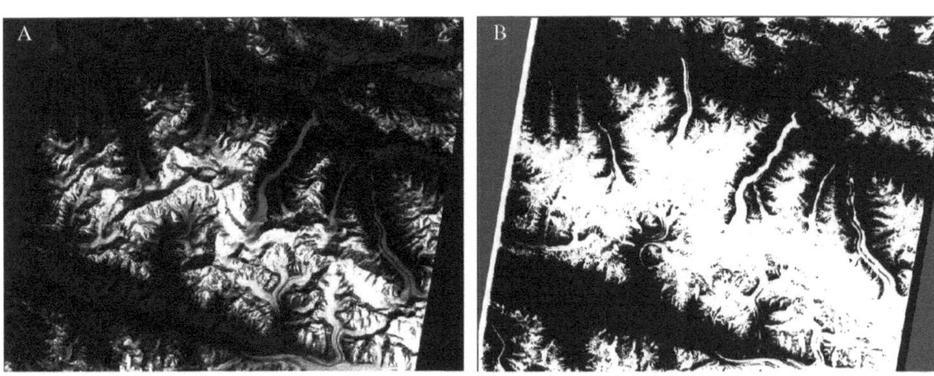

Figure 25.5. ASTER false-color composite of the Hispar Glacier region in the Karakoram Himalaya (A). Neural network technology can be utilized to accurately map snow and ice distributions in the Himalaya (B).

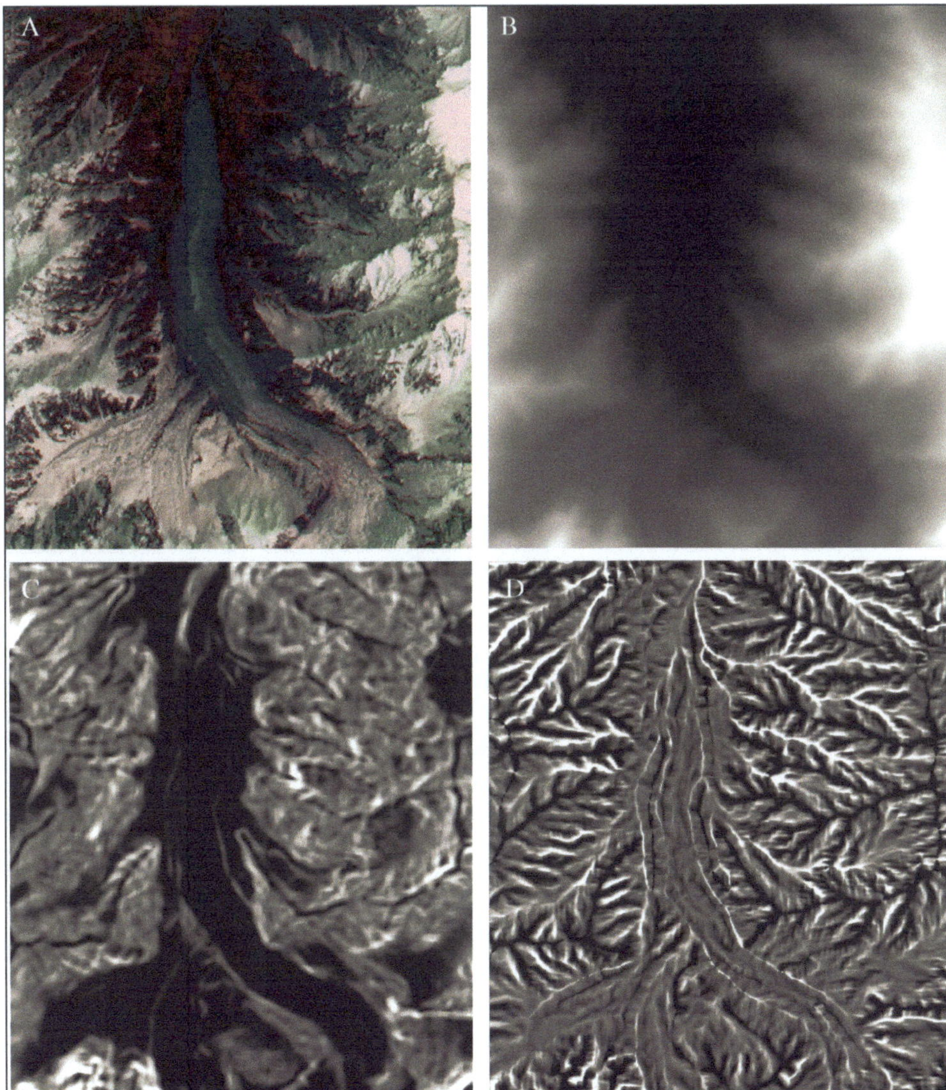

Figure 25.6. Raikot Glacier on the north side of Nanga Parbat. A SPOT false-color composite acquired in June 1996 clearly depicts the debris-covered terminus (A). A stereo-pair of SPOT panchromatic images were used to generate a DEM (B). DEM-derived slope (C) and curvature information (D) can be used to better delineate glacier boundaries.

clearly reveals that the Raikot Glacier surface in the terminus region can be easily differentiated from valley wall surfaces, as glaciers in the Karakoram and Nanga Parbat Himalaya exhibit low slope angles across relative altitude ranges.

In addition to mapping the spatial extent and temporal fluctuations of alpine glaciers, it is important to estimate surface altitude changes caused by radiative

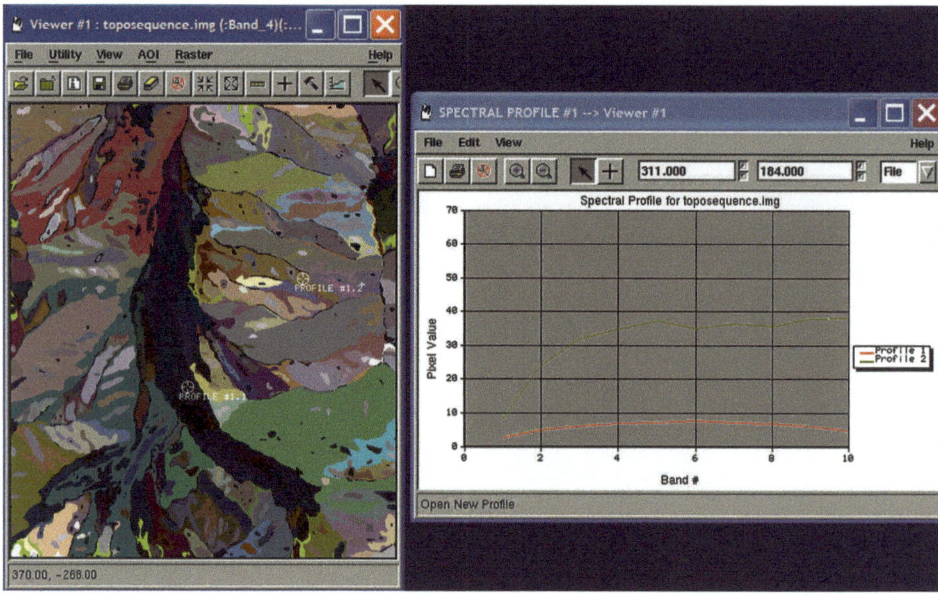

Figure 25.7. Toposequence information for the Raikot Glacier generated from object-oriented analysis of altitude, slope and slope azimuth parameters. Note the difference in the slope-altitude curve for an object representing the glacier surface, versus an object representing part of the valley wall.

forcing, temperature changes, and ice-flow dynamics. DEMs generated from multi-temporal ASTER imagery permit statistical and geomorphometric analysis of glacier surface changes. For example, topographic profiles can be used to ascertain the magnitude of surface altitude change over a particular time period. Changes may be the result of increased ablation or changes in mass balance and ice-flow dynamics. Information can be generated along the flow-line of the glacier, or it is possible to compute the altitude-distance function by comparing every pixel representing the surface of the glacier to a terminus position, and plotting the average altitude in a distance bin (e.g., Fig. 25.8). Such an analysis for the Baltoro Glacier in the Karakoram suggests that the glacier surface has downwasted from 2000 to 2004. In addition, the analysis suggests that the downwasting rate varies with altitude. This pattern can be explained due to variations in the debris-cover depth with altitude, where thick debris protects ice and thin debris enhances ablation. It is essential, however, that systematic biases be removed before analysis and interpretation of results.

Another approach to characterizing the entire surface is to perform glacier hypsometric analysis. Hysometric analysis is essentially a frequency analysis, which characterizes the altitude/area distribution property of the glacier. A hypsometric curve represents the altitude/area function (e.g., Fig. 25.9). An analysis of the Baltoro Glacier surface (without biases removed) suggests downwasting, as the hypsometric integral for 2000 versus 2004 is 0.6319 and 0.6019, respectively. This would represent an average decrease in surface elevation from 4064 to 4009 m for the two aforementioned dates.

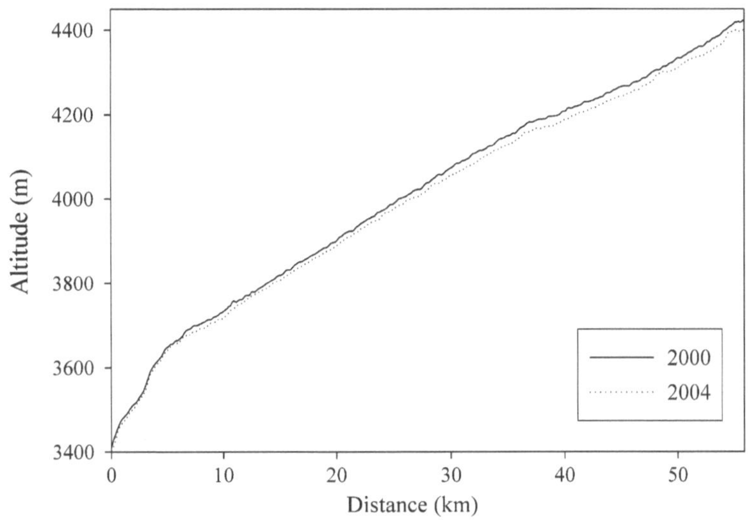

Figure 25.8. Glacier surface profiles of the Baltoro Glacier from ASTER-derived DEMS from 2000 and 2004. Note the temporal difference in the surface altitude with increasing altitude (systematic biases not removed). Changes in the slope of the profiles are due to the method of computation and therefore glacier geometry, not the slope of the glacier surface.

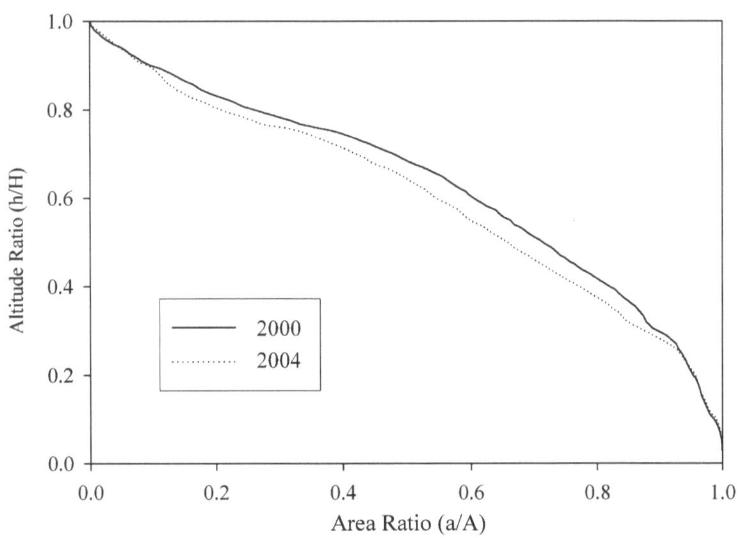

Figure 25.9. Hypsometric curves for the Baltoro Glacier generated from ASTER-derived DEMs from 2000 and 2004 imagery (systematic biases not removed).

4. Numerical modeling

The integration of GIScience and numerical modeling represents a more sophisticated approach to the study of glacier dynamics and the interpretation of causal mechanisms. Information generated from satellite imagery and DEMs can be used to establish boundary conditions, estimate input parameters, and constrain simulations. We briefly demonstrate the significance of numerical models for understanding glacier fluctuations.

4.1. Climate modeling

While mapping and monitoring the present conditions of Himalayan glaciers is a fundamental and necessary task, predictions of future mass balances must rely on knowledge of what the climate will likely be sometime in the future. Such knowledge can only come from numerical modeling of the climate system. In addition, general circulation models (GCMs) can be used for paleo-climate studies in which direct comparisons to proxy climate data recovered from the Himalaya may be made. For example, inferred spatio-temporal fluctuations in glacier advance/retreat across the Himalaya (Phillips et al., 2000; Richards et al., 2000) correspond to simulated changes in the south Asian monsoon system associated with changing orbital parameters, sea level, and atmospheric carbon dioxide (Bush, 2002).

Climate simulations suggest that the glaciers in the Himalaya will respond to greenhouse-gas forcing (Fig. 25.10). We know that solar radiative forcing increased the strength of the monsoon in Holocene time, thereby enhancing orographic precipitation and snow accumulation. With higher temperatures, an increase in ablation is expected at lower altitudes. With higher temperatures there may be an increase in evaporation and greater moisture flux into the Himalaya via a strengthening of the monsoon.

Greater precipitation at higher altitudes may cause glaciers to reach an equilibrium, and perhaps exhibit positive mass balance. An increase in latent heat and high altitude glacier area is thought to enhance the strength of the monsoon, such that positive feedback mechanisms might facilitate more enhancement and snow

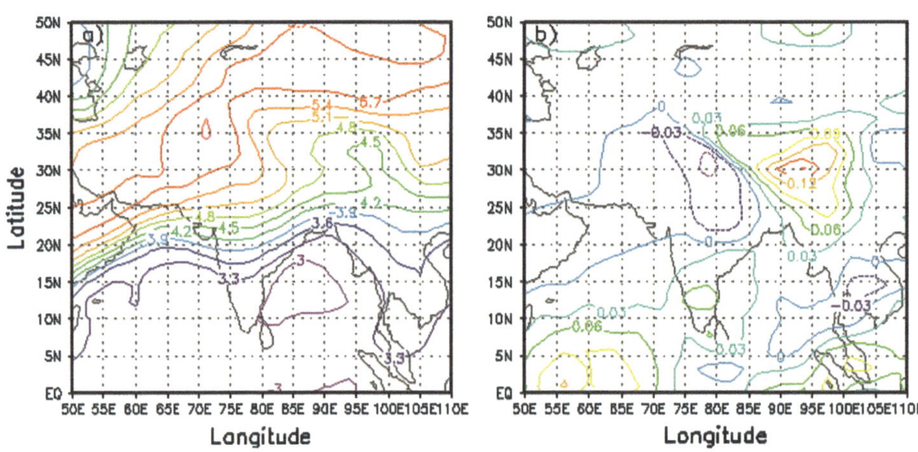

Figure 25.10. Simulated (a) annual mean temperature difference and (b) change in precipitation (cm/day) over the Himalaya in a double CO_2 climate compared to today.

accumulation. This prediction is opposite to our current understanding of higher temperatures causing an increasing negative mass-balance trend with time. Nevertheless, we might expect modern-day scientific investigations to support the notion of negative mass balance in the region due to lower altitude empirical studies that do not systematically account for glacier dynamics and glacier-climate feedback mechanisms. Consequently, remote sensing and GIS studies need to be interpreted in the context of regional and global climate simulation results, as climate time-series data are not systematically measured across altitude and latitude in the Himalaya. Furthermore, the situation is complicated by spatial variations in numerous forcings that could result in different mass balance trends across the Himalaya, as suggested in Fig. 25.10.

4.2. *Energy-balance modeling*

The glaciers in the Himalaya receive both summer and winter snow accumulation. Together with accumulation, the overall factor that governs the mass balance of a glacier is the surface energy balance. The surface energy budget is composed of a variety of processes that determine the magnitude of ablation. It has been demonstrated that the spatial and temporal variability of ablation is high in the Himalaya and other mountain regions given altitude ranges, topographic conditions, and debris-cover variability (Bishop et al., 1995; Arnold et al., 2006).

It is generally acknowledged that three-dimensional numerical modeling of the energy budget is required to characterize the spatio-temporal complexity of numerous processes that effect ablation. Consequently, input of spatio-temporal data from remote sensing and GIS is needed to reduce the uncertainties associated with key parameters and processes such as albedo, irradiance flux, and turbulent fluxes.

A physically based approach for estimating ablation requires an assessment of the energy fluxes. The components of the energy fluxes can be represented as

$$Q_N + Q_H + Q_L + Q_G + Q_R + Q_M = 0 \tag{3}$$

where Q_N is the net radiation, Q_H is the sensible heat flux, Q_L is the latent heat flux (Q_H and Q_L are referred to as the turbulent heat fluxes), Q_G is the ground heat flux, Q_R is the sensible heat flux supplied by rain, and Q_M is the energy used for melting snow and ice. Positive quantities represent energy gains while negative magnitudes indicate an energy loss at the surface.

Ablation rate (M) estimates generated from energy-balance modeling produce reliable results (Hock, 2005). Ablation is computed as

$$M = \frac{Q_M}{\rho_w L_f}, \tag{4}$$

where ρ_w is the density of liquid water and L_f is the latent heat of fusion.

4.2.1. *Net radiation*

The net radiation component represents the balance between the incoming and outgoing energy absorbed and emitted by the surface. It is characterized and modeled as short-wave (0.15–4.0 μm) and long-wave (4.0–120 μm)) radiation fluxes, and it is the dominant energy source in the Himalaya. The net radiation is regulated by numerous processes and key parameters that represent atmospheric, topographic, and surface properties.

Radiation transfer processes are highly complex and wavelength (λ) dependent. The radiation balance can be computed as

$$Q_n = E(1-\alpha) + L_s^{\downarrow} + L_t^{\downarrow} + L^{\uparrow} \tag{5}$$

where E is the total surface irradiance, α is the surface albedo, $L_s^{\downarrow}$ is the long-wave sky radiation, $L_t^{\downarrow}$ is the long-wave radiation from the surrounding terrain, and $L^{\uparrow}$ is the emitted long-wave radiation.

The total surface irradiance can be highly variable depending upon atmospheric, topographic, and surface biophysical characteristics. It is computed as

$$E(\lambda) = E^{b}(\lambda) + E^{d}(\lambda) + E^{t}(\lambda), \tag{6}$$

where E^b is the direct/beam irradiance, E^d is the diffuse-skylight irradiance, and E^t is the adjacent-terrain irradiance.

The direct irradiance varies due to solar activity, orbital parameter and solar geometry variations, and atmospheric and topographic conditions. Atmospheric models and DEMs are required to accurate estimate the direct irradiance.

Similarly, the diffuse-skylight irradiance component is governed by atmospheric conditions, topography, and land cover. Atmospheric models are required to account for Rayleigh-scattering and aerosol-scattering subcomponents, DEMs permit the modeling of the sky-view factor, and land-cover information derived from satellite imagery permits the backscattered subcomponent to be characterized more realistically.

Another important component is the hemispherical adjacent-terrain irradiance. Topographic and land cover information is required to model it effectively. This component is not frequently utilized in energy-budget modeling, although its magnitude can be significant when vegetation and/or snow cover is present in basins. It is difficult to accurately model because of DEM errors and lack of information regarding the bidirectional reflectance distribution function (BRDF) for a multitude of land-cover classes.

The surface albedo is a key parameter that regulates the net radiation. It represents the spectrally and hemispherically integrated BRDF and is represented as

$$\alpha = \int_{\lambda} \int_{0}^{2\pi} \int_{0}^{\pi/2} \mathrm{BRDF}(\theta_i, \phi_i, \theta_r, \phi_r, \lambda)\, d\theta_r d\phi_r d\lambda \tag{7}$$

where θ_i is the zenith incident angle, ϕ_i is the azimuth incident angle, θ_r is the zenith reflected angle, and ϕ_r is the azimuth reflected angle. In practice, the spatial variability of albedo can be estimated using satellite imagery. It is also possible to measure albedo in the field and empirically model the relationship between measured albedo and surface radiance. This parameter, however, does vary over time, and numerous approaches to modeling albedo have been developed (Arendt,1999).

The long-wave radiation can be characterized using numerous parameterization schemes. In general, the incoming long-wave irradiance ($L^{\downarrow}$) can be represented as

$$L^{\downarrow} = L_{\mathrm{s}}^{\downarrow} + L_{\mathrm{t}}^{\downarrow} = (\varepsilon_{\mathrm{a}}\sigma T_{\mathrm{a}}^{4})V_{\mathrm{f}} + (\varepsilon_{\mathrm{s}}\sigma T_{\mathrm{s}}^{4})(1 - V_{\mathrm{f}}) \tag{8}$$

where σ is the Stefan–Boltzmann constant, ε_{a} and ε_{s} are the emissivities of the air and surrounding surface, T_{a} and T_{s} are the temperature of the air and surface, and V_{f} is the sky-view factor which represents the angular portion of the sky that is not blocked by the topography. The sky-view factor is best estimated using a DEM such that

$$V_{\mathrm{f}} = \sum_{\phi=0}^{360} \cos^{2}\theta \frac{\Delta\phi}{360} \tag{9}$$

where θ is the local horizon angle at a given azimuth, ϕ.

Finally, the outgoing long-wave radiation can be calculated from

$$L^{\uparrow} = \varepsilon_{\mathrm{s}}\sigma T_{\mathrm{s}}^{4} + (1 - \varepsilon_{\mathrm{s}})L^{\downarrow}. \tag{10}$$

Satellite thermal imagery can be used to estimate the surface temperature at image acquisition times. This data can be used to validate and constrain energy-balance modeling of surface temperature.

4.3. *Turbulent heat fluxes*

Sensible and latent heat fluxes vary due to temperature and moisture gradients and atmospheric turbulence. They do not dominate over longer time periods, although the latent heat flux can be of major importance for short-term ablation rates.

Modeling these components is usually done using the bulk aerodynamic expressions that characterize the turbulent fluxes as

$$Q_{\mathrm{H}} = \rho_{\mathrm{a}} c_{\mathrm{p}} C_{\mathrm{H}} \bar{u}(\bar{\theta}_{\mathrm{z}} - \bar{\theta}_{\mathrm{s}}) \tag{11}$$

$$Q_{\mathrm{E}} = \rho_{\mathrm{a}} L_{\mathrm{v}} C_{\mathrm{E}} \bar{u}(\bar{q}_{\mathrm{z}} - \bar{q}_{\mathrm{s}}) \tag{12}$$

where ρ_{a} is the air density, c_{p} is the specific heat capacity of the air, L_{v} is the latent heat of evaporation, $\bar{u}$ is the mean wind speed, $\bar{\theta}_{\mathrm{z}}$ and $\bar{\theta}_{\mathrm{s}}$ are mean potential temperatures, $\bar{q}_{\mathrm{z}}$ and $\bar{q}_{\mathrm{s}}$ are mean specific humidities, and C_{H} and C_{E} are the exchange coefficients for heat and vapor pressure, respectively. Radiation transfer and atmospheric modeling are required to estimate the spatio-temporal variability in these parameters, however, field-based measurements from point weather stations are typically used.

4.4. *Ground heat*

For temperate glacial ice, the surface temperature is essentially zero with nocturnal freezing. For cold glacier ice, the surface layer can be an energy sink. The surface temperature is regulated by heat conduction from the englacial heat flux where the temperature gradient is a function of depth, advection and thermal conductivity.

Sources for this energy component include the geothermal heat flux, friction caused by basal sliding, and internal ice deformation. The flux across the surface can be represented as

$$Q_G = \int_0^z \rho c_p \frac{\partial T}{\partial t} dz, \tag{13}$$

where $\partial T/\partial t$ is the rate of change in ice temperature, ρ is ice density, and c_p is the specific heat capacity of ice.

4.5. Glacier mass-balance modeling

Estimating glacier mass balance is complicated because ice-mass gain or loss takes place at the glacier surface, within the glacier, or at the base of the glacier. At the base, loss occurs as a result of geothermal heat and sliding friction, although the surface loss is most significant. Internally, mass can accumulate as meltwater refreezes and losses occur due to thermal erosion. Surface accumulation includes snowfall, wind drift, avalanches, resublimation and condensation. Surface ablation includes melting, sublimation, ice avalanches, and snowdrift.

Mass balance estimates have been traditionally generated using the glaciological method, which involves inferring the mass balance from point measurements taken from ablation stakes and snow pits. Measurements such as these are sparse in the Himalaya (Bishop et al., 2004). Furthermore, energy-balance modeling has demonstrated the significant variability in ablation caused by glacier topography and localized topographic parameters that govern short-wave and long-wave energy fluxes. In the Himalaya, glacier topographic variation is considerable, and relief and topographic shielding significantly varies over relatively short distances. Consequently, the spatial complexity in ablation highlights the difficulty of using systematic stake measurements for accurately inferring mass balance gradients (Arnold et al., 2006).

Remote sensing and GIS permit the use of the geodetic method, wherein changes in glacier topography over a given period of time can be used to estimate volume change. Satellite-derived DEMs can be used to accomplish this, and the total mass balance of a glacier (B) is

$$B = \int_t \int_{A_s} b(x, y) dA_s dt = \rho dV, \tag{14}$$

where t is time, A_s is the surface area of the glacier, b is the mass balance at a particular point (x, y), ρ is the glacier density, and dV is the volume change. This approach can be used in a GIS to compute the mean specific mass balance ($\bar{b}$), which is useful for comparing glaciers. It is computed as

$$\bar{b} = \frac{\Sigma_x \Sigma_y \rho_{xy} dV_{xy}}{A_s}. \tag{15}$$

The geodetic method, however, has its limitations including (1) DEM accuracy and effective removal of biases; (2) Volume estimates need to be corrected for density of the surface materials; and (3) It generates an average mass balance and does not characterize the spatial pattern of mass balance. Generation procedures for DEMs permit altitude characterization in mountain environments on the order of 15–30 m (Eckert et al., 2005; Fujisada et al., 2005). Density corrections below the equilibrium line altitude (ELA) are reasonably accurate (~830–917 kg m^{-3}), although above the ELA density can vary from 50 to 830 kg m^{-3}. Finally, the point mass balance, b, cannot be determined from the geodetic method because glacier elevation is also a function of mass-flux divergence and convergence due to ice flow. Consequently, numerical flow modeling must also be utilized.

Hubbard et al. (2000) proposed to use the mass-continuity equation to compute mass balance such that

$$b(x, y) = \frac{\partial h(x, y)}{\partial t} + \nabla \cdot [H(x, y)\bar{v}(x, y)], \tag{16}$$

where, h is the ice elevation of the glacier surface, H is the ice thickness, $\bar{v}$ is the vertically averaged velocity vector, and ∇ is the horizontal difference operator. The term, $\partial h/\partial t$, can be determined from the geodetic method, while the local mass flux term is derived from numerical three-dimensional ice-flow modeling. Surface ice-velocity fields generated using feature-tracking and multi-temporal satellite imagery can also be used to constrain ice-flow simulations. Their results demonstrated that this approach has potential for deriving high spatial and temporal-resolution estimates of mass balance.

4.6. Glacier erosion

Glacier erosion also plays an important role in glacier fluctuations. It is difficult to assess because of the intractability of directly observing and measuring erosion at the ice-bed contact. Sediment yield studies of glaciated basins, suggest that glaciers are exceptionally efficient at eroding and transporting sediment (Hallet et al., 1996; Shuster et al., 2005). Empirical studies that make use of DEMs also support the interpretation that glaciers are very effective erosional agents (Bishop et al., 2003, 2004).

Glacial erosion partially controls ice-surface elevation, and simplistic erosion (abrasion) parameterization schemes assume that erosion is proportional to the basal-sliding velocity (Harbor, 1992; MacGregor et al., 2000). The basal shear stress (τ_b) can be computed as

$$\tau_b = \rho_i g H_i \sin \beta, \tag{17}$$

where ρ_i is the ice density (917 kg m^{-3}), g is gravitational acceleration (9.81 m s^{-1}), H_i is the thickness/depth of ice, and β represents the ice-surface slope ($\partial z/\partial x$) which can be initially obtained from a DEM. The basal sliding velocity (U_b) is estimated as

$$U_b = \frac{k\tau_b^2}{N}, \tag{18}$$

where k is a sliding coefficient (0.0012) and N is the effective normal stress. The later represents the difference in the stress caused by ice and water such that

$$N = \rho_i g H_i - \rho_w g H_w, \tag{19}$$

where ρ_w is the density of water (1000 kg m^{-3}) and H_w is the depth of water. Frequently, simulation studies assume H_w that is 80 percent the depth of the ice, which is reasonable if one is attempting to model temperate glaciers in the mid-latitudes. The glacier erosion rate E, is proportional to U_b by

$$E = 0.0001\, U_b, \tag{20}$$

where the erosion coefficient is presumably related to rock resistance to erosion, influenced by rock strength and stress conditions. Glacier erosion (incision) occurs normal to valley slopes which generates distinctive U-shaped valleys over time in cross section (Harbor, 1992), and parabolic topographic profiles in longitudinal profile (MacGregor et al., 2000).

Glacial erosion should fluctuate with ice discharge (Q_i) and topography, as radiative and atmospheric forcing will govern Q_i and the net mass balance, b. Consequently, it is possible to account for fluctuations in both parameters using the continuity equation for a one-dimensional flow model such that

$$\frac{\partial Q_i}{\partial x} + \frac{\partial A}{\partial t} = bW, \tag{21}$$

where A is the cross-sectional area, t is time, and W is the surface width of the glacier initially obtained from satellite imagery. The surface net mass balance can be simplistically estimated as

$$b(z) = k(z - z_{ela}), \tag{22}$$

where z is altitude, k is a scaling constant, and z_{ela} is the equilibrium-line-altitude. Finally, the change in ice depth can be estimated as

$$\frac{\partial H_i}{\partial t} = b + \frac{\partial Q_i}{\partial x}. \tag{23}$$

We stress that this approach to estimating the mass balance gradient does not effectively address variations in ablation due to multi-scale topographic influences on the surface irradiant flux, other energy fluxes, or debris-cover variations, but rather represents a flexible approach to account for atmospheric temperature change with altitude. Although it is recognized that an increase in erosion and relief production facilitates topographic shielding over geologic time, thereby enhancing erosion and reducing ablation, many models do not effectively account for this positive feedback, as modeling this effectively requires three-dimensional representations and the integration of climate, erosion and ice-flow models.

Boundary conditions and constraints for modeling come from glacier outlines, satellite-derived velocity fields, and DEM-derived profiles and cross sections. Specifically, ice-velocity fields can be generated from multi-temporal imagery (Scambos et al., 1992). Statistical scaling relations of ice width to depth can be used (Bahr et al., 1997), as ice-depth information from radio-echo sounding is not readily available for many glaciers in the Himalaya. Similarly, satellite-derived debris-cover maps can be used to account for the first-order spatial variation in debris depths.

5. Discussion

The potential use of satellite imagery for glacier change-detection studies is widely recognized, however, accurate mapping of alpine glaciers represents a unique problem. Based upon our experience, and the work of others, visual interpretation of satellite imagery does not produce reliable or repeatable results. It is extremely difficult to correctly delineate the accumulation areas and terminus position for many debris-covered glaciers in the Himalaya. It is widely believed that experience in the field is required in order to do an adequate job, although this does not guarantee accurate results. Glacier size, topographic conditions, spectral saturation, and lithological/supraglacial debris conditions affect interpretation. Similar problems with this approach have been obtained by the international Global Land Ice Measurements from Space (GLIMS) consortium in numerous experiments. Research on glacier mapping indicates that the integration of topographic information is very important in order to produce reliable results (Bishop et al., 2001), however, methodologies significantly vary. At its best, visual interpretation and digitization should be viewed as a first-order approximation.

Results from such change-detection studies suggest that glaciers in the Himalaya may be reacting differently to climate forcing. For example, we have found that many glaciers in Pakistan have not significantly retreated over the last decade, although a large number of glacier surges have been identified in the Karakoram, and this is significantly more than the number currently reported in the literature. Similarly, Smiraglia et al. (2007) indicates that the Baltoro Glacier has not significantly retreated in recent years. It should be noted, however, that high-quality long-term records do not exist for many glaciers in Pakistan. In contrast, research in India, Nepal, and Bhutan indicates that many glaciers exhibit systematic glacial retreat (Nakawo et al., 1999; Naito et al., 2000; Kulkarni and Bahuguna, 2002; Karma et al., 2003; World Wildlife Fund, 2005). It is tempting to infer that the downwasting rate might be greater in these countries, although higher precipitation rates in the east might support greater mass through-put for some glaciers that could balance ablation rates.

The geodetic method can be used to assess the downwasting rate and produce estimates of mass balance. This work, however, has yet to be systematically conducted in the Himalaya. The first published results for the Himalaya are from Berthier et al. (in press). They conducted a study to estimate regional mass balance in the Himachal Pradesh, Himalaya in India. They compared coarse resolution DEMS from 2000 to 2004 and found significant glacier downwasting. They indicate that the rate of ice loss is twice as large as the long-term mass balance record for the Himalaya. This type of analysis is required throughout the Himalaya to better understand how the glaciers are responding to climate forcing. Special care should be taken in interpreting the results, however, given the uncertainty in estimating the magnitude of error caused by method of data acquisition and removal of systematic biases.

Similarly, research involving the integration of remote sensing/GIS and numerical modeling for glacier parameter estimation and system coupling is in its early stages. For example, regional climate modeling is sorely needed to better understand the influence of the monsoon and the partitioning of monsoon and westerly precipitation. Furthermore, more information regarding the magnitude and spatial variability of ablation is required throughout the Himalaya via rigorous radiation-transfer and energy-budget modeling. Finally, more glacier-flow modeling and ice-flow velocity information is required to better understand glacier erosion and regional and glacier mass balance conditions.

Numerical modeling offers numerous possibilities, but there are also interesting challenges too. It is essential to develop parameterization schemes that can address spatial and temporal fluctuations. For example, many glacier simulations are implemented on a yearly basis and do not account for diurnal and seasonal ablation, ice-velocity, mass balance, or erosion fluctuations (e.g., MacGregor et al., 2000). Furthermore, the influence of topography on numerous processes and feedback mechanisms has not been adequately characterized or studied. Computer-assisted analysis of satellite imagery and DEMs can greatly assist in numerical modeling efforts to better understand the causal mechanisms associated with glacier fluctuations detected by space-based change-detection studies.

Finally, field-based observations support the notion that the glaciers across the Himalaya could exhibit different mass-balance trends. Valley climate stations in the central Karakoram in Pakistan indicate increases in precipitation over the past 50 years with small declines in summer temperatures (Roohi, 2007). In theory, this would support positive mass balance conditions in the extreme western Himalaya. Interestingly enough, Hewitt (2005) reports on glacier expansion in high altitude regions, and our preliminary results in the Nanga Parbat Himalaya, which is strongly influenced by the monsoon, indicate that we cannot detect significant thinning for some glaciers, although we know that the ablation rates are relatively high. This suggests significant mass through-put.

Climate modeling scenarios also indicate the possibility of a strengthening of the monsoon and an increase in precipitation given carbon-dioxide forcing. The system dynamics in this region are extremely complicated and the glaciers will also respond to other atmospheric forcings. For example, Lau and Kim (2007) discuss the role of aerosol forcing over India using climate modeling. They indicate that increased levels of aerosols over the Himalaya could reduce solar radiation reaching the surface that would cause surface cooling in the lower atmosphere. In addition, more aerosols increases the number of cloud condensation nuclei that prolong cloud cover, thereby further cooling the surface, but weakening the water cycle. They also describe how aerosols absorb solar radiation that may lead to an intensification of monsoon rainfall during the early monsoon season. Consequently, there is scientific debate regarding aerosol forcing, although it is not clear how multiple forcings will effect temperature and precipitation distributions in the Himalaya. Empirical data from field-based and remote-sensing studies suggest that the glaciers in the eastern Himalaya may be responding faster to climate change. More systematic analyses using remote sensing, GIS and numerical modeling are required in order to characterize glacier sensitivity and mass-balance trends in the Himalaya.

6. Conclusions

Alpine glacier fluctuations are directly and indirectly related to climate change. Consequently, it is essential to be able to assess glacier fluctuations and determine the causal mechanisms responsible for such change. Space-based change-detection studies can be effectively utilized to map and delineate alpine glaciers, although the traditional human-interpretation approach does not produce reliable results. More sophisticated GIS-based mapping models are required that incorporate spectral, texture, topographic and topological information from spatial data sets.

Interpreting the causal mechanisms responsible for glacier fluctuations requires additional analysis that integrates remote sensing and GIS analyses with numerical modeling. Climate, energy-budget, ice-flow, and glacier erosion modeling are required to assess glacier dynamics and mass balance. Only in this way can a multitude of forcing factors and interrelated processes be evaluated in an objective way to quantitatively ascertain glacier response to climate change. We can then begin to generate reliable estimates of the regional mass-balance trend and determine the contribution of Himalayan glaciers to rising sea level.

Acknowledgements

The field and laboratory work for this paper were supported by grants from the U.S. National Aeronautics and Space Administration under the NASA OES-02 program (Award NNG04GL84G), the U.S. National Science Foundation (Award BCS-0242339), the National Geographic Society, and the U.S. Geological Survey. We also thank the EV-K^2-CNR Committee of the Technical and Executive Secretariat of the Government of Italy for the opportunity to present our research on glacier fluctuations in Afghanistan and Pakistan. We also thank Dr. Andrew A.B.G. Bush at the University of Alberta, Canada for GCM simulation results.

References

Arendt, A., 1999. Approaches to modelling the surface albedo of a high Arctic glacier. Geografiska Annaler 81 (4), 477–488 Series A.

Arnold, N.S., Rees, W.G., Hodson, A.J., and Kohler, J., 2006. Topographic controls on the surface energy balance of a high Arctic valley glacier. Journal of Geophysical Research 111 .

Avouac, J.P. and Burov, E.B., 1996. Erosion as a driving mechanism of intracontinental mountain growth. Journal of Geophysical Research 101, 747–769.

Bahr, D.B., Meier, M.F., and Peckham, S.D., 1997. The physical basis of glacier volume-area scaling. Journal of Geophysical Research 102, 20,355–20,362.

Beniston, M. (Ed.), 1994, Mountain Environments in Changing Climates. Routledge, London, .

Berthier, E., Arnaud, Y., Kumar, R., et al., in press. Remote sensing estimates of glacier mass balances in the Himachal Pradesh (Western Himalaya, India). Remote Sensing of Environment.

Bishop, M.P., Barry, R.G., Bush, A.B.G., et al., 2004. Global land-ice measurements from space (GLIMS): Remote sensing and GIS investigations of the Earth's cryosphere. Geocarto International 19 (2), 57–84.

Bishop, M.P., Bonk, R., Kamp, U. Jr., and Shroder, J.F. Jr., 2001. Terrain analysis and data modeling for alpine glacier mapping. Polar Geography 24 (4), 257–276.
Bishop, M.P. and Colby, J.D., 2002. Anisotropic reflectance correction of SPOT-3 HRV imagery. International Journal of Remote Sensing 23 (10), 219–222.
Bishop, M.P. and Shroder, J.F. Jr., 2000. Remote sensing and geomorphometric assessment of topographic complexity and erosion dynamics in the Nanga Parbat massif. In: Khan, M.A., Treloar, P.J., Searle, M.P., and Jan, M.Q. (Eds), Tectonics of the Nanga Parbat Syntaxis and the Western Himalaya. Geological Society, London, Vol. 170, pp. 181–200.
Bishop, M.P. and Shroder, J.F., Jr., (Eds), 2004. Geographic Information Science and Mountain Geomorphology. Springer-Praxis, Chichester.
Bishop, M.P., Shroder, J.F. Jr., Bonk, R., and Olsenholler, J., 2002. Geomorphic change in high mountains: A western Himalayan perspective. Global and Planetary Change 32, 311–329.
Bishop, M.P., Shroder, J.F. Jr., and Colby, J.D., 2003. Remote sensing and geomorphometry for studying relief production in high mountains. Geomorphology 55, 345–361.
Bishop, M.P., Shroder, J.F. Jr., and Hickman, B.L., 1999. High resolution satellite imagery and neural networks for information extraction in a complex mountain environment. Geocarto International 14 (2), 17–26.
Bishop, M.P., Shroder, J.F. Jr., Hickman, B.L., and Copland, L., 1998a. Scale dependent analysis of satellite imagery for characterization of glacier surfaces in the Karakoram Himalaya. Geomorphology 21, 217–232.
Bishop, M.P., Shroder, J.F. Jr., Sloan, V.F., et al., 1998b. Remote sensing and GIS technology for studying lithospheric processes in a mountain environment. Geocarto International 13 (4), 75–87.
Bishop, M.P., Shroder, J.F. Jr., and Ward, J.L., 1995. SPOT multispectral analysis for producing supraglacial debris-load estimates for Batura Glacier, Pakistan. Geocarto International 10 (4), 81–90.
Brozovik, N., Burbank, D.W., and Meigs, A.J., 1997. Climatic limits on landscape development in the northwestern Himalaya. Science 276, 571–574.
Burbank, D., Leland, J., Fielding, E., et al., 1996. Bedrock incision, rock uplift and threshold hillslopes in the northwestern Himalaya. Nature 379, 505–510.
Bush, A.B.G., 2000. A positive feedback mechanism for Himalayan glaciation. Quaternary International 65-6, 3–13.
Bush, A.B.G., 2002. A comparison of simulated monsoon circulations and snow accumulation in Asia during the mid-Holocene and at the Last Glacial Maximum. Global and Planetary Change 32, 331–347.
Chavez, P.S. Jr., 1989. Radiometric calibration of landsat thematic mapper multispectral images. Photogrammetric Engineering and Remote Sensing 55 (9), 1285–1294.
Chavez, P.S. Jr., 1996. Image-based atmospheric corrections – revisited and improved. Photogrammetric Engineering and Remote Sensing 62 (9), 1025–1036.
Dyurgerov, M.B. and Meier, M.F., 2000. Twentieth century climate change: evidence from small glaciers. Proceedings of the National Academy of Science 97 (4), 1406–1411.
Eckert, S., Kellenberger, T., and Itten, K., 2005. Accuracy assessment of automatically derived digital elevation models from ASTER data in mountainous terrain. International Journal of Remote Sensing 26 (9), 1943–1957.
Fujisada, H., Bailey, G.B., Kelly, G.G., et al., 2005. ASTER DEM performance. IEEE Transactions on Geoscience and Remote Sensing 43 (12), 2707–2714.
Haeberli, W., 1998. Historical evolution and operational aspects of worldwide glacier monitoring. In: Haeberli, et al. (1998), Into the Second Century of World Glacier Monitoring – Prospects and Strategies, UNESCO Publishing, Paris, pp. 35–47.
Haeberli, W. and Beniston, M., 1998. Climate change and its impacts on glaciers and permafrost in the Alps. Ambio 27 (4), 258–265.
Haeberli, W., Hoelzle, M., and Suter, S., (Eds), 1998. Into the Second Century of Worldwide Glacier Monitoring: Prospects and Strategies. UNESCO, Paris.
Hallet, B., Hunter, L., and Bogen, J., 1996. Rates of erosion and sediment evacuation by glaciers: A review of field data and their implications. Global and Planetary Change 12, 213–235.

Harbor, J., 1992. Numerical modeling of the development of U-shaped valleys by glacial erosion. Geological Society of America Bulletin 104, 1364–1375.
Harbor, J. and Warburton, J., 1992. Glaciation and denudation rates. Nature 356, 751.
Harbor, J. and Warburton, J., 1993. Relative rates of glacial and nonglacial erosion in alpine environments. Arctic and Alpine Research 25, 1–7.
Henderson-Sellers, A. and Pitman, A., 1992. Land surface schemes for future climate models: Specification, aggregation, and heterogeneity. Journal of Geophysical Research 97, 2687–2696.
Hewitt, K., 2005. The Karakoram anomaly? Glacier expansion and the 'elevation effect', Karakoram Himalaya. Mountain Research and Development 25 (4), 332–340.
Hock, R., 2005. Glacier melt: A review of processes and their modelling. Progress in Physical Geography 29 (3), 362–391.
Hubbard, A., Willis, I., Sharp, M., et al., 2000. Glacier mass-balance determination by remote sensing and high-resolution modelling. Journal of Glaciology 46 (154), 491–498.
Kargel, J.S., Abrams, M.J., Bishop, M.P., et al., 2005. Multispectral imaging contributions to global land ice measurements from space. Remote Sensing of Environment 99, 187–219.
Karma, A., Ageta, Y., Naito, N., et al., 2003. Glacier distribution in the Himalayas and glacier shrinkage from 1963 to 1993 in the Bhutan Himalayas. Bulletin of Glaciological Research 20, 29–40.
Kaser, G., 2001. Glacier-climate interaction at low latitudes. Journal of Glaciology 47, 195–204.
Kimes, D.S. and Kirchner, J.A., 1981. Modeling the effects of various radiant transfers in mountainous terrain on sensor response. IEEE Transaction on Geoscience and Remote Sensing 19 (2), 100–108.
Koons, P.O., 1995. Modeling the topographic evolution of collisional belts. Annual Review Earth and Planetary Science 23, 375–408.
Kotlyakov, V.M., Serebrjanny, L.R., and Solomina, O.N., 1991. Climate change and glacier fluctuations during the last 1,000 years in the southern mountains of the USSR. Mountain Research and Development 11, 1–12.
Kulkarni, A.V. and Bahuguna, I.M., 2002. Glacial retreat in the Baspa Basin, Himalayas, monitored with satellite stereo data. Journal of Glaciology 48, 171–172.
Lambeck, K. and Chappell, J., 2001. Sea level change through the last glacial cycle. Science 292, 679–686.
Lau, K.M. and Kim, K.M., 2007. Does aerosol weaken or strengthen the Asian monsoon water cycle? In: Shroder, J.F. Jr. (Ed.), Mountains, Witnesses of Global Change. Developments in Earth Surface Processes Book Series. Elsevier Publishing, Amsterdam, Vol. 10, pp. 11–20.
MacGregor, K.R., Anderson, R.S., Anderson, S.P., and Waddington, E.D., 2000. Numerical simulations of glacial-valley longitudinal profile evolution. Geology 28, 1031–1034.
Maisch, M., 2000. The longterm signal of climate change in the Swiss Alps: Glacier retreat since the end of the Little Ice Age and future ice decay scenarios. Geografia Fisica e Dinamica Quaternaria 23 (2), 139–151.
Mattson, L.E., 2000. The influence of a debris cover on the mid-summer discharge of Dome Glacier, Canadian Rocky Mountains. In: Nakawa, M., Raymond, C.F., and Fountain, A. (Eds), Debris covered glaciers. Proceedings and Reports. IAHS Press, Wallingford, UK, Vol. 264, pp. 25–33.
McClung, D.M. and Armstrong, R.L., 1993. Temperate glacier time response from field data. Journal of Glaciology 39 (132), 323–326.
Meier, M.F., 1984. Contribution of small glaciers to global sea level. Science 226, 1418–1421.
Meier, M.F. and Dyurgerov, M.B., 2002. How Alaska affects the world. Science 297, 350–351.
Meier, M.F. and Wahr, J.M., 2002. Sea level is rising: Do we know why?. Proceedings of the National Academy of Science 99 (10), 6524–6526.
Molnar, P. and England, P., 1990. Late Cenozoic uplift of mountain ranges and global climate change: Chicken or egg?. Nature 346, 29–34.
Montgomery, D.R., 1994. Valley incision and the uplift of mountain peaks. Journal of Geophysical Research 99, 913–921.
Naito, N., Nakawo, M., Kadota, T., and Raymond, C.F., 2000. Numerical simulation of recent shrinkage of Khumbu Glacier, Nepal Himalayas. In: Nakawa, M., Raymond, C.F., and Fountain, A. (Eds), Debris covered glaciers. Proceedings and Reports. IAHS Press, Wallingford, UK, Vol. 264, pp. 245–254.

Nakawo, M., Fujita, K., Ageta, U., et al., 1997. Basic studies for assessing the impacts of the global warming on the Himalayan cryosphere. Bulletin of Glacier Research 15, 53–58.

Nakawo, M., Yabuki, H., and Sakai, A., 1999. Characteristics of Khumbu Glacier, Nepal Himalaya: Recent change in the debris covered area. Annals of Glaciology 28, 118–122.

Phillips, W.M., Sloan, V.F., Shroder, J.F. Jr., et al., 2000. Asynchronous glaciation at Nanga Parbat, northwestern Himalaya Mountains, Pakistan. Geology 28, 431–434.

Raymo, M.E. and Ruddiman, W.F., 1992. Tectonic forcing of the late Cenozoic climate. Nature 359, 117–122.

Raymo, M.E., Ruddiman, W.F., and Froelich, P.N., 1988. Influence of late Cenozoic mountain building on geochemical cycles. Geology 16, 649–653.

Richards, B.W., Owen, L.A., and Rhodes, E.J., 2000. Timing of late Quaternary glaciations in the Himalayas of northern Pakistan. Journal of Quaternary Science 15, 283–297.

Rivera, A., Casassa, G., Bamber, J., and Kääb, A., 2005. Ice-elevation changes of Glaciar Chico, southern Patagonia, using ASTER DEMs, aerial photographs and GPS data. Journal of Glaciology 51 (172), 105–112.

Roohi, R., 2007. Research on global changes in Pakistan. In: Shroder, J.F. Jr. (Ed.), Mountains, Witnesses of Global Change. Developments in Earth Surface Processes Book Series. Elsevier Publishing, Amsterdam, Vol. 10, pp. 329–340.

Ruddiman, W.F. (Ed.), 1997, Tectonic Uplift and Climate Change. Plenum Press, New York, .

Scambos, T.A., Dutkiewicz, M.J., Wilson, J.C., and Bindschadler, R.A., 1992. Application of image cross-correlation to the measurement of glacier velocity using satellite image data. Remote Sensing of Environment 42 (3), 177–186.

Schaper, J., Martinec, J., and Seidel, K., 1999. Distributed mapping of snow and glaciers for improved runoff modelling. Hydrological Processes 13 (12-13), 2023–2031.

Seltzer, G.O., 1993. Late-quaternary glaciation as a proxy for climate change in the central Andes. Mountain Research and Development 13, 129–138.

Shroder, J.F. Jr. and Bishop, M.P., 1998. Mass movement in the Himalaya: New insights and research directions. Geomorphology 26, 13–35.

Shroder, J.F. Jr. and Bishop, M.P., 2000. Unroofing of the Nanga Parbat Himalaya. In: Khan, M.A., Treloar, P.J., Searle, M.P., and Jan, M.Q. (Eds), Tectonics of the Nanga Parbat Syntaxis and the Western Himalaya. Geological Society, London, Vol. 170, pp. 163–179.

Shroder, J.F., Jr. and Bishop, M.P., in press. Satellite-image analysis of glaciers of northern Pakistan. In: Williams, R S. and Ferrigno, J.G. (Eds), Satellite Image Atlas of Glaciers. Professional Paper 1386-F. United States Geological Survey, Reston, pp. 1–24.

Shuster, D.L., Ehlers, T.A., Rusmore, M.E., and Farley, K.A., 2005. Rapid glacial erosion at 1.8 Ma revealed by $^{4}He/^{3}He$ thermochronometry. Science 310, 1668–1669.

Smiraglia, C., Mayer, C., Mihalcea, C., et al., 2007. Ongoing variations of Himalayan and Karakoram glaciers as witnesses of global changes: Recent studies of some glaciers. In: Shroder, J.F. Jr. (Ed.), Mountains, Witnesses of Global Change. Developments in Earth Surface Processes Book Series. Elsevier Publishing, Amsterdam, Vol. 10, pp. 235–247.

Thorne, K., Palluconi, F., Takashima, T., and Masuda, K., 1998. Atmospheric correction of ASTER. IEEE Transactions on Geoscience and Remote Sensing 36 (4), 1199–1211.

Whipple, K.X. and Tucker, G.E., 1999. Dynamics of the stream-power river incision model: Implications for height limits of mountain ranges, landscape response timescales, and research needs. Journal of Geophysical Research 104, 17661–17674.

World Wildlife Fund, 2005. An Overview of Glaciers, Glacier Retreat and Subsequent Impacts in Nepal, India and China. World Wildlife Fund Nepal Program, Kathmandu.

Zeitler, P.K., Koons, P.O., Bishop, M.P., et al., 2001a. Crustal reworking at Nanga Parbat, Pakistan: Metamorphic consequences of thermal-mechanical coupling facilitated by erosion. Tectonics 20 (5), 712–728.

Zeitler, P.K., Meltzer, A.S., Koons, P.O., et al., 2001b. Erosion, Himalayan geodynamics, and the geomorphology of metamorphism. GSA Today 11, 4–8.

Mountains Witnesses of Global Changes
R. Baudo, G. Tartari, and E. Vuillermoz (Editors)

26

Ongoing variations of Himalayan and Karakoram glaciers as witnesses of global changes: recent studies on selected glaciers

Claudio Smiraglia, Christoph Mayer, Claudia Mihalcea, Guglielmina Diolaiuti, Marco Belò and Giorgio Vassena

Abstract

The glaciers located on the highest Asian mountain ranges (Pamir, Karakoram and Himalaya) represent the largest alpine glaciations in the world, outside the polar regions. Runoff from these glaciers feeds rivers (e.g. Indus, Ganges, etc.) that provide much-needed water supply for several hundred million people. In the Himalayan and Karakoram regions debris-covered glaciers (DCG) are the most common glacier type, which seems currently to be spreading in the world's mountain regions due to the feedback between ice thinning and rock-wall downwasting. To contribute to a better understanding of the complex relations between DCG and climate and to forecast on a decadal scale the glacier response to climate change and its impact on the runoff of high mountain regions, different studies have been carried out on some selected glaciers located in the Karakoram (Baltoro and Liligo in the K2 region) and in the Nepal Himalayas (Changri Nup in the Everest region). To quantify the recent and ongoing fluctuations of the ice masses, historical sources, e.g. maps and photographs, and satellite images have been processed and analysed. In addition, direct field measurements on selected glaciers were performed in order to validate the remotely sensed data and to investigate the variability and magnitude of surface ablation on the DCG tongues.

1. Introduction

The Himalayan and Karakoram glaciers represent the largest ice masses in the world apart from high latitude glaciation. They cover a surface area of ~58.000 km^2. Especially during hot and dry periods, runoff from these glaciers is a major contribution to the large river systems (e.g. Indus, Ganges and Brahmaputra). Especially with growing population and more intense agriculture, melt water from the Asian glaciers becomes an increasingly important natural resource for several hundred million people (Hewitt et al. 1989; Hagg and Braun, 2005).

ISSN: 0928-2025 DOI: 10.1016/S0928-2025(06)10026-7

During recent decades the scientific community is debating about global glacier shrinkage and its consequences on sea level rise and the mountain ecosystem. In the Himalaya and Karakoram the question is not only if ice masses are reducing, but also how fast, how much, and for how long melt water can be produced.

To quantify the recent and historical variations of glaciers in the Himalaya and Karakoram ranges, the analysis of remote sensing data and of historical sources (Bishop et al. 2000, 2005) is necessary due to the large size of the glacier systems (e.g. Siachen, Biafo, Hispar and Baltoro glaciers are more than 60 km long), and to their remote location which makes it difficult and expensive to manage direct and accurate field surveys. Nevertheless, direct field measurements on a selection of glaciers are performed to improve and test remote sensing techniques and also to understand the specific processes of glaciers in these regions.

Several authors reported a strong phase of retreat affecting Asian high elevation glaciers since the second half of the 19th century. According to Ageta (2001) the shrinkage of DCG as well as debris-free glaciers is accelerating in the recent years in Nepal and Bhutan Himalayas, although, response time scales and sensitivities to climate change are quite different. Ageta and Fujita (1996) mentioned a drastic glacier terminus retreat of Yala Glacier in Langtang Valley starting from 1990s.

Hewitt (2005) found that a dozen glaciers he surveyed at the end of the 1990s and which he revisited in 2000 and 2001, respectively in the Upper Champursan, Chalt, Naltar, Karambar and Darkot valleys in the Karakoram continued to thin and retreat. Moreover, he emphasised the strong limitation of remote sensing techniques (e.g. Landsat imagery) for evaluating the changes of these debris-covered glaciers (DCG), mainly due to image accuracy and resolution, and he underlined the need for field surveys.

According to Ageta and Kadota (1992) and Naito et al. (2001), glaciers in the Himalaya are to be considered more sensitive to the recent global warming due to their characteristic of peak accumulation in the summer months. Nevertheless, not all Asian high elevation glaciers showed terminus retreats and an exceptional number of glacier surges was reported by Hewitt (1998a), in which 6 of a possible 9 Karakoram glaciers showed surge type behaviour between 1986 and 1996. The same author found in his surveys between 1997 and 2002, 13 advancing glaciers of intermediate size (10 to 20 km in length). Moreover, some of the larger glaciers (40 to 70 km in length) exhibited 5 to 15 m of thickening over substantial areas of the ablation zone, locally more than 20 m (Hewitt, 2005).

As a contribution to these efforts in understanding glacier change in the Asian high mountain ranges, data from several glaciers located in the Karakoram (Baltoro and Liligo) and Himalaya (Changri Nup) are presented. The data have been collected from 1994 up to the more recent scientific expeditions in the frame of the projects managed by the Ev-K^2-CNRcommittee.

2. Geographical setting of the studied glaciers

Baltoro Glacier is one of the world's largest valley glaciers, draining an area of about 1500 km^2 and having a length extension of more than 60 km (Fig. 26.1). The basin of

Figure 26.1. Location map and locations of photos presented herein.

Baltoro Glacier is situated on the watershed between the Indus Basin, of which Baltoro Glacier is a tributary, and the Tarim Basin (Desio et al., 1961) on the south side of the Karakoram Range. Its elevation extends from 3370 m a.s.l. at the terminus, to K2 (8611 m a.s.l.) the highest peak of the Karakoram. The erratic block, "Desio Boulder", just in front of the glacier at 3,365.6 m a.s.l., was named after A. Desio, who in 1954 used it as a reference point close to the glacier snout.

The glacier is debris-covered below about 5000 m a.s.l., initially by medial moraines then gradually spreading to a uniform cover across the entire surface. Near the terminus, the debris thickness exceeds 1 m in many places. The average width of Baltoro Glacier is about 2.1 km, reaching up to 3.1 km at Gore 1 just upstream of the confluence of Yermandu Glacier with the main glacier tongue. With a mean slope of 3.9%, it is a rather flat glacier compared with other glaciers of the Karakoram Range. In some areas (e.g. south of Concordia) the mean glacier surface rises only 21 m over a distance of 2 km. On the other hand, the glacier surface is very rough perpendicular to the flow direction. Surface undulations of 25 m over a horizontal distance of 140 m are common for many parts of the lower glacier. These undulations partly exist due to the abundance of melt water, forming large superficial melt-water streams. But also a rather high number of large tributary glaciers form a bundle of parallel ice bulges after merging.

Liligo Glacier is a smaller debris-covered glacier (DCG) located in a transversal valley, which enters on the left or south side of Baltoro Glacier. The glacier measures about 15 km in length and has a surface area of about 17 km^2. At the moment the glacier snout calves into an ice-contact lake that formed recently during the last retreat of the glacier terminus.

Baltoro and Liligo glaciers provide a long sequence of investigations. They have been explored and scientifically investigated since the observations of Conway in the 19th century (Conway, 1894). Many studies have been carried out on its geographical and morphological characteristics (Dainelli and Marinelli, 1928; De Filippi, 1912; Savoia-Aosta and Desio, 1936; Desio et al., 1961). Also they have been analysed in more detail because of the surge-type phenomena that have occurred on Liligo and other Baltoro tributaries (Hewitt, 1969; 1998b&c; Pecci and Smiraglia 2000; Diolaiuti et al., 2003). Attention has been paid to the debris cover and its effects on ablation (Mayer et al., 2006; Mihalcea et al., 2006).

In Nepal, the Changri Nup Glacier is one of the tributaries of the larger Khumbu Glacier. It is located in the Sagarmatha National Park, in the Nepal Himalaya. The glacier is oriented E–W and it is partially debris covered from the terminus until it reaches an altitude of about 5350 m a.s.l. From that elevation to the highest parts (5700 m a.s.l.), its surface is debris free. The debris cover on the glacier tongue varies in thickness between a few cm to 1–2 m. The glacier terminus is at an elevation of about 5200 m a.s.l. not far from the tongue of Khumbu Glacier on its right lateral side. The glacier area of about 8 km^2 is considerably smaller than the area of the two other glaciers, whereas the mean glacier width is approximately 800 m. The accumulation area starts at an elevation of around 5400 m, which leads to two separate accumulation basins. Only the northern one is nowadays still contributing to the glacier flow.

On the debris-covered tongue of Changri Nup Glacier several supraglacial lakes exist and the roughness is rather high (with elevation changes of 30–40 m on small distances due to differential ablation). Changri Nup Glacier is located not far away from the Pyramid Laboratory (5050 m a.s.l.), which allows frequent investigations and the rather easy implementation of different experiments.

2.1. *Recent investigations on Baltoro and Liligo glaciers (Karakoram)*

2.1.1. *Baltoro Glacier*

During the scientific expedition to Baltoro Glacier in 2004 (in commemoration of the 50th anniversary of the first ascent to K2 peak) comprehensive data sets on glacier dynamics, geometry, morphology and on its surface energy and mass exchanges have been collected by an Italian–German scientific team. Field experiments have been carried out on the debris-covered sector in order to evaluate the present dynamics of the main glacier tongue, to quantify the volume, surface and morphological changes that have occurred during the last 50 years, and to investigate the ablation of buried ice and its relations with debris thickness and debris features. The glacier tongue of Liligo Glacier was also visited and investigated.

The evaluation of the different ablation rates depending on debris thickness is important not only as a topic of high scientific interest, but also for water management. In fact, the large glacier system of Baltoro provides an important part of the water utilised in Pakistan during the dry season. Therefore the quantification of glacier ablation and of glacier energy exchanges in relation to different climate conditions can be very important for forecasting seasonal runoff.

The tongue of Baltoro Glacier today is basically formed by two contributions: Trango and Uli Biaho glaciers, which occupy the northern part of the valley; and the main flowband of the Baltoro Glacier, which flows down from Concordia and covers the southern part of the valley. Over the last fifty years (1954–2004) the glacier terminus has retreated and the snout has become flatter. Notably, the snout morphology shows small changes, whereas the main glacial stream varied its position. The glacier portal is now located in the northern sector of the glacier terminus. However, in a picture from A. Desio in 1929 (reported in Mayer et al., 2006) the entire glacier snout looks almost identical to the modern one. An analysis of the measurements carried out during the last 91 years shows that there is no clear trend of recession or advance. The terminus variations (Fig. 26.2), which have been derived from a number of measurements since the beginning of the last century, show only limited fluctuations of the glacier terminus. The accumulated value is −65 m over the period 1913–2004 which reflects a quite stable snout position.

By comparing an image of the Concordia area, the central confluence at about 4650 m, taken in 1909 by Sella with the one taken in 2004 by Mayer, it is possible to observe small changes that have occurred over the last 95 years (Mayer et al., 2006). A careful analysis of the images has shown that the mean elevation difference of the glacier surface in this time span is less than 40 m. Along almost the entire glacier margin that is identifiable on the images, the ice is more or less at the same level as 95 years ago. Some minor changes are discernible but their magnitude is minor

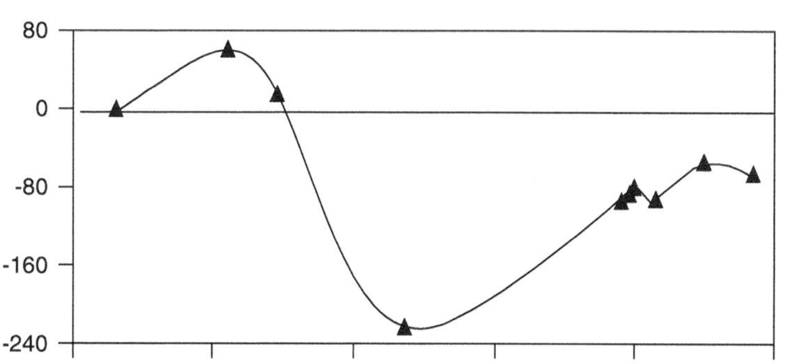

Figure 26.2. Fluctuations in meters (retreat in the positive direction of the y-axis) of the Baltoro Glacier terminus, 1913–2004.

compared to other mountain glaciers especially in the Alps, where glaciers in the same time span lost about 30–40% of their surface area (Haeberli et al., 1999; Beniston, 2003; Haeberli, 2005). The reason for such limited changes on Baltoro is surely due to the debris cover present on its surface. It reduces the surface ablation and thus slows down the glacier recession (Fujii, 1977; Bozhinskiy et al., 1986; Conway and Rasmussen, 2000; Fukita and Sakai, 2000). The observations of changes in the snout position and variations in the glacier elevation at Concordia lead to the conclusion that the extent of Baltoro Glacier showed only minor changes during the last century.

On Baltoro Glacier the debris cover has been present on the ablation zone for at least more than a century (Conway, 1894) and the complex relations between debris and ice ablation have been active for a long time. An exact calculation of debris cover is rather difficult, but the analysis of the available satellite images enabled a statistical evaluation of the glacier-surface conditions. Stretches of bare ice exposed at the glacier surface are usually short, because high debris loads and intensive melt lead to a fast coverage with debris. On the main glacier, a few kilometres from the equilibrium line (~5500–5600 m) the mean debris cover across the glacier is more than 50%. Downstream from about 5000 m the debris cover rapidly increases to 70–90%. In the area of Urdukas and further down almost the entire glacier is covered by debris. Here, melt-water ponds, superficial streams, and bare ice only account for a few percent of the glacier area. In total about 38% of the glacier area is debris covered. As a consequence, ~33% of the glacier surface consists of bare ice. The accuracy of this analysis is ±5% (Mayer et al., 2006). The quantification of the debris thickness is even more difficult, due to the number of sample points and the large thickness variations even over short distances. Thickness measurements have been carried out at more than 60 locations. In general, the debris thickness is larger close to the snout and lower in higher areas. There is, however, no clear trend in the thickness distribution. Even in higher areas patches with rather thick debris cover occur. Generally, debris cover is thick between 0.3 m and 0.5 m even if in some places more than 1.5 m have been measured. Also the grain size shows a very large variation all over the glacier, from sub-millimetre silt to boulders of a few metres diameter.

Using all the available topographic maps and the most recent satellite images, the catchment area of Baltoro and the total glaciated surface was calculated. The area is ~524 km^2 (considering all the glacier tributaries nourishing the Baltoro tongue), while the whole catchment area is ~1500 km^2 (Mayer et al., 2006).

Analysis of the glacier flux was difficult due to the presence of several large tributaries entering the main glacier below its ELA (Uli Biaho, Trango, Muztagh, Younghusband, Mundu, Yurmandu, Biarchedi, Godwin Austen, Vigne and other glaciers). The measurements were referred to a transversal section located just downstream from the confluence of Baltoro South and Godwin Austen glaciers at Concordia. There, the two major glaciers join and at the same time changing their flow direction towards the west. An analysis of ASTER images from 2001 has shown that the flowband of these two glaciers together still represents 60% of the glacier width at Urdukas and 40% close to the glacier snout. There, Trango and Uli Biaho glaciers, entering the main glacier about 5 km and 2 km upstream, respectively, form another 45% of the glacier width. Observed variations of the glacier snout, thus, only to some extent reflect changes in mass balance and ice dynamics of the main glacier (Mayer et al., 2006).

Baltoro glacier is a firn-kettle-type glacier (Wissmann, 1959) nourished from short high elevation basins and avalanches from the adjacent steep valley slopes. The firn areas of the glacier (without slopes above the bergschrund) comprise 29% of the total glacier area. Snow accumulation in these areas, together with avalanche snow, is the only source of mass input for the subsequent formation of the extensive glacier tongue. In contrast to the 29% of firn area for the glacier itself, the accumulation area ratio (AAR) is 40%, if the high slopes contributing avalanche snow are included in the analysis.

As mentioned above, the cross section at Concordia is a natural location for the computation of mass fluxes. The area upstream of Concordia represents about 56% of the entire glacier area. Information about ice velocities and estimates of ice thickness are also available at this confluence (Desio et al., 1961). Balance calculations have been carried out for the part of Baltoro Glacier upstream of Concordia. An estimated vertical precipitation profile (Mayer et al., 2006), the area/elevation distribution and the prevailing climatic conditions have been used for the determination of the net accumulation above the equilibrium line, which resulted in a mass flux of 0.511 km^3/yr for this part of the glacier. Ablation measurements (Mihalcea et al., 2006) at Concordia and further upstream provided the basis for calculating the net ablation. Using a local lapse rate determined from air temperature data collected at two automatic weather stations located on the glacier surface at different elevations (Mihalcea et al., 2006) and extrapolating the temperature data over the entire ablation period by using long-term temperature data from Srinagar, Kashmir, the estimated net ablation for the specified area resulted in $-0.167\,km^3/yr$. Obviously, errors are large due to the lack of long-term local data. Nevertheless, the possible variations should not exceed 30%, which results in a net balance estimate above Concordia of $0.344 \pm 0.103\,km^3/yr$ (Mayer et al., 2006).

2.1.2. Liligo Glacier

The Liligo Glacier was investigated within the context of the occurrence of surge-type phenomena (Desio, 1954; Hewitt, 1969, 1998a, 1998b; Weake and Searle, 1993; Pecci and Smiraglia 2000; Diolaiuti et al., 2003). Terminus variations of Liligo Glacier starting from 1892 were reconstructed by means of various methods and sources, e.g. historical documents, cartography, photographs, satellite images and field surveys (Diolaiuti et al., 2003).

The actions of the glacier were characterised by a phase of advance until 1909, followed by a strong retreat until 1985 and then by another significant advance phase in the last 10 years of the 20th century. The large rate of the glacier advances, together with some observed ice-surface features, such as the heavily crevassed surface at certain times, are considered to be indicative of a surge-type glacier. Liligo Glacier retreated by about 1300 m between 1929 and 1954 and advanced by about 350 m between 1954 and 1986. Between 1986 and 1997 a surge-type advance took place and the distance of the snout movement was about 1400 m (Diolaiuti et al., 2003).

During the last 5 years a strong retreat of the glacier terminus was observed, followed by the formation of an ice contact lake. The surge-type reaction that affected Liligo at least twice during the past century seems to have ended. The glacier snout is now downwasting and losing mass.

3. The recent investigations on Changri Nup Glacier

In contrast to Baltoro and Liligo Glaciers of Pakistan, Changri Nup Glacier in Nepal was investigated only during the past few decades. Historical photographs as for the Karakoram glaciers are not available. The oldest map showing the entire glacier surface was the one (1957) at 1:25,000 scale produced by the Austrian Alpenverein; in addition there are maps of the "Arbeitsgemeinschaft für vergleichende Hochgebirgsforschung," first edition 1965 and subsequent editions 1978, 1985, 1988, 2005, at a scale of 1:50,000. The more recent (1988) large-scale map was the one at a 1:25,000 scale produced by the National Geographic Society and the ETH in Zurich in co-operation with the Swiss Federal Office of Topography. This last map does not cover the whole glacier surface, which makes it difficult to evaluate the volume changes derived from DTM comparisons.

The Changri Nup Glacier has been investigated several times since 1994 to quantify the change in position of its debris-free sector (Fig. 26.3), which resulted between 1994 and 2005 in more than 140 m of retreat. In contrast it was not possible to measure the exact glacier terminus on the debris-covered tongue due to the large thickness of debris cover on the lower sector. It seemed interesting, however, to analyse the behaviour and the evolution of the debris-free sector, which flows from the upper basin partly overlapping the debris-covered tongue (Fig. 26.4).

In addition, starting from 1998, a network of topographic benchmarks was positioned on the debris-covered tongue to measure glacier velocity and elevation

Figure 26.3. The white terminus of Changri Nup Glacier.

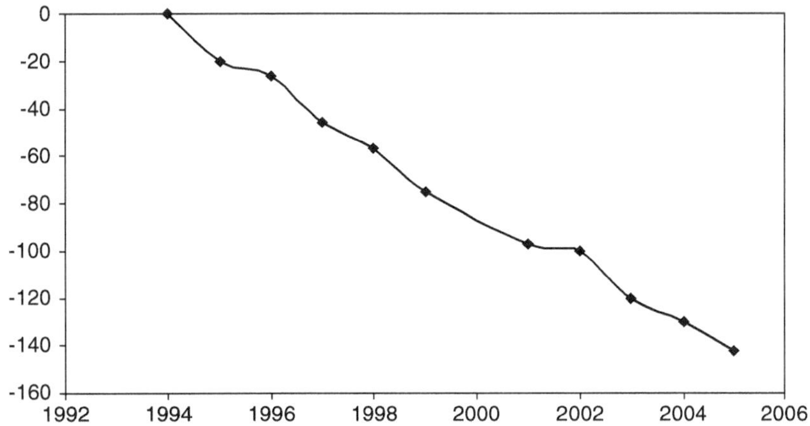

Figure 26.4. Terminus fluctuations in meters of the debris free sector Changri Nup Glacier.

changes by DGPS measurements. The surveys were performed in a fast static technique with reference to a base station installed at 4993 m a.s.l. outside the glacier surface on a stable rock. On the glacier surface the rover station was installed for measuring position data at nine points located on big stable boulders. The acquisition time was 30 min for each position, using a cut-off angle of 15° and maximum GDOP of 6. The data accuracy was evaluated to ~2 cm in planimetry and ~4 cm in elevation. The DGPS surveys were performed from 1998 until 2001 (Table 26.1). Due to the fast changes affecting the glacier tongue it was difficult to preserve the benchmarks, which, every year had to be substituted.

The planimetric differences between the measurements (E and N) in the period 1998–2001 are meaningful for V400 and V600. In the case of V700 they are in the same order as the accuracy. The vertical movements detected show smaller values as the measurement accuracy for the points located at lower elevations. A meaningful change has only been computed for the highest benchmark.

4. Discussion

The terminus variations in Fig. 26.2 indicate that the terminus of Baltoro Glacier (~3370 m a.s.l.) experienced limited fluctuations (cumulated value of −65 m for the period 1913–2004, yearly average ~−0.7 m) and reflect a quite stable snout position. This behaviour is representative for DCG not only in high elevation regions but also in the Alps (D'Agata et al., 2005). During the last century the mass balance of the DCGs was less negative (or even positive in some cases) and the position of the termini was almost the same as during the LIA. Debris-free glaciers, on the contrary, showed for the largest part of the world a strong negative balance and their tongues have experienced a rather large retreat. This retreat can also be observed at Changri Nup Glacier, where the terminus fluctuations of its debris-free sector showed a strong retreat rate (~−10 m /y), mainly due to the supraglacial conditions that permit a

Table 26.1. Average of the position changes (in directions East: ΔE, North: ΔN and vertical: ΔH) measured on three representative benchmarks located on the debris covered tongue of the Changri Nup Glacier between 1998 and 2001.

Benchmark code	Benchmark Long	Benchmark Lat	Benchmark elevation (m)	−ΔE (m)	−ΔN (m)	−ΔH (m)
V400	E86° 48′ 36.88956″	N27° 58′ 31.13254″	5161.989	+0.12	−0.50	+0.05
V600	E86° 46′ 51.85882″	N27° 59′ 17.44940″	5438.334	+2.25	−5.03	−1.30
V700	E86° 48′ 23.40417″	N27° 58′ 29.16390″	5163.024	−0.02	+0.00	−0.02

Note: The coordinates reported are referred to the WGS84 system. The differences reflect the entire period between 1998 and 2001. The lat/long positions are given for 1999; the elevation is elipsoidic.

large amount of ablation occurring even at an altitude of 5350 m a.s.l. The recent variations of Changri Nup Glacier show a fast terminus retreat, comparable to recent shrinkage of glaciers in the Alps. Both Baltoro and Changri Nup are high elevation glaciers with significant accumulation.

The climate of Karakoram is characterised by a dry season and a wet season. The annual mass balance is critical for the position of the ELA. With the observed climate warming, the altitudinal limit of solid precipitation occurs at higher elevations, which results in a reduction of the accumulation area, whereas the ablation area expands. Down-glacier below 5000 m a.s.l. the latent heat flux of rain contributes additionally to ablation. Since 1931 in the Karakoram region the extreme temperatures in the dry environments have increased in general, while the extreme annual rainfall is decreasing (Roohi, 2005). By analysing separately the available meteorological data (in Srinagar, Kashmir, 30 years of daily temperature data, from 1973 to 2003 and the mean monthly temperature values for the decades 1931–1961) in different periods, an increasing air temperature can be observed, which could cause higher melt-water production. Nevertheless the changes of Baltoro Glacier during the last century as described above do not suggest too strong an influence by these climate variations. At Baltoro glacier the ELA is situated close to 5700 m a.s.l. During the expedition in 2004, the accumulation area of Baltoro was not investigated and information on local accumulation is not available. The ELA was calculated using the degree-day factor and the air temperature from Srinagar, Kashmir (Braithwaite, 1984).

If we consider the long-term fluctuation analysed by indirect sources some more can be added to the surge-type events that have occurred. Surging glaciers have been identified only in some particular areas on the Earth (Paterson, 1994), although they exist in a wide range of climatic and morphological situations: from polar to subtropical areas and from maritime to mid-continental areas (Hewitt, 1998a). As Hewitt (1969) suggested, about 90% of known surging glaciers are located in Alaska and in the Karakoram. But, there is definitely more information available on events in North America than there is on events in Asia. Yet, some of the largest advances of this type of glacier have taken place in the Karakoram. In 1953, Kutiah Glacier advanced 12 km in 2 months (Desio, 1954; Hewitt, 1969), whereas Chiring Glacier advanced 15.5 km between 1994 and 1996 (Hewitt, 1998a). While also taking into account the practical consequences of such phenomena, for example blocking off access to yak pastures or to trekking routes as was the case in 1989 with Phumari Chhish Glacier, a tributary of Hispar Glacier (Searle, 1991), the presentation of details for the recent history of a small glacier in the Karakoram, Liligo Glacier, is useful.

The yearly frontal advance calculated for Liligo Glacier during 1986–1997 ($\sim$140 m/y) is lower than the rate of the other known surging glaciers in the Karakoram. Nevertheless, its ice-surface features indicate surge mechanisms. The lack of folded and looped moraines, which are clearly visible in the satellite images of other Karakoram surging glaciers, may be due to the fact that Liligo advances in an ice-free valley and has not flowed into Baltoro for at least a century.

5. Conclusion

The response of DCG to the current warming climate is a long-term negative mass balance. The magnitude, however, is decidedly smaller than for debris-free glaciers. The debris cover has an insulating function once a critical debris thickness is exceeded (Nakawo and Takahashi, 1982; Nakawo and Young, 1981). Baltoro glacier, for instance, the largest DCG chosen as a main field test site, seems to be only slightly changed in respect to its past dimensions. Liligo was advancing, but it might be misleading to interpret this advance in term of direct response to regional climate change, rather than due to an oscillatory change (i.e. surge). On the contrary the signs of a strong shrinkage of the Changri Nup debris-free terminus, connect to the ongoing warming, are indisputable (more than 10 m per year of retreat).

Acknowledgements

This study was carried out within the framework of the Ev-K^2-CNR "Scientific and Technological Research in Himalaya and Karakorum" project and in the framework of the scientific-mountaineering project "K2 2004-50 years later" (Glaciology research group, leader C. Smiraglia). The research conducted was also made possible thanks to the contributions from the IMONT (Italian Mountain Institute) and from the 2005 MIUR project (National leader C. Smiraglia). The mobility of the Italian–German researchers involved in the project has been supported by the Vigoni-DAAD project 2005.

References

Ageta, Y., 2001. Study project on the recent shrinkage of summer accumulation type glaciers in the Himalayas, 1997–1999. Bulletin of Glaciological Research 18, 45–49.

Ageta, Y. and Fujita, K., 1996. Characteristics of mass balance of summer-accumulation type glaciers in the Himalayas and Tibetan Plateau. Zeitschrift für Gletcherkunde und Glazialgeologie 32, 61–65.

Ageta, Y. and Kadota, T., 1992. Predictions of changes of glacier mass-balance in the Nepal Himalaya and Tibetan Plateau: a case study of air temperature increases for three glaciers. Annals of Glaciology 16, 89–94.

Beniston, M., 2003. Climatic change in mountain regions: a review of possible impacts. Climatic Change 59, 5–31.

Bishop, M.P., Kargel, J.S., Kieffer, H.H., et al., 2000. Remote sensing science and technology for studying glacier processes in High Asia. Annals of Glaciology 31, 164–170.

Bishop, M.P., Barry, R.G., Bush, A.B.G., et al., 2005. Global land ice measurements from space (GLIMS): remote sensing and GIS investigations of the Earth's cryosphere. Geocarto International 19 (2), 57–84.

Bozhinskiy, A.N., Krass, M.S., and Popovnin, V.V., 1986. Role of debris cover in thermal physics of glaciers. Journal of Glaciology 32 (111), 255–266.

Braithwaite, R., 1984. Calculation of degree days for glacier climate research. Zeitschrift für Gletscherk. Glazialgeol. 20, 1–8.

Conway, W.M., 1894. Climbing and Exploration in the Himalayas. Fisher Unwin, London.

Conway, H. and Rasmussen, L.A., 2000. Summer temperature profiles within supraglacial debris on Khumbu Glacier, Nepal. In: Nakawo, M., Raymond, C. F., and Fountain, A. (Eds), Debris Covered Glaciers, IAHS 264, pp 89–97.

Dainelli, C. and Marinelli, O., 1928. Spedizione Italiana De Filippi nell'Himalaya,Karacorum e Turchestan Cinese (1913–1914). Serie II—Risultati geologici e geografici. Vol. IV, Zanichelli, Bologna.

De Filippi, F., 1912. La Spedizione di S.A.R. il Principe Luigi Amedeo di Savoia Duca degli Abruzzi nel Karakorum e nell'Himalaya occidentale (1909). Zanichelli, Bologna.

Desio, A., 1954. An exceptional glacier advance in the Karakoram–Ladakh region. Journal of Glaciolio-logy 2 (16), 383–385.

Desio, A., Marussi, A., and Caputo, M., 1961. Glaciological Research of the Italian Karakorum Expedition 1953–1955, International Association of Scientific Hydrology, Publ. 54 (General Assembly of Helsinki 1960—Snow and Ice), 224–232.

Diolaiuti, G., Smiraglia, C., and Pecci, M., 2003. Liligo Glacier (Karakoram): a reconstruction of the recent history of a surge-type glacier. Annals of Glaciology 36, 20–33.

Fujii, Y., 1977. Field experiments on glacier ablation under a layer of debris cover. Journal of Japanese Society of Snow and Ice (Seppyo) 39, 20–21.

Fukita, K. and Sakai, A., 2000. Air temperature environment on the debris-covered area of Lirung Glacier, Langtang Valley, Nepal Himalayas. Debris-Covered Glacier IAHS 264, 83–88.

Haeberli, W., Frauenfelder, R., Hoelzle, M., and Maisch, M., 1999. On rates and acceleration trends of global glacier mass changes. Geografiska Annaler 81a (4), 585–591.

Haeberli, W., 2005. Climate change and glacial/periglacial geomorphodynamics in the Alps: a challenge of historical dimensions. Geografia Fisica e Dinamica Quaternaria VII (Suppl.), 9–14.

Hagg, W. and Braun, L., 2005. The influence of glacier retreat on water yield from high mountain areas: comparison of Alps and Central Asia. In: De Jong, C., Collins, D., and Ranzi, R. (Eds), Climate and Hydrology in Mountain Areas. Wiley, Chichester, pp. 263–275.

Hewitt, K., 1969. Glacier surges in the Karakoram Himalaya (Central Asia). Canadian Journal of Earth Sciences 6, 1009–1017.

Hewitt, K., Wake, C.P., Young, G.J., and David, C., 1989. Hydrological investigations at Biafo Glacier, Karakoram Himalaya: an important source of water for the Indus River. Annals of Glaciology 13, 103–108.

Hewitt, K., 1998a. Glaciers receive a surge of attention in the Karakoram Himalaya. Eos Transactions, American Geophysical Union 79 (8), 104–105.

Hewitt, K., 1998b. Recent Glacier Surges in the Karakoram Himalaya, South Central Asia. American Geophysical Union, http://www.agu.org/eos_elec/97016e.html.

Hewitt, K., 2005. The Karakoram anomaly? Glacier expansion and the 'elevation effect', Karakoram Himalaya. Mountain Research and Development 25 (4), 332–340.

Mayer, A., Lambrecht, M., Belò, C., et al. (2006). Glaciological characteristics of the ablation zone of Baltoro Glacier, Karakorum. Annals of Glacioliology 43, 123–131.

Mihalcea, C., Mayer, C., Diolaiuti, G., et al. (2006). Ice ablation and meteorological conditions on the debris covered area of Baltoro Glacier (Karakoram, Pakistan). Annals of Glaciology 43, 292–300.

Nakawo, M. and Takahashi, S., 1982. A simplified model for estimating glacier ablation under a debris layer. In: Glen, J.W. (Ed.) Hydrological Aspect of Alpine and High Mountain Areas (Proc. Kathmandu Symp., November 1992; Proc. Exeter Symp., July 1982). IAHS 138, 137–145.

Nakawo, M. and Young, G.J., 1981. Field experiments to determinate the effect of a debris layer on ablation of glacier ice. Annals of Glaciology 2, 85–91.

Naito, N., Ageta, Y., Nakawo, M., Waddington, E.D., Raymond, C.F., and Conway, H., 2001. Response sensitivities of a summer-accumulation type glacier to climate changes indicated with a glacier fluctuation model. Bulletin of Glaciological Research 18, 1–8.

Paterson, W.S.B., 1994. The Physics of Glaciers. 3rd edn. Pergamon Press, 480 pp.

Pecci, M. and Smiraglia, C., 2000. Advance and retreat phases of the Karakorum Glaciers during the 20th century: case studies in Braldo Valley (Pakistan). Geogr. Fis. Dinam. Quat. 23, 73–85.

Roohi, R., 2005. Research on global changes in Pakistan. Mountains, Witnesses of Global Changes. Research in the Himalaya and Karakoram: SHARE-Asia Project (abstract book). Rome, 16–17 November 2005, 72–73.

Savoia-Aosta, A. and Desio, A., 1936. La Spedizione geografica italiana al Karakoram (1929)—Storia del viaggio e risultati geografici. Bertarelli, Milano.

Searle, M.P., 1991. Geology and Tectonics of the Karakoram Mountains. Wiley, Chichester.

Wissmann, H.Von, 1959. Die heutige Vergletscherung und Schneegrenze in Hochasien. Abh. Math. Naturwiss. Kl., Akad. Wiss. Lit., Mainz 14, 1101–1407.

Mountains Witnesses of Global Changes
R. Baudo, G. Tartari, and E. Vuillermoz (Editors)

27

Changing climates, changing lives: strengthening adaptive response capacities to climate change in the Huascarán Biosphere Reserve, Peru, and Sagarmatha (Mt. Everest) National Park, Nepal

Alton C. Byers

Abstract

Mountains are particularly sensitive to changes in climate because of their slope, aspect, verticality, mass, and altitude. Global climate change over the past century has resulted in the dramatic recession of glaciers throughout the world, particularly in the subtropical Andean ranges of South America. The rapid melting of snow and ice has resulted in an increase in the formation of high-altitude glacial lakes, sometimes too fast to monitor accurately, accelerating the potential for catastrophic down-valley floods which can destroy everything in their paths. The likelihood of other high-magnitude/low-frequency events such as debris flows and landslides has increased, exacerbated by the very nature of the dynamic mountain environment that is naturally predisposed to earthquakes and mass movement processes. Less understood, however, are the human dimensions of climate change that are already impacting water supplies (irrigation, drinking, power), agriculture systems, high-altitude vegetation dynamics, conflicts over irrigation rights, local economies, adventure tourism (climbing, trekking), and other highland/lowland interactions. Likewise, an understanding of how institutional capacities and partnerships can be strengthened to effectively deal with, and adapt to, these changes is lacking. The mission of The Mountain Institute (TMI) is to conserve high-priority mountain ecosystems, improve mountain livelihoods, and promote the well being of mountain people through research, education, and outreach. This paper discusses an evolving program within TMI that plans to conduct a detailed analysis of the human dimensions of climate change within two of its work regions: the Huascarán Biosphere Reserve, Peru, and Sagarmatha (Mt. Everest) National Park, Nepal. The goal of the project is to better manage natural resources and improve human livelihoods within these regions through:

(1) better understanding of climate change impacts in the Huascarán Biosphere Reserve and Sagarmatha National Park on peoples lives, livelihoods, safety, and environments,

ISSN: 0928-2025 DOI: 10.1016/S0928-2025(06)10027-9

(2) strengthened integration of the social and physical sciences within the climate change research, analysis, project design, and implementation processes, and
(3) strengthened stakeholder capacities to access and use scientific information to better adapt and respond to the risks and vulnerabilities associated with contemporary changes in climate, land use patterns, and tropical high mountain environments.

Mountains Witnesses of Global Changes
R. Baudo, G. Tartari, and E. Vuillermoz (Editors)

28

Chemical composition of fresh snow in the Himalaya and Karakoram

Stefano Polesello, Michele Comi, Licia Guzzella, Angela Marinoni, Massimo Pecci, Claudio Roscioli, Claudio Smiraglia, Gianni Tartari, Paola Teti, Sara Valsecchi and Elisa Vuillermoz

Abstract

The interpretation of firn and ice cores in high-altitude sites in central Asia requires a detailed knowledge of fresh-snow chemistry, especially in the extra-monsoon season. Since 1992 the Water Research Institute of CNR has been involved in sampling and analysis of wet and snow deposition in the Himalayan area. The first campaigns, which were based at the EV-K^2-CNR Pyramid site in Khumbu Valley, were focused on the southern side of the Everest group in monsoon season with the aim of evaluating the long-range transport of inorganic pollutants from the Indian subcontinent to the Himalayan range. In the following years, involving climbing expeditions, we extended our research to other regions outside the monsoon season. This approach allowed us to get a better knowledge of spatial and temporal distribution of major ions in snow deposition of the Himalayan region. Furthermore, our results show that nitrate and ammonium concentrations can be biased by post-depositional gas absorption. In fact the interpretation of nitrate values in glaciochemistry is rather difficult because nitrate concentrations in snow are affected by post-exchange with the atmosphere over a broad range of environmental conditions.

1. Introduction

In the last few years a great deal of research has been devoted to evaluating the impact of anthropogenic pollution on remote areas. The Himalayan region represents an ideal environment for studying the chemistry of remote areas because it is located far from industrialised zones and, at the same time, strongly influences the global atmospheric circulation.

The glaciochemical record contained in the Himalayan glaciers represents a valuable resource that can be used to document the atmospheric deposition of this region and reconstruct its past climate (Wake et al., 1990; Seko and Takahashi, 1991;

ISSN: 0928-2025 DOI: 10.1016/S0928-2025(06)10028-0

Nijampukar et al., 1993; Thompson et al., 1997; Kang et al., 1999; Thompson et al., 2000). Furthermore, the ion content of Himalayan snow packs either provides background measurements of deposition unaffected by local sources of pollution (Mayewski et al., 1983, 1986; Jenkins et al., 1987), or shows the impact of long-range transport of anthropogenic emissions (Steinegger et al., 1993; Xie et al., 1999, 2000). Nevertheless, recovering representative samples requires an understanding of the processes that control the chemical content of snow, the post-depositional effects and the local-to-regional scale variations.

Studies carried out in the eighties were devoted to defining the spatial and temporal distribution of snow chemistry in central Asia (Mayewski et al., 1983, 1986; Wake et al., 1990; Williams et al., 1992), in order to identify different sources of ions in high-altitude areas. These findings concluded that snow chemistry over the Tibetan plateau was dominated by desert dust from the vast arid regions of central Asia, whereas snow from the southern slopes of the eastern Himalaya was characterised by very low ion burdens representative of relatively clean, free tropospheric air (Wake et al., 1993). More recent studies were focused on how well snow chemistry represents the atmospheric chemistry, in order to evaluate the feasibility of using snow pack core data to reconstruct the past atmospheric changes (Kang et al., 1999). This approach required a comparison between aerosol and snow chemistry (Shrestha et al., 1997; Sun et al., 1998), or wet, dry and snow depositions (Valsecchi et al., 1999). These studies showed that snow chemistry was strongly influenced by post-depositional effects such as absorption and scavenging of gaseous species (HCl, HNO_3, NH_3), and it was often not representative of weaker monsoon and local circulation. Furthermore, our previous work (Valsecchi et al., 1999) concluded that the monsoon air influence did not overcome the Ev-K^2-CNR Pyramid Research Laboratory in the Khumbu valley (5050 m a.s.l.), and that background tropospheric levels could be measured above 5500 m altitude, according to findings of Mayewski et al. (1983) in the Ladakh Himalaya.

The third phase of glaciochemical studies in Himalaya was the drilling of ice cores (Thompson et al., 1997, 2000; Kang et al., 2000, 2001; Kreutz et al., 2001). A strong seasonal alternation (with higher ionic concentrations in winter–spring layers) was found in ice layers (Kang et al., 2000) and also confirmed our results on snow chemistry (Marinoni et al., 2001).

According to this international scientific framework, our research group, coordinated by IRSA–CNR, carried out studies on fresh snow in central Asia since the beginning of the nineties. First we carried out spatially extensive and temporal intensive sampling of wet, dry (Valsecchi et al., 1999) and snow deposition in Khumbu valley, at the Pyramid station (Marinoni et al., 2001). The second approach was to use climbing expeditions to collect samples at different altitudes on different mountains, which allows collection of samples from both slopes of some mountains (Balerna et al., 2003) and culminated in the participation of the Italian project for the ascent of K2 and Everest in 2004 (Teti et al., in press).

In the course of the different sampling campaigns, different research groups have been involved and the snow concentrations of different variables have been determined, such as ionic composition, nutrients, light carboxylic acids, trace metals, persistent organic pollutants and in situ radioactivity.

The present work is a short review of our activities in this field carried out in high-altitude regions of central Asia with the aim of deepening the knowledge of the spatial and temporal distribution of snow deposition in relation also to atmospheric circulation.

2. Experimental part

The comparability of different sampling campaigns has been assured by the use of a standardised protocol carried out by trained climbers. When possible, pH and conductivity were measured directly in situ to avoid contamination or carbon dioxide dissolution. Frozen samples were sent to Italy by using dedicated freezers with a continuous record of internal temperature.

Ionic variables and carboxylic acids were analysed by ion chromatography, while total nitrogen and total phosphorus were determined by molecular spectrometry after persulphate digestion, according to Valderrama procedure (Valderrama, 1981). Analysis of organochlorine compounds (OCs) and polycyclic aromatic compounds (PAHs) was carried out by concentrating melted snow on C18 Speedisks (J.T. Baker, USA) and the eluted extract was analysed by GC–MS. Detailed information on analytical procedures can be found elsewhere (Balerna et al., 2003; Teti et al., in press)

3. Results and discussion

3.1. Spatial variability and seasonal effects in fresh-snow deposition

Detailed investigation on snow chemistry has been carried in the Khumbu valley and Everest region, thanks to logistical support from the Ev-K^2-CNR Pyramid research station. Statistical analysis of the full data set collected in this area by principal component analysis clearly showed the presence of two distinct contributions (Marinoni et al., 2001). Looking at the factor loading plot, the first component includes crustal ions (Ca^{2+} and Mg^{2+}) and acidifying species (NO_3^-, SO_4^{2-}, NH_4^+), while the second axis is connected to ion species related to a marine source (Na^+, K^+ and Cl^-). Plotting samples with respect to these two new axes (score plot) shows that samples collected in the summer monsoon period are located on the second axis (i.e. the marine axis), while for extra-monsoon samples the crustal source prevails. Concentration differences between these two periods are evident plotting the mean for selected ions. Concentrations in the summer monsoon are of the same order of magnitude (if not lower) than internal Antarctic snowfall, while in the rest of the year concentrations are similar (if not higher as in the case of sulphate) to those measured in the Italian Alps (Fig. 28.1).

We tried to verify whether concentrations on southern or northern slopes are different. During the summer monsoon there are no differences apart from calcium that prevails on northern slope. The same result was obtained also for the

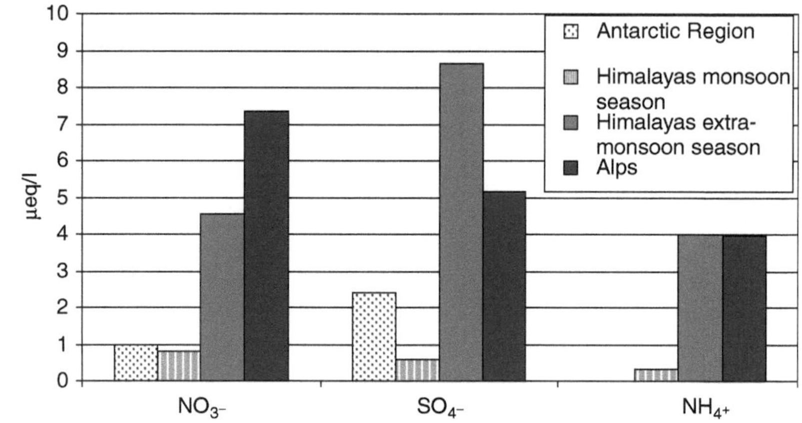

Figure 28.1. Comparison between mean values of the Everest region, during monsoon and extra-monsoon season, Alps and internal Antarctica data sources: *Alps*: Nickus et al., 1998; Polesello et al., unpublished results. *Himalaya* : Marinoni et al., 2001; Balerna et al., 2003; Kang et al., 2004. *Antarctica*: Whitlow et al., 1992; Legrand, 1987.

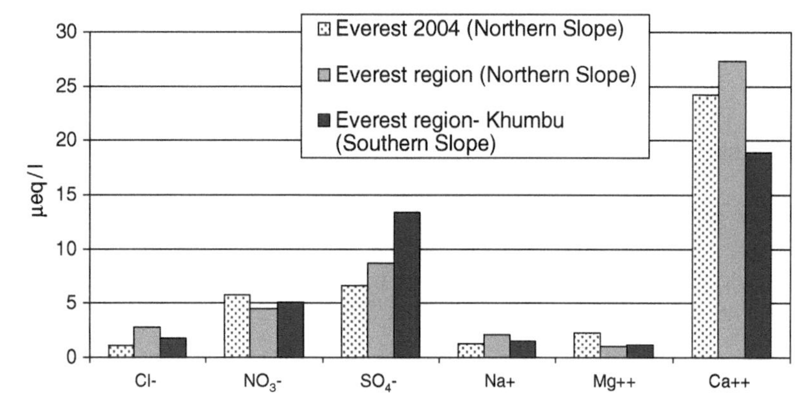

Figure 28.2. Comparison between Everest 2004 campaign and previous campaigns in the same area during extra-monsoon season. Data sources: Jenkins et al., 1983; Marinoni et al., 2001; Balerna et al., 2003.

extra-monsoon period (Fig. 28.2), though on a different concentration scale (up to 30 $\mu eq\,l^{-1}$ for calcium).

From this preliminary data analysis we conclude that there is a substantial homogeneity of main ion concentrations in fresh snow, which does not depend on the geographical location but only on seasonality; and that in pre-monsoon season, the Himalaya is not an effective barrier for dust-storm transportation. In a similar way, the monsoon and late monsoon ionic concentrations, which are one order of magnitude lower than the pre-monsoon ones, are similar on the two slopes of Himalaya, and comparable to those measured in internal Antarctic snowfalls. Nevertheless these conclusions needed to be enforced by collecting data on fresh-snow chemistry on the northern slope of the central Himalaya, which are very scarce in the literature, as well as from other region of the central Asia mountain ranges. This result has also been confirmed by data collected in 2004 on the ascent of Mt. Everest (Fig. 28.2), and it is an interesting confirmation that the inter-annual variability is low. On the other hand, analysis of samples collected almost simultaneously on both slopes of Mt. K2 showed that there are significant differences for all ions, except nitrate and ammonium, between the two slopes.

3.2. Relationship between altitude and concentrations

The use of climbing expeditions allowed us to investigate the relationship between concentrations and altitude. During two ascents a significant number of samples have been collected at different altitudes: (1) on October 2000, samples were collected from Mt. Pumori at seven different altitudes from 5800 to 7160 m a.s.l.; (2) on April–May 2004 snow from six sites at altitudes from 5200 to 7300 m a.s.l. was sampled on the northern slope of Mt. Everest.

Generally, a clear relationship between altitude and concentrations can never be evidenced, because it can be concealed from a widespread of factors, such as analytical and sampling uncertainty, post-depositional effects and other factors. Nevertheless interesting information on atmospheric chemical processes can be inferred.

As an example, in the case of the Pumori ascent, there is no correlation between concentrations and altitude, as shown also by a correlation matrix. But looking at this matrix, we can observe that all analysed compounds, except for oxalate, are perfectly correlated with each other, suggesting that the ions had a common source or had been transported by the same air mass.

On the other hand, samples collected at different altitudes on the northern slope of Everest seem to be correlated with altitude, except from one site at about 7100 m a.s.l., which behaves as an outlier.

The correlation matrix (Fig. 28.3) shows that ions can be divided into two groups, as in the case of the Khumbu valley study (Marinoni et al., 2001). Sodium, chloride and, partially, potassium, connected with a marine source, are inversely but significantly correlated with altitude; while calcium, magnesium and sulphate, which derive from dust transport, showed a less significant but direct correlation with altitude. The concentration increase with altitude of sulphate (Fig. 28.3) can be attributed to the increasing impact of aerosol deposition at higher altitude, possibly connected with Atmospheric Brown Cloud (ABC) phenomena in Central Asia. The global circulation, witnessed by the presence of jet streams throughout the sampling period (Pecci and Mortara, 2005), influenced the chemical composition of snow during this campaign. Ammonium and nitrate are not significantly correlated with other chemical species or altitude, suggesting a prevalent origin by local transport or post-depositional effects on snow. An interesting correlation is observed also between ammonium and oxalate, which can be attributed to a common burning source. This preliminary evidence underlines the importance of the determination of organic acids in fresh snow for understanding the sources and the transport mechanisms of pollutants in remote and high-altitude areas (Chebbi and Carlier, 1996).

EVEREST 27 April- 18 May 2004:
6 sites from 5200 to 7300 m a.s.l.

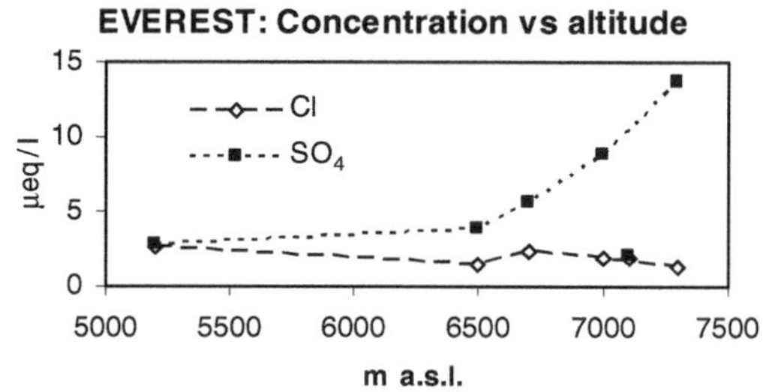

	Cl	NO_3	SO_4	C_2O_4	Na	NH_4	K	Mg	Ca	Altitude
Cl	**1.00**									
NO_3	0.56	**1.00**								
SO_4	0.58	0.55	**1.00**							
C_2O_4	0.13	-0.30	0.22	**1.00**						
Na	**0.97**	0.46	0.68	0.28	**1.00**					
NH_4	0.53	0.37	0.61	0.74	0.60	**1.00**				
K	0.79	0.36	0.36	-0.35	0.74	-0.09	**1.00**			
Mg	0.61	0.60	1.00	0.21	0.70	0.65	0.36	**1.00**		
Ca	0.55	0.68	**0.92**	-0.17	0.59	0.37	0.49	**0.92**	**1.00**	
Altitude	**0.72**	0.31	0.42	0.02	**0.73**	0.10	**0.79**	0.40	0.34	**1.00**

Figure 28.3. Analysis of correlation data for ion concentrations and altitude for the ascent of northern slope of Everest in 2004.

3.3. *Organic compounds and aerosols*

3.3.1. *Carboxylic acids*

Concentrations of carboxylic acids in deposition give information about photochemical reactivity in the atmosphere. Since the nineties we determined them in wet and dry deposition at the Pyramid site (Valsecchi et al., 1999). Mean carboxylic acid concentrations in wet and bulk deposition were about one order higher than concentrations measured in the French Alps, located in densely industrialised areas (Maupetit and Delmas, 1994). High levels of carboxylic acids can be produced by intensive use of wood burning (Chebbi and Carlier, 1996), which is the most relevant energy source for Nepalese people in high-altitude valleys (Davidson et al., 1986). The amount of organic acids in rain can also be increased by an efficient scavenging mechanism because of the neutral to alkaline pH of the samples (Valsecchi et al., 1999). The high level of formate measured in campaigns at the Pyramid site showed that an intense oxidative photochemical activity was present in this remote high-altitude area.

The technology progress in ion chromatographic systems, such as the introduction of an electrolytic eluent generator, allowed us to routinely determine organic acids in fresh snow in the last campaign on Mts. Pumori, Everest and K2. Data on organic acids in this area are available only for an ice core (Kang et al., 2001; Lee et al., 2002a, b, 2003) and are of the same order of magnitude as our preliminary results measured in fresh snow (Fig. 28.4).

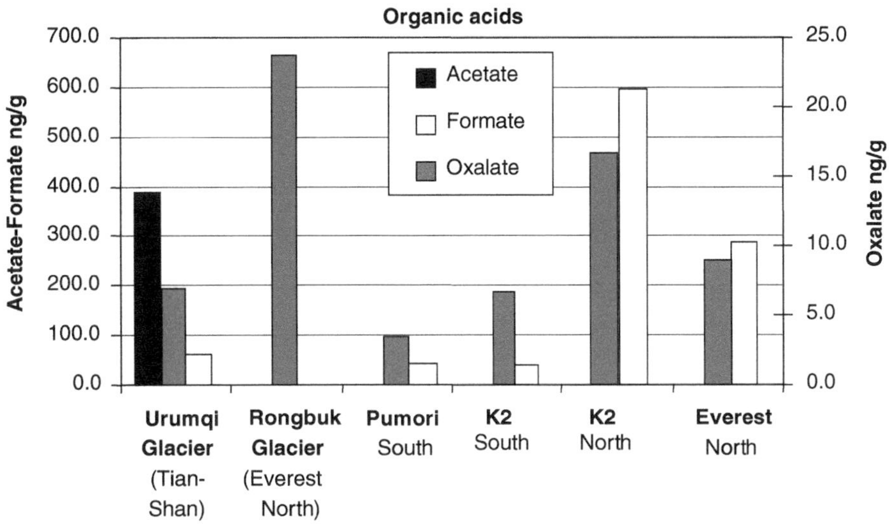

Figure 28.4. Comparison between mean concentrations of organic acids in fresh snow samples and ice cores in central Asia. Data sources for ice cores: Lee et al., 2002a, b, 2003; Kang et al., 2001.

A further important bit of evidence is that concentrations measured on northern slopes are significantly higher than those measured on southern slopes, especially for that which concerns formate. The presence of high levels of organic acids in snow, confirming our previous data on wet deposition (Valsecchi et al., 1999), is correlated with the photochemical activity in the aerosol because the organic acids can be considered the last terminal of oxidation of organic compounds in the atmosphere. These oxidative processes take place both in the hydrometeors or on the surface of air transported dust; but the final products, the organic acids, dissolve in the wet fraction of the aerosol and sink by deposition.

3.3.2. *Organic compounds*

During the last campaign on Everest and K2, we were able to measure also the concentrations of selected persistent organic micropollutants (POPs) in fresh snow at different altitudes. The data are shown in Teti et al. (in press). Two different classes were analysed: OCs, such as p,p′-DDE, a metabolite of the insecticide p,p′-DDT, the insecticides Lindane (γ-HCH) and Methoxychlor, the fungicide HCB whose presence is connected with present or past crop protection use in Asiatic countries; and selected polycyclic aromatic hydrocarbons (PAHs) that are directly related with combustion processes and aerosol transport mechanisms.

The OCs concentrations showed a decreasing trend with increasing altitude, probably caused by volatilisation process; therefore the air dispersion of OC compounds probably prevailed on the recondensation process at an altitude greater than 6500 m a.s.l.. The concentrations of OCs in the K2 northern slope samples were all under the detection limit of the analytical method adopted ($<0.05\,ng\,L^{-1}$), showing that northern K2 slope is less influenced by anthropogenic activities.

Table 28.1. OC and PAH concentrations (ng L^{-1}) in the samples from the Everest northern slope and K2 southern slope.

		EVEREST northern slope (m a.s.l)				K2 southern slope (m a.s.l)		
		6489	6700	7000	7300	5000	6100	6470
OCs	p,p′-DDE	–	–	–	–	0.02	0.01	0.01
	γ-HCH	0.01	0.03	–	–	–	–	–
	HCB	–	0.01	–	–	0.03	–	–
	Methoxychlor	–	–	–	–	1.00	0.02	0.03
	Total	0.01	0.04	–	–	1.05	0.03	0.03
2 r PAH	Naphthalene	2.13	70.0	1.71	24.6	9.77	41.0	92.6
	Acenaphthtene	–	–	–	–	0.54	–	0.07
3 r PAH	Anthracene	–	–	–	–	25.7	–	0.05
	Fluorene	0.07	0.07	0.04	–	0.37	–	–
4 r PAH	Fluoranthene	–	0.10	–	–	–	–	–
	Benzo[a]anthracene	–	–	–	–	–	–	–
	Benzo[k]fluoranthene	98.8	0.48	5.18	–	8.20	–	1.63
	Pyrene	0.04	0.10	0.08	–	–	–	–
PAH	Total	2.25	0.75	7.01	–	18.9	41.00	1.75

As regards PAHs (Table 28.1), the lighter compounds prevailed over the heavier ones. Naphthalene was the main detected pollutant. Naphthalene is commonly found in atmospheric depositions because of its highest volatility among PAHs. As for OCs PAHs with two to three rings had a decreasing trend with increasing altitude, in agreement with the process of their volatilisation (Sonnefeld et al., 1983).

The PAHs with five rings, the heaviest ones, showed a different trend. The high concentrations of benzo[k]fluoranthene in the advanced base camp samples of the Everest's northern slope (6489 and 7000 m a.s.l) and of K2 southern slope (5000 m a.s.l.) can be probably attributed to local pollution sources such as the presence of electricity generator, and the use of fuel combustion. This is also confirmed by the analysis of PAH percentage distribution (Fig. 28.5) that showed a different shape for 6489 and 7000 m a.s.l. snow samples, in comparison with the snow sample collected in a remote area at 7300 m a.s.l. As for OCs, the PAH concentrations in the K2's northern slope samples were all under the detection limit of the analytical method (< 1 ng L^{-1}).

4. Conclusions

The survey on chemical composition of fresh snow in high-altitude regions of central Asia shows that there is a substantial homogeneity of main ion concentrations in the fresh snow, not dependent on geographical location but only on seasonality. This experimental evidence suggests that in pre-monsoon season the Himalayas are not an

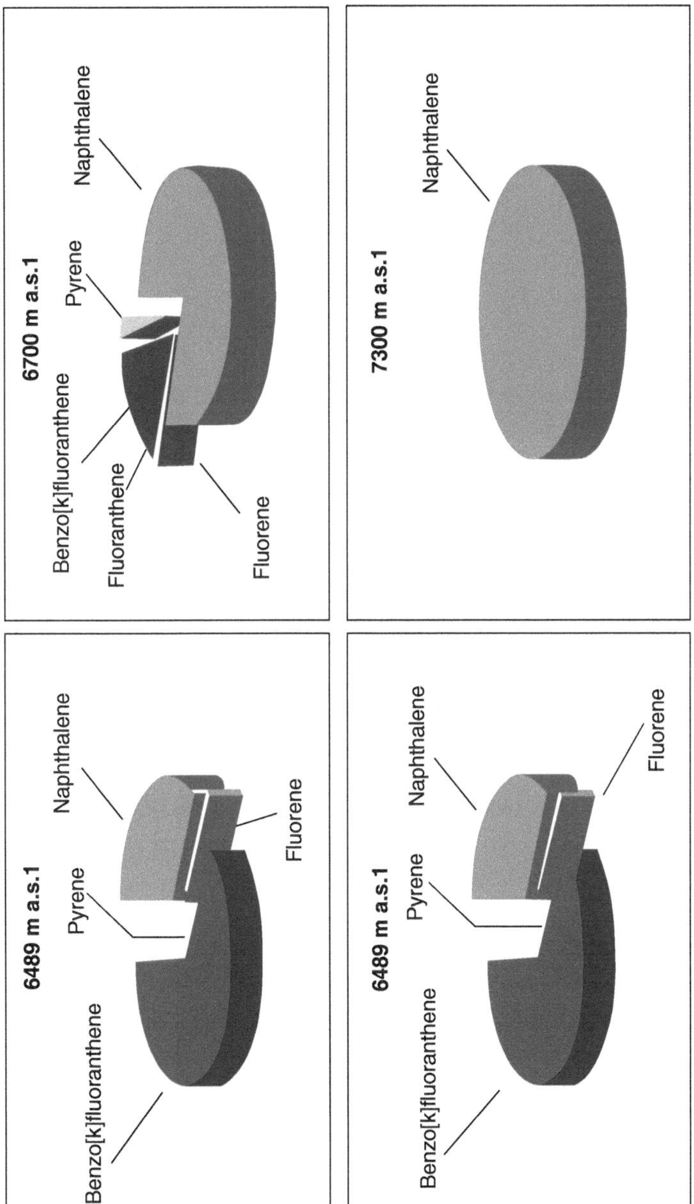

Figure 28.5. PAH distribution in the Everest samples during the 2004 campaign.

effective barrier for dust storm transportation. In a similar way, the monsoon and late monsoon concentrations, which are one order of magnitude lower than the pre-monsoon one, are similar on the two slopes of Himalaya, and comparable to those measured in internal Antarctic snowfall. This comparison proves that summer depositions in high elevation sites in central Asia are not substantially influenced by anthropogenic inputs and may be useful for the investigation of the composition of the remote continental troposphere.

On the other hand in winter and spring seasons the highest concentrations of crustal elements and acidifying species, especially sulphate, are determined. This alternation of concentration levels is confirmed by ice core studies and it is used to date accumulation layers (Kang et al., 2000).

The open issues that need further investigation are the geographical origin of crustal particles and calcium and the identification of the sulphate source, particularly in winter-spring time. The sulphate issue, which was shown to increase with altitude in Everest region, is directly connected to the atmospheric processes taking place in Atmospheric Brown Clouds. As is known, sulphate and organics represents more than 50% of the total aerosol composition in central Asia (UNEP and C4, 2002). Our future aim will be to integrate deposition chemistry with aerosol chemistry, giving complementary information on atmospheric processes involving organic compounds. In fact the study of organic acids can give information on pollutants oxidation pathways and pollution, and help to understand the relationship between aerosol chemistry and snow chemistry.

Acknowledgements

The sampling and analytical campaigns have been carried out in collaboration with Italian research Institutes such as the University of Venezia, Istituto di Ricerca sulla Montagna and other CNR Institutes, and with the essential logistic support of Ev-K^2-CNR Committee. We thank Prof. Laj of the Universitè "Blaise Pascal" of Clermont Ferrand for having given us the opportunity to perform determination of organic acids and formaldehyde in fresh-snow samples.

References

Balerna, A., Bernieri, E., Pecci, M., et al., 2003. Chemical and radio-chemical composition of fresh-snow samples from northern slopes of Himalayas (Cho Oyu range, Tibet). Atmospheric Environment 37, 1573–1581.

Chebbi, A. and Carlier, P., 1996. Carboxylic acids in the troposphere; occurrence, sources, and sinks: a review. Atmospheric Environment 30, 4233–4249.

Davidson, C.I., Lin, S., Osborn, J.F., et al., 1986. Indoor and outdoor air pollution in the Himalayas. Environmental Science and Technology 20, 561–567.

Jenkins, M.D., Driver, J.I., Reider, R.G., and Buchanan, T., 1987. Chemical composition of fresh snow on Mount Everest. Journal of Geophysical Research 92, 10,999–11,002.

Kang, S.C., Qin, D.H., Mayewski, P.A., et al., 2001. Climatic and environmental records from the Far East Rongbuk ice core, Mt. Qomolangma (Mt. Everest). Episodes 24, 176–181.

Kang, S.C., Qin, D.H., Yao, T.D., et al., 1999. Summer monsoon and dust signals recorded in the Dasuopu firn core, central Himalayas. Chinese Science Bulletin 44, 2010–2015.

Kang, S.C., Mayewski, P.A., Qin, D., et al., 2004. Seasonal difference in snow chemistry from the vicinity of Mt. Everest, central Himalayas. Atmospheric Environment 38, 2819–2829.

Kang, S.C., Wake, C.P., Qin, D.H., et al., 2000. Monsoon and dust signals recorded in Dasuopu glacier, Tibetan Plateau. Journal of Glaciology 46, 222–226.

Kreutz, K.J., Aizen, V.B., Cecil, L.D., and Wake, C.P., 2001. Oxygen isotopic and soluble ionic composition of a shallow firn core, Inilchek glacier, central Tien Shan. Journal of Glaciology 47, 548–554.

Lee, X.Q., Qin, D.H., Jiang, G.B., et al., 2003. Atmospheric pollution of a remote area of Tianshan Mountain: ice core record. Journal of Geophysical Research 108 (D14), 4406.

Lee, X.Q., Qin, D.H., Hou, S.G., et al., 2002a. Changes in chemical and isotopic properties near infiltrated cracks in an ice core from Urumqi glacier No. 1, Tien Shan, China. Annals of Glaciology 35, 162–166.

Lee, X.Q., Qin, D.H., Ren, J.W., et al., 2002b. Past 43 year oxalate record: Urumqi glacier No. 1, Tien Shan, China, and its link with far East Rongbuk glacier, Qomolangma (Mount Everest). Annals of Glaciology 35, 285–290.

Legrand, M., 1987. Chemistry of Antarctic snow and ice. Journal de Physique 48, C1,77–C1,86.

Marinoni, A., Polesello, S., Smiraglia, C., and Valsecchi, S., 2001. Chemical composition of fresh snow samples from the southern slope of Mt. Everest region (Khumbu-Himal region, Nepal). Atmospheric Environment 35, 3183–3190.

Maupetit, F. and Delmas, R.J., 1994. Carboxylic acids in high-altitude Alpine glacier snow. Journal of Geophysical Research 99, 16,491–16,500.

Mayewski, P.A., Lyons, W.B., and Ahmad, N., 1983. Chemical composition of a high altitude fresh snowfall in the Ladakh Himalayas. Geophysical Research Letters 10, 105–108.

Mayewski, P.A., Lyons, W.B., Spencer, M.J., and Clayton, J.L., 1986. Snow chemistry from Xixabangma Peak, Tibet. Journal of Glaciology 32, 542–543.

Nickus, U., Kuhn, M., Novo, A., and Rossi, G.C., 1998. Major element chemistry in Alpine snow along a North-South transect in the eastern Alps. Atmospheric Environment 23, 4053–4060.

Nijampukar, V.N., Sarin, M.M., and Rao, D.K., 1993. Chemical composition of snow and ice from Chhota Shigri glacier, central Himalaya. Journal of Hydrology 151, 19–34.

Pecci, M. and Mortara, G., 2005. Everest versante nord: neve e ghiaccio dell'alta quota himalayana. Rivista neve e valanghe, 56, dicembre 2005.

Seko, K. and Takahashi, S., 1991. Characteristics of winter precipitation and its effect on glaciers in the Nepal Himalaya. Bulletin of Glacier Research 9, 9–16.

Shrestha, A.B., Wake, C.P., and Dibb, J.E., 1997. Chemical composition of aerosol and snow in the high Himalaya during the summer monsoon season. Atmospheric Environment 31, 2815–2826.

Sonnefeld, W.J., Zoller, W.H., and May, W.E., 1983. Dynamic coupled/column liquid chromatographic determination of ambient temperature vapour pressures of polynuclear aromatic hydrocarbons. Analytical Chemistry 55, 275–280.

Steinegger, U., Braun, L.N., Kappenberger, G., and Tartari, G., 1993. Assessment of annual snow accumulation over past 10 years at high elevations in the Langtang region, in Snow and Glacier Hydrology. Proceedings of the Kathmandu Symposium, November 1992, IAHS Publ. no. 218, 155–165.

Sun, J.Y., Qin, D.H., Mayewski, P.A., et al., 1998. Soluble species in aerosol and snow and their relationship at Glacier 1, Tien Shan, China. Journal of Geophysical Research 103 (D21), 28,021–28,028.

Teti, P., Vuillermoz, E., Polesello, S., et al., in press. Chemical composition of fresh snow samples synoptically collected on Mount Everest (northern slope) and K2 (northern and southern slopes) during May–July 2004. Proceedings of International Karakorum Conference, Islamabad, Pakistan, 25–27 April 2005.

Thompson, L.G., Yao, T., Davis, M.E., et al., 1997. Tropical climate instability: the last glacial cycle from a Qinghai–Tibetan ice core. Science 276, 1821–1825.

Thompson, L.G., Yao, T., Mosley-Thompson, E., et al., 2000. A high resolution millennial record of the South Asian Monsoon from Himalayan ice cores. Science 289, 1916–1919.

UNEP, C^4, 2002. The Asian Brown Cloud: climate and other environmental impacts. Environmental Science and Pollution Research 9, 289–295.

Valderrama, J.C., 1981. The simultaneous analysis of total phosphorus in natural waters. Marine Chemistry 10, 109–122.
Valsecchi, S., Smiraglia, C., Tartari, G., and Polesello, S., 1999. Chemical composition of Monsoon deposition in the Everest region. Science of the Total Environment 226, 187–199.
Wake, C.P., Mayewski, P.A., and Spencer, M.J., 1990. A review of central Asian glaciochemical data. Annals of Glaciology 14, 301–306.
Wake, C.P., Mayewski, P.A., Zichu, X., et al., 1993. Regional distribution of monsoon and desert dust signals recorded in Asian glaciers. Geophysical Research Letters 20, 1411–1414.
Whitlow, S., Mayewski, P.A., and Dibb, J.E., 1992. A comparison of major chemical species seasonal concentration and accumulation at the South Pole and the Summit, Greenland. Atmospheric Environment 26A, 2045–2054.
Williams, M.W., Tonnesen, K.A., Melack, J.M., and Yang, D., 1992. Sources and spatial variation of the chemical composition of snow in the Tien Shan, China. Annals of Glaciology 16, 25–32.
Xie, S.C., Yao, T.D., Kang, S.C., et al., 1999. Climatic and environmental implications from organic matter in Dasuopu glacier in Xixiabangma in Qinghai-Tibetan Plateau. Science in China Series D – Earth Sciences 42, 383–391.
Xie, S.C., Yao, T.D., Kang, S.C., et al., 2000. Geochemical analyses of a Himalayan snowpit profile: implications for atmospheric pollution and climate. Organic Geochemistry 31, 15–23.

Mountains Witnesses of Global Changes
R. Baudo, G. Tartari, and E. Vuillermoz (Editors)

29

Shrinking cryosphere in South Asia

Syed Iqbal Hasnain

Abstract

In the Indian subcontinent of South Asia, the cryosphere (glaciers and snow cover) provides up to 80% of the lowland dry season flows of the Indus, Ganges and Brahmaputra river system through their vast irrigation networks. Deglaciation is considered to be a global problem; there is a particular concern at the alarming rate of retreat of Himalayan glaciers. The cryosphere retreat is likely to lead a temporary increase followed by reduction in river flows, but the quantity, timing and consequences are unknown. The IPCC report (1995) estimated that a doubled CO_2 level, which was expected sometime around the mid 21st century, would raise average temperature somewhere between 1.5 and 4.5°C. A mass-balance study was started by the initiative of the former ICSI, presently known as the Commission on Cryospheric Sciences, in 2002 at the Chhota Shigri (benchmark) Glacier (32°11′–32°17′ N and 77°30′–77°32′ E), located in the Lahaul and Spiti valley, Himachal Pradesh, India. This glacier is unique as it receives nourishment both by the southwest summer monsoon, as well as the westerlies. This work is being conducted in collaboration with the Great Ice Research Unit of the *Institut de Recherche pour le Développement* from France, and field measurements were carried out during the years 2002–2005. The study based on the 2 years of observations shows a negative mass balance of –1.06 and –1.20 m water equivalent (w.e.) for the year 2002–2003 and 2003–2004 respectively. This has increased slightly in hydrological year 2003–2004 as compared to hydrological year 2002–2003. The glacier is likely to become thinner at lower altitudes, as there is increased negative net mass balance per year. The equilibrium line altitude (ELA) is also shifting upward and stays at an interval of 4800–5100 m, thus reducing the accumulation area. The results clearly indicate the impact of warming in the region. The work is still continuing and long-term mass balance measurements will show the climate signals affecting the cryosphere in South Asia.

1. Introduction

The Hindu Kush–Himalaya represents a critical region in terms of glacial melt-water contribution to irrigation network for millions of South Asians. The region lacks

ISSN: 0928-2025 DOI: 10.1016/S0928-2025(06)10029-2

reliable quantitative estimates of ice volume. The regional runoff model developed under the Sagarmatha Project shows distinct differences in the potential impacts of deglaciation in the east–west direction along the Himalayan arc, and within individual catchments (Rees et al., 2005).

As global climate changes, there is an accordant trend towards global recession and wasting of glaciers. Several analyses show that it is not only increased temperature and/or decreased precipitation that are responsible for these retreats, but also changes in air humidity (Kaser et al., 2002).

Glacier mass balance is the link between climate and glacier dynamics. Mass losses affect the local hydrology because mass is lost generally through melt-water runoff. Thus the prediction of mass balance changes is also a prediction of their hydrological effect, which is important for regional water supplies. In the present study an attempt has been made to check climate signals by monitoring mass changes. The sensitivity of the equilibrium-line altitude (ELA) is also discussed.

2. Area

Chhota Shigri Glacier lies on the Chandra–Bhaga river basin on the northern ridge of Pir Panjal range in the Lahaul–Spiti valley of Himachal Pradesh, India. It is included in the upper basin of the Chandra River, contributing to the Chenab River, one of the components of the Indus basin. It is located about 3 km (glacier terminus) south of Chhota Darra across the left bank of Chandra River (the main river of this region), is trending about N–S to NNE to SSW. This glacier extends between 32°11′–32°17′ N and 77°30′–77°32′ E and occupies an elevation of 4100 to more than 6000 m a. s. l. It extends for about 9 km up to Sara Umga Pass and the width varying between a few metres to about half a kilometre in central portion as shown in Fig. 29.1.

Alternatively influenced by the southwest monsoon during the summer, and subject to the westerlies in the winter, the glacier offers a complex accumulation/ablation regime. The glacier falls in the monsoon-arid transition zone; therefore this glacier is considered to be a potential indicator of the northern limits of the intensity of the monsoon (Krenek and Bhawan, 1945).

The total drainage area of the Chhota Shigri basin is about 45 km^2, with a glacierised area of about 10 km^2 and the glacier occupying about 20% of the drainage area (Dobhal et al., 1995). It is a valley glacier and the shape is complex; including two main flows in the central part and several small-suspended sub-glaciers in the accumulation zone. The lower portion of the glacier is covered by debris due to the maximum secondary weathered (mechanical) effects caused by sudden variations of climatic changes. The thickness of the glacier ice varies from 5 to 80 m between the snout and accumulation zone; its average thickness is about 55 m (Nijampurkar and Rao, 1992).

3. Climate

The climatic records of the region are not available but the nearest weather station at Keylong records a maximum temperature of not exceeding 24°C. The region is

Figure 29.1. Aster 123 image of Chandra Valley's glacierised basins showing the location of Chotta Shigiri (benchmark) Glacier, Indian Himalaya.

mainly characterised by the cold winter extending from October to April. The short-term meteorological observations (July to September) on the glacier during 1987–1989 showed a temperature ranging between 10.5°C and −5.2°C at an elevation of 4600 m a.s.l., while 16–4°C near the snout (Dobhal et al., 1995). The main valley (3000–7000 m deep), in which the glacier is situated, is dry. The annual precipitation on the glacier is 150–200 cm of snow (~600 kg m^2 year^{-1}). The average environmental lapse rate on Chhota Shigri Glacier remained pseudo-adiabatic on most days during the summer and varied from 0.38–0.67°C/100 m (Bhutiyani and Sharma, 1989).

4. Methodology

The mass balance study was started by the initiative of former International Commission of Snow and Ice (ICSI), presently known as Commission on Cryospheric Sciences, in 2002 (September–October) at the Chhota Shigri (benchmark) Glacier. Since then the work is being continued in collaboration with *Directeur de Recherche* (IRD), France and the Society for Himalayan Glaciology, Hydrology, Ice, Climate and Environment (HIGHICE–India).

In this respect the glacier has been revisited and measurements have been carried out for the old stakes, and new stakes were installed every year. During the field expedition from September 18 to October 10, 2003, the stake positions of 2002 were re-measured with the help of differential GPS and melting was observed through the exposed-stake measurements. The new stakes were installed on the glacier and the number has been increased to 22 from 14 in order to increase the stake density for better accuracy of the mass balance. Accumulation pits were dug to know the yearly accumulation. The hydrological measurements were also performed down stream from the snout.

A field campaign during September 14–28, 2004 was organised to revisit the glacier for the measurement of 2003 stakes and to put the new stakes on the glacier. The density of stakes has been increased further to 27 as compared to 22 in the year 2003. Three accumulation pits were dug to know the yearly accumulation at nearly the same place where the last year's observations were performed.

The current status of stakes installed in September 2004 is presented on the topographic map of the Chhota Shigri Glacier. This clearly shows the distribution of stakes on the glacier as shown in Fig. 29.2.

5. Results and analyses

After the re-measurement of the 2003 stakes in 2004 with the help of differential GPS in both years, we are able to calculate the horizontal shift of the glacier. The result so obtained is represented graphically (Fig. 29.3). It is observed that the upper ablation stakes (01–09) of the glacier shows a displacement of more than 35 (35–47) m per year while the lower ablation stake shows a displacement between 25 and 30 m per year. In the accumulation area, three snow/firn pits were dug in order to obtain information on yearly accumulation of snow. The only recognisable changes in the stratigraphy were a few slight differences in the size of the firn grains and few thin ice layers. The ice layers were clear at some places but with dirt at other places.

Density measurements are performed to get the net annual accumulation of the site. Density distributions for all the three pits are presented for the years 2003 and 2004 in Figs. 29.4 and 29.5 respectively. Very little change in the density profile was observed for the first two pits at the elevation of 5200 and 5405 m a.s.l. in the year 2003. The density first increases and then decreases with depth. The third pit (year 2003) at an elevation of 5500 m shows a large variation in the density profile increasing with depth and reaching 0.86 g cm^{-3} (glacier ice) at the depth of 300–338 cm.

In the year 2004, the pit at 5180 m a.s.l. shows clean ice with bubbles at the depth between 25 to 165 m showing a density of 0.79 g cm^{-3} and thereafter decreased to 0.58 g cm^{-3} up to the depth of 220 m. The two upper elevation pits (5400 and 5500 m a.s.l.) are showing the same trend and density reaches up to 0.8 g cm^{-3} at several depths as the ice layers were observed.

At every point on a glacier, there is a specific net mass balance. In general, this is positive at higher altitudes; whereas accumulation is more and negative at lower altitudes where the ablation is greater than the accumulation. The net balance of a particular glacier may vary sharply from one year to another. In order to obtain an accurate picture of glacier changes, mass balance programmes therefore need to be continued for many years. The shape, or geometry, of a glacier may have a significant effect on its mass balance.

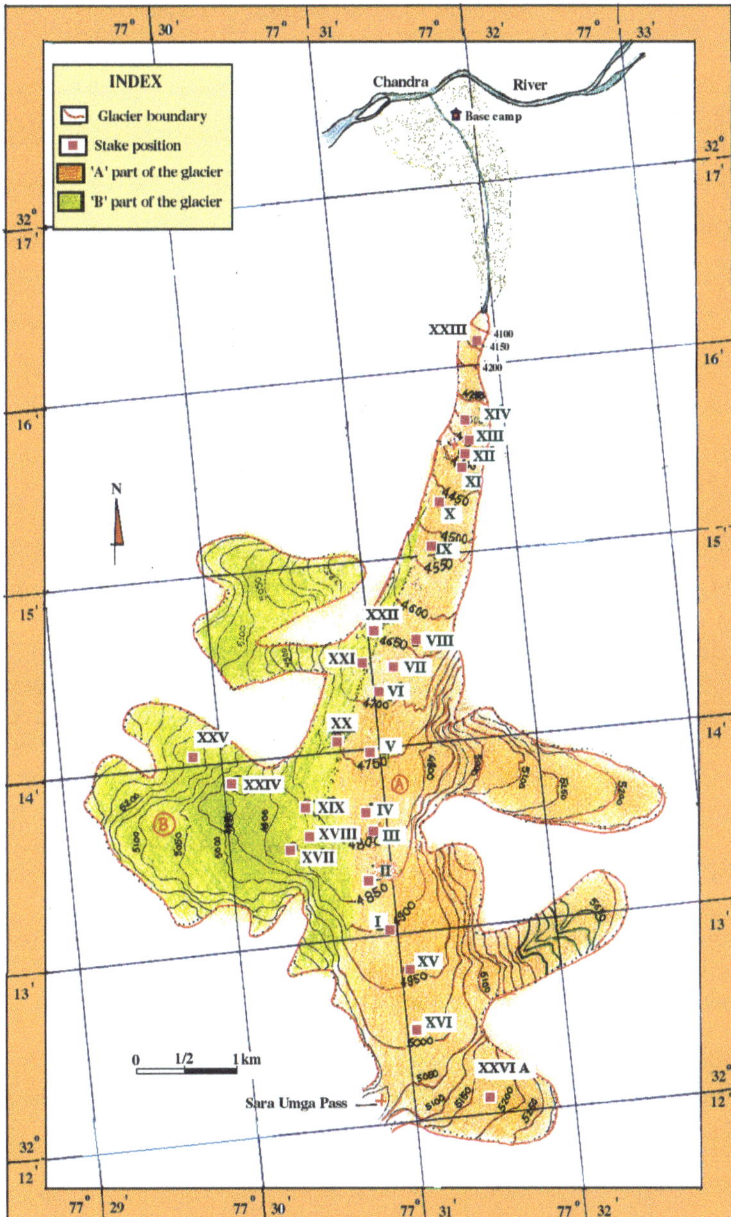

Figure 29.2. Location of stakes installed on the Chhota Shigri Glacier during September 2004.

The glacier's annual net mass balance is calculated from the results of measurements of the winter and summer balances. The surface area of the glacier is divided into altitudinal zones based on the contours of 50 m difference. The average accumulation and ablation in each zone are calculated from the differential exposure of

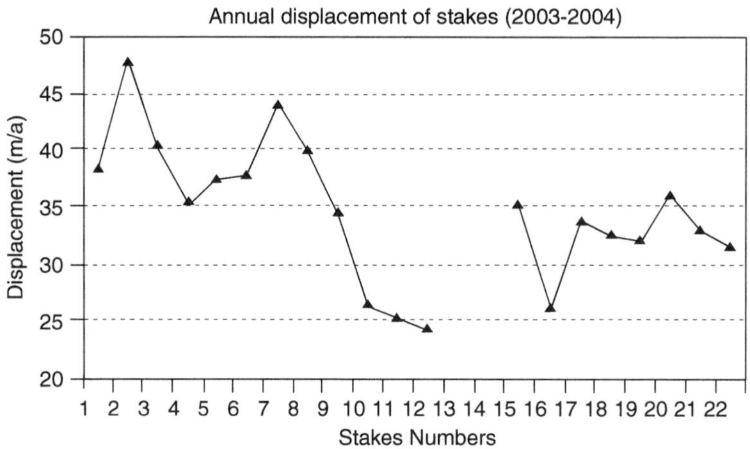

Figure 29.3. Annual displacement of stakes on Chhota Shigri Glacier.

the stakes in one year. The surface area of each altitudinal zone used in mass balance calculations is based on a survey of the map of India.

On the basis of the 2 years of the stake measurements (2002–2003 and 2003–2004), the specific net mass balance is calculated at every point of measurement and presented in a graphical form (Figs. 29.6 and 29.7) against the representative elevations of the stakes. The glacier melt increases from higher altitude towards lower altitude from 4917 to 4370 m a.s.l. and abruptly the melting reduces for two stakes in the lower ablation zone. This sudden decrease of melting may be due to the reduction in sunshine hours at this region because of narrowing of the valley observed during the field campaign. This reduced ablation may also be a result of the debris-cover zone just above the snout region.

The various positions of the snout line monitored during 1987–1989 (campaign of the Department of Science and Technology, New Delhi) are an indication of the annual climatic variation on the glacier and have been accompanied by three main episodes of advance and retreat. Fluctuations of the equilibrium line, observed during the same period, support the above observations. The snout line of the glacier continued to recede at a rate of 18.7 m year^{-1} during 1986–1988. This retreat was accompanied by a negative mass balance observed during 1987–1988 (Nijampurkar and Rao, 1992).

6. Discussion and conclusions

Our study based on the 2 years of observations (2002–2003 and 2003–2004) shows a negative specific mass balance of −1.06 and −1.20 m w.e. for 2002–2003 and 2003–2004 respectively. This has increased slightly in hydrological year 2003–2004 as compared to hydrological year 2002–2003. The glacier is likely to become thinner at lower altitudes, as there is increased negative net mass balance year to year.

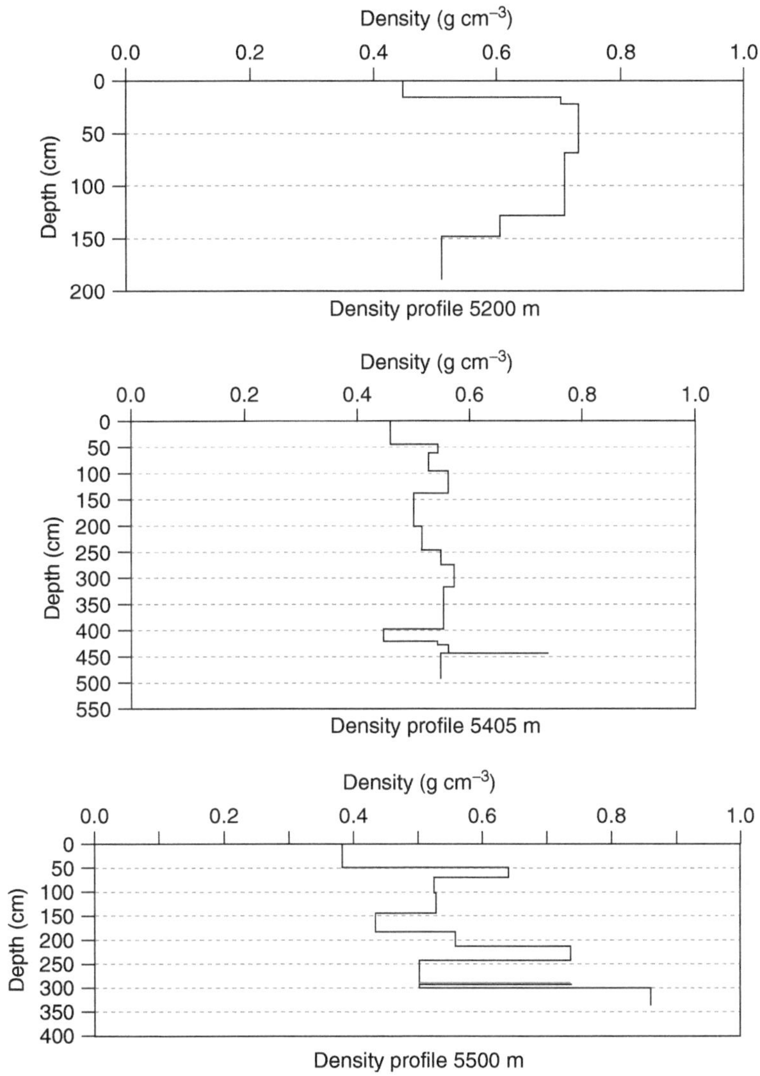

Figure 29.4. Density distribution in the pits on October 7, 2003 at different elevations.

At the end of the summer, the equilibrium line marks the position at which summer ablation just equals the accumulation of the preceding winter. The equilibrium line altitude (ELA) is the most suitable parameter for interpreting glacier responses to climate variations. The sensitivity of the ELA to climate variations depends on the gradients of the mass balance terms at the respective altitudes. In order to study the sensitivity of the equilibrium lines under different conditions, their mean positions in the respective vertical profiles of the specific mass balances have to be determined. Frequently, there is a clear relationship between the ELA and the net mass balance of the glacier.

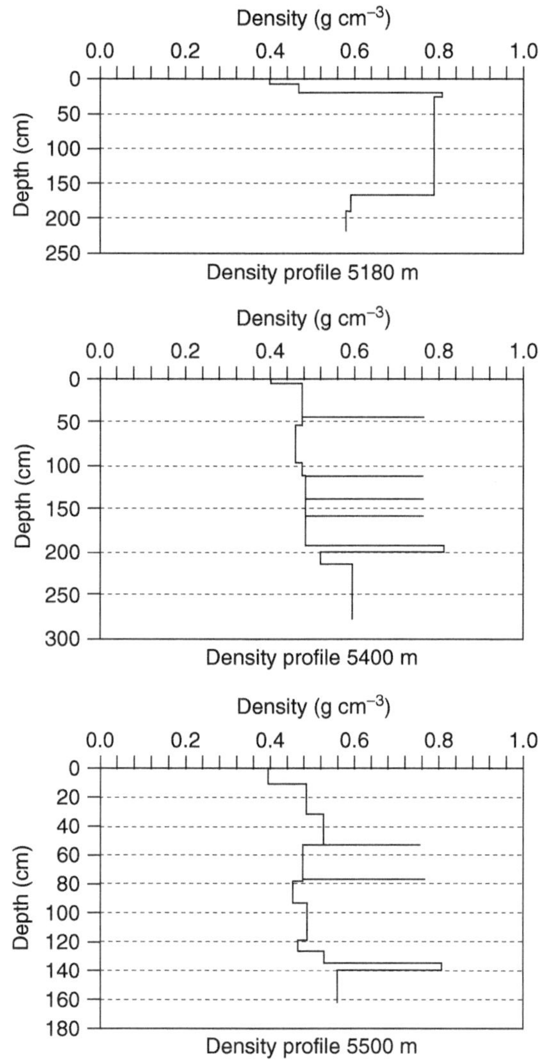

Figure 29.5. Density distribution in the pits on September 22, 2004 at different elevations.

The ELA in year 2003 was 5050 m a.s.l, whereas it stayed a little higher at 5100 m a.s.l in 2004. The accumulation area ratio (AAR) has also decreased from 0.40 in 2003 to 0.31 in 2004 indicating that the glacier is melting more, which is also evident from the calculated net mass balance of the glacier for the years 2002–2003 and 2003–2004.

Global-warming impacts are clearly visible on the benchmark glacier as evident through the increase in the net loss of mass year by year. The glacier is likely to become thinner at lower altitudes, as there is increased negative net mass balance. The ELA in 2003 was 5050 m a.s.l and has gone up to 5100 m in 2004. The decrease in

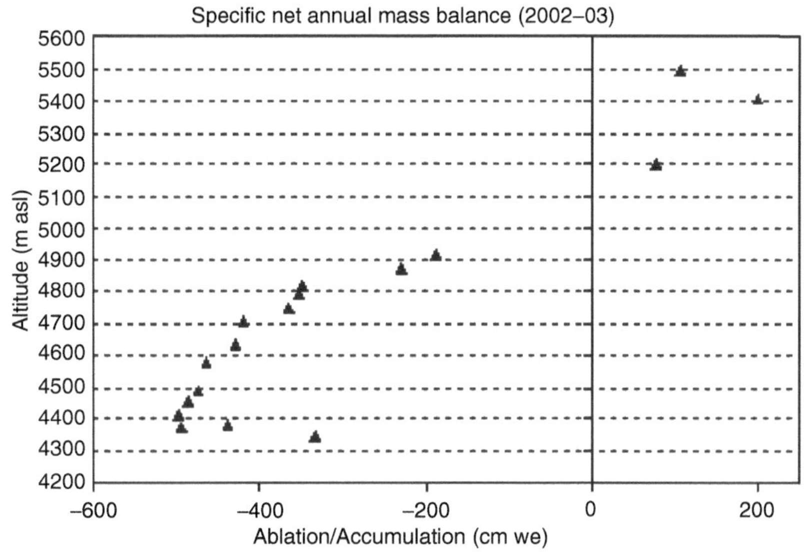

Figure 29.6. Specific net annual mass-balance (2002–2003) near measurement points averaged against elevation on the Chhota Shigri Glacier.

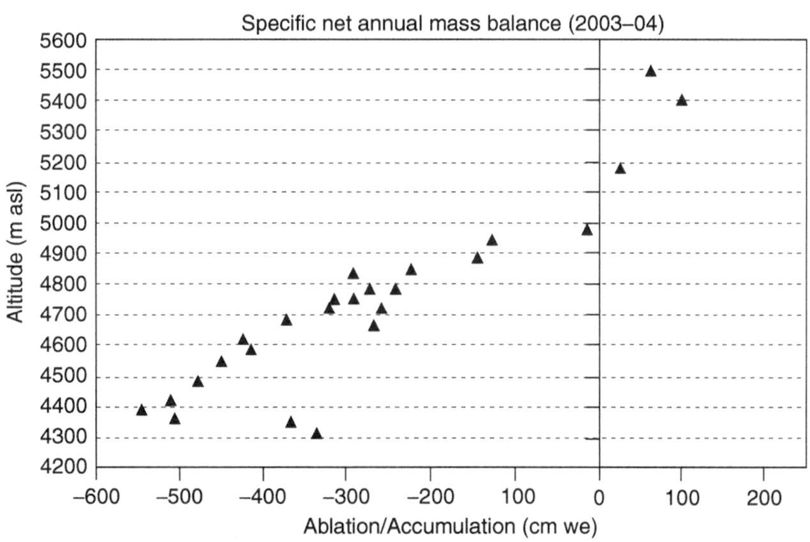

Figure 29.7. Specific net annual mass-balance (2003–2004) near measurement points averaged against elevation on the Chhota Shigri Glacier.

AAR value in 2 years gave the glacier a further threat for more melting as the accumulation area has reduced and the ablation area has increased in 2 years of observation. This is also evident from the calculated net mass balance of the glacier.

Acknowledgement

The author is grateful to Mountaineering and Allied Sports, Manali, Himachal Pradesh for their support during the field campaigns.

References

Bhutiyani, M. R. and Sharma, M. C., 1989. A report on the glaciological studies carried out on Chhota Shigri glacier. Technical Report of the Multi-disciplinary Glacier Expedition to Chhota Shigri Glacier, Department of Science and Technology, New Delhi, Report No. 3, pp. 203–236

Dobhal, D.P., Kumar, S., and Mundepi, A.K., 1995. Morphology and glacier dynamics studies in monsoon-arid transition zone: an example from Chhota Shigri glacier, Himachal-Himalaya, India. Current Science 68 (9), 936–944.

Kaser, G., Fountain, A., and Jansson, P., 2002. A manual for monitoring the mass balance of mountain glaciers. Technical Document in Hydrology, IHP-VI, No. 59, 107 pp.

Krenek, L. and Bhawan, V., 1945. Recent and past glaciation of Lahaul. Indian Geography Journal 3, 93–102.

Nijampurkar, V.N. and Rao, D.K., 1992. Accumulation and flow rates of ice on Chhota Shigri glacier, central Himalaya, using radio-active and stable isotopes. Journal of Glaciology 38 (128), 43–50.

Rees, H. G., Collins, D. N., Shrestha, A. B., et al., 2005. An assessment of the potential impacts of climate change induced glacier retreat on Himalayan river flows. Technical Report (unpublished), Snow and Glaciers Aspects of Water Resources Management in the Himalayas, DFID KAR Project No. R7980, U.K., 54 pp.

SHARE-Asia Scientific Partners: Commitments to High-Altitude Environmental Monitoring in Asia

Mountains Witnesses of Global Changes
R. Baudo, G. Tartari, and E. Vuillermoz (Editors)

30

The third pole of the planet: the Mountain Research Initiative

Gregory B. Greenwood

Abstract

The Mountain Research Initiative (MRI) is funded by the Swiss National Science Foundation to promote global change research in mountain regions around the world. Mountains are very sensitive to global change, especially climate change. A great proportion of humanity depends on resources from mountain regions, especially water. MRI established a research agenda for mountains through its Global Change in Mountain Regions (GLOCHAMORE) Project funded by the European Union and focused on Man and the Biosphere (MAB) Biosphere Reserves nominated by UNESCO throughout the world. The perspectives of MRI and SHARE-Asia are complementary and may intersect in the inclusion of Mountain Biosphere Reserves (MBRs) in the SHARE-Asia network.

1. Introduction

The Mountain Research Initiative (MRI) is a joint network of the International Human Dimensions Programme (IHDP) and the International Geosphere–Biosphere Programme (IGBP) funded by the Swiss National Science Foundation, but global in scope. The MRI focuses on Global Change in Mountain Regions (GLOCHAMORE) around the world. While not all of these regions have nival zones, many do, such as the American Cordillera and the Hindu Kush–Himalaya. For this reason, MRI can be considered as focusing on the Third Pole of the planet, a pole with all the sensitivities on the other two, but unlike the others, located among, or near, a significant faction of humanity.

2. The MRI itself

While the role and importance of mountains in global change research was manifested with IGBP from the early 1990s, the MRI took concrete conceptual form with the publication in 2001 of Report 49 of the IGBP (Becker and Bugmann, 2001). MRI differs from many (but not all) of the global change projects supported by the Earth Systems Science Partnership in that it focuses on a place, as opposed to an

ISSN: 0928-2025 DOI: 10.1016/S0928-2025(06)10030-9

element or a system. As such, the MRI incorporates both biophysical and socio-economic aspects of global change. MRI is in a certain sense the new Global Lands Project for mountains, as GLP, the successor to the Land Use and Cover Change Project, has been specifically constructed as a project that takes a whole system view of the land system, including both human and biophysical aspects. This linkage led as well to endorsements by the Global Terrestrial Observing System (GTOS), with its emphasis on observations, and the Man and the Biosphere Program (MAB), with its emphasis of the interaction between people and the biosphere.

The Coordination Office of MRI in Bern, Switzerland, has been supported by a grant from the Swiss National Science Foundation running from 2004 to October 2007. This grant is an expression of both Swiss scientific interest in global change and in mountain research, and of Swiss foreign policy in support of international scientific cooperation and of mountain regions throughout the world.

The MRI has been involved in scientific synthesis, exemplified by the recent Springer publication "Global Change and Mountain Regions: An Overview of Current Knowledge" (Huber et al., 2005). MRI has also been involved in promoting new research, most particularly in conjunction with the MAB program through the implementation of the GLOCHAMORE project, funded by the European Union (EU), which establishes a research strategy on global change for MAB Biosphere Reserves around the world. Biosphere Reserves are not areas off-limit to human use but are in fact designed as sites for the exploration of sustainable development. The project started in 2003 and ended in December 2005 with the creation of its research strategy (GLOCHAMORE, 2006).

3. The importance of mountains in global change research

Mountains constitute an important nexus in global change research. First, both model output and observations show that climatic changes will be amplified in mountain regions (Bradley et al., 2004). Mountains are marginal environments, highly sensitive to change. Changes wrought in mountains often affect areas far beyond the mountain regions themselves. Significant gaps in our knowledge, however, hinder our ability to assess changes in mountain regions and to develop adaptation and mitigation strategies.

Water supply from mountains illustrates these points. During the International Year of the Mountain, mountains were correctly touted as the "water towers of the planet." Viviroli et al. (2003) looked at the proportions of flow in different river systems that are due to discharge from mountains (Fig. 30.1). They classified river basins into four classes. In the first class are rivers in which essentially all of the flow is attributable to discharge from mountains. This class includes some major world rivers: the Colorado in the western North America, the Amu Darya in Central Asia, and the Nile in Africa.

In the second set, discharge from mountains accounts for 50–85% of annual discharge, and during periods of low water, mountain discharge accounts for 100% of the flow in the river. This class includes the rivers in Mesopotamia, the Indus in South Asia, the Senegal and the Niger in Africa. Even in the third class, which

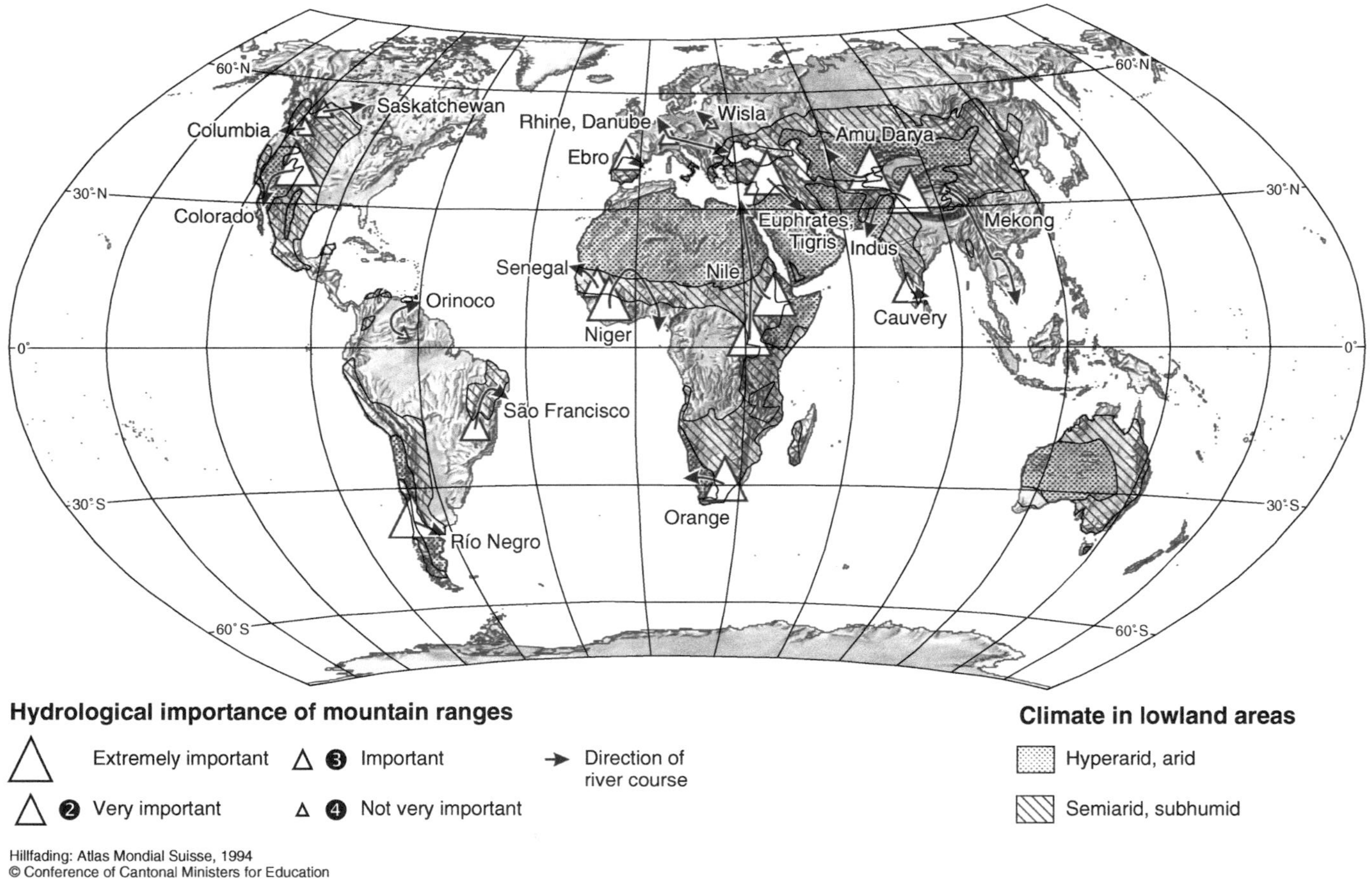

Figure 30.1. Mean annual proportion of total discharge attributable to mountains and proportion of total catchment area covered by mountains (*, arid and semiarid areas; °, humid areas). The vertical lines denote the maximum and minimum monthly amounts of discharge attributable to mountains. From Viviroli et al. (2003) and reproduced with permission of IMS and UNU.

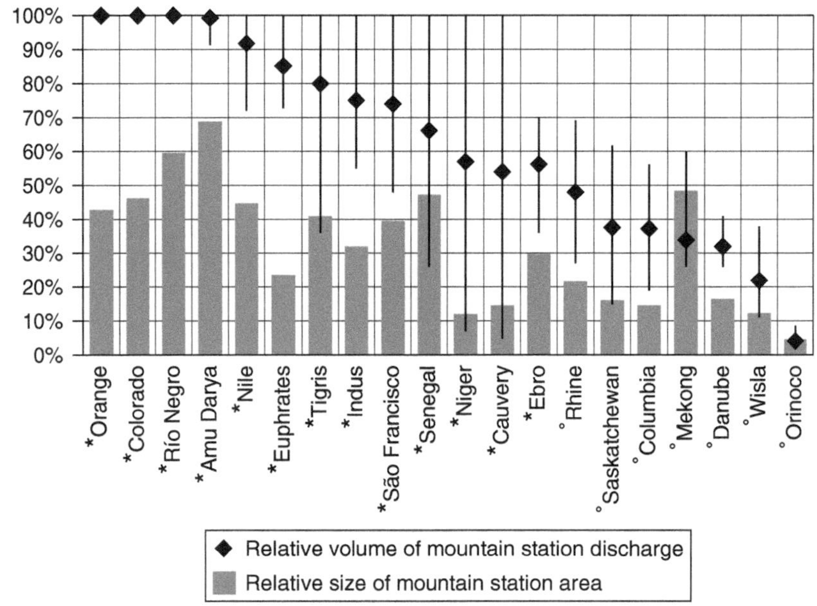

Figure 30.2. Hydrologic significance of mountain ranges for lowland river basins as studied by Viviroli et al. (2003). From Viviroli et al. (2003) and reproduced with permission of IMS and UNU.

includes the Rhine in Europe and the Columbia in North American, mountains contribute between 30 and 60% of the total flow, in many cases nearly double that of the area occupied by the mountains within the watershed. At a global level, and outside of the tropics, mountains generate nearly double the amount of water that one would expect based on the area they occupy within the watershed.

The fate of a large fraction of humanity is tied to these rivers. Many of the rivers in the first and second classes flow through arid regions between 30° N and 30° S, where roughly 70% of the world's population lives (Fig. 30.2).

3.1. Global change in mountain regions (GLOCHAMORE)

The GLOCHAMORE project, funded by the EU 6th Framework Programme, strove to establish a framework for long-term research efforts by taking advantage of the infrastructure and ongoing research activities in UNESCO MAB's Mountain Biosphere Reserves (MBRs) in European countries with the explicit goal of implementing the strategy in mountain Biosphere Reserves around the world, in both developed and developing countries.

The GLOCHAMORE project focused on 27 MBRs located on 6 continents. All UNESCO MBRs are protected areas, where core zones of more strict preservation are buffered by zones in which sustainable land uses are promoted. UNESCO Biosphere Reserves are structured to fulfil three functions: (1) conservation, (2) development, and (3) logistical support function. The latter is critical and is designed "to provide support for research, monitoring, education, and information exchange related to local, national and global issues of conservation and development."

To make realistic, feasible proposals for global change research in mountain regions, GLOCHAMORE queried the managers of the MBRs on their perspective of global change and its likely impact on their reserves. Specifically GLOCHAMORE asked

- Which different groups care about the reserves and why?
- Which resources and which interest groups are likely to be affected by global change, especially climate change as it affects the reserve?
- Will climate change exacerbated any existing or created any new natural hazards?
- What are the contentious scientific issues associated with your reserve?

The responses of the managers (Greenwood, 2005) complemented the understanding of the research scientists invited to participate in GLOCHAMORE workshops. While scientists had already identified water supply and forests as important issues, the reserves added active recreation, tourism, and grazing as equally important issues to be addressed by research. Among the second tier of issues, managers added fires, governance, and endangered species issues to the issues already identified by scientists: geologic hazards, floods, and snow cover. This inquiry led not only to a more intellectually complete research strategy, but also to one that addressed issues important to managers, thereby gaining their support for implementation.

4. Relevance to SHARE-Asia

The Mountain Research Initiative is focused principally on the development of global change research programs that can, in the first instance, be implemented through existing infrastructure, such as that offered by MBRs. SHARE-Asia appears to focus principally on the enhancement of infrastructure, specifically, high-elevation observatories in the Hindu Kush–Himalaya. The perspectives of the two programs are thus complementary: some of the MBRs may be good sites for the implantation of high-elevation observatories (e.g., Nandi Devi BR in India) while the sustainability and value of the observatories may be enhanced by anchoring the network within a research program that certainly includes observations but also considers deeply the needs of planning for impact and for ultimately addressing needs of sustainable development.

References

Becker, A. and Bugmann, H., 2001. Global change in mountain regions: the Mountain Research Initiative. International Geosphere–Biosphere Program Report 49, Stockholm.

Bradley, R.S., Keimig, F.T., and Diaz, H.F., 2004. Projected temperature changes along the American Cordillera and the planned GCOS network. Geophysical Research Letters 31, L16210, doi:10.1029/2004GL020229.

GLOCHAMORE, 2006. Global change and mountain regions research strategy. Mountain Research Initiative, Zürich.

Greenwood, G., 2005. What are the important global change themes and issues in mountain biosphere reserves? Projecting Global Change Impacts, and Sustainable Land Use and Natural Resources Management in Mountain Biosphere Reserves. UNESCO, Paris, pp. 179–194.
Huber, U.M., Bugmann, H.K.M., and Reasoner, M.A. (Eds.), 2005. Global change and mountain regions: an overview of current knowledge. Springer, Dordrecht, pp. 650.
Viviroli, D., Weingartner, R., and Messerli, B., 2003. Assessing the hydrological significance of the world's mountains. Mountain Research and Development 23 (1), 32–40.

Mountains Witnesses of Global Changes
R. Baudo, G. Tartari, and E. Vuillermoz (Editors)

31

Global changes and sustainable development in the Hindu Kush–Karakoram–Himalaya

Bidya Banmali Pradhan and Basanta Shrestha

Abstract

The Hindu Kush–Karakoram–Himalayan (HKKH) region, stretching 3500 km over eight countries, from Afghanistan in the west to Mynamar in the east, is home to more than 150 million people and affects the lives of three times as many in the plains and river basins below. The region is not only the world's highest mountain region, but also its most populous. The wealth of the HKKH lies in an immense diversity of flora and fauna, as well as ethnic groups and languages. It is also an important source of water, energy, and biological diversity. Yet despite this rich diversity, in reality, the HKKH regions are exceedingly fragile. Each day, climate change, pollution, as well as exploitative mining and unsound agriculture practices, take a toll on mountain environments and the most vulnerable to these changes are the inhabitants of this region. Already, they are among the world's poorest, hungriest and most marginalized people. Sustainable development of these mountain regions is a challenging task because these areas have highly diverse and fragile ecosystems. The HKKH region is geologically the youngest mountain range giving rise to a high degree of natural hazards. Specific information on ecology, natural resource potential, and socio-economic activities is essential for sustainable development of this region. There is, however, a lack of sufficient knowledge of mountain ecosystems for the reasons of understanding of mountain specificities, the effort of International Center for Integrated Mountain Development (ICIMOD) has been to establish itself as a hub for mountain-specific knowledge, which will help in sustainable development. Among several programs ICIMOD is contributing to two important programs related to climate change: (1) glaciers and glacial lakes; and (2) trans-boundary air pollution. Glaciers and glacial lakes are the repositories of information for exploring Quaternary climate changes, as they remain sensitive to global temperature conditions. Rising temperature trends with climate change are much more pronounced in the higher altitudes causing the net shrinkage and retreat of glaciers, and the increase in size and number of glacial lakes. A number of glacial lake outburst floods (GLOFs) have been reported in the region in the last few decades and many potential threats of more GLOF in the HKKH region have been identified, which may pose high death tolls in downstream populations, as well as destruction of property and infrastructure.

ISSN: 0928-2025 DOI: 10.1016/S0928-2025(06)10031-0

Though mountains are often associated with clean air, they can receive contamination due to transport of pollution by winds. The Malè Declaration on Control and Prevention of Air Pollution and its likely Trans-boundary Effects for South Asia, and Project Atmospheric Brown Cloud (ABC) seek to find the nature of transport of pollution for informed decision-making to tackle trans-boundary air pollution through regional cooperation. Initial study has shown that aerosols have led to a large reduction of surface solar radiation during winter. This may affect agriculture, health, and the hydrological cycle; eventually contribute to climate change.

1. Background

The Hindu Kush–Karakoram–Himalayan (HKKH) region sustains approximately 140–150 million people and has an impact on the lives of three times as many people living on the plains and in the river basins below. The region is not only the world's highest mountain region, but also most populous. It extends over 3500 km from Afghanistan in the west to Myanmar in the east, and ranges from the plateau regions of Tibet and other mountain areas of China in the North to the Ganges Basin of India in the South (Fig. 31.1). As a macro-region, it contains the upland watersheds of major river systems: the Indus, the Ganges, the Tsangpo–Brahmaputra, the Nu–Salween, the Lacang–Mekong, and the Yangtze (Jinsha).

The wealth of the HKKH lies in an immense diversity of flora, fauna, and biological diversity than most other ecoregions on the planet, including lowland rainforests. Furthermore, these mountains are a source of such key resources as water, minerals, forest and agricultural products, and recreation. They are also rich repositories of cultural diversity, traditions, ethnic groups, and languages. Yet as diverse as mountain ecosystems are, and as strong and powerful as the image may be, in reality, mountains are exceedingly fragile. Each day, climate change, pollution, as well as exploitative mining and unsound agriculture practices, take a toll on mountain environments. Mountain people – the guardians of mountain ecosystems – are most vulnerable to these changes.

Sustainable development of mountain regions is a challenging task. Among the mountain areas of the world, the HKKH region offers the greatest challenge due to diverse and fragile ecosystem. This region is geologically the youngest mountain range giving rise to a high degree of natural hazards. It is the most densely populated mountain range with majority of the population living below the poverty line. The main cause of poverty of these people is due to a poor productive base, isolation, social and political exclusion, and above all, the severity of the constraints of unfavorable geographical situation. Besides, the life and property of these people are highly vulnerable to natural calamities such as earthquake, flash floods and landslides.

2. Global environment change

Global environment change can be categorized into two types: systemic changes and cumulative changes. Broadly speaking, a systemic change is one that, while taking

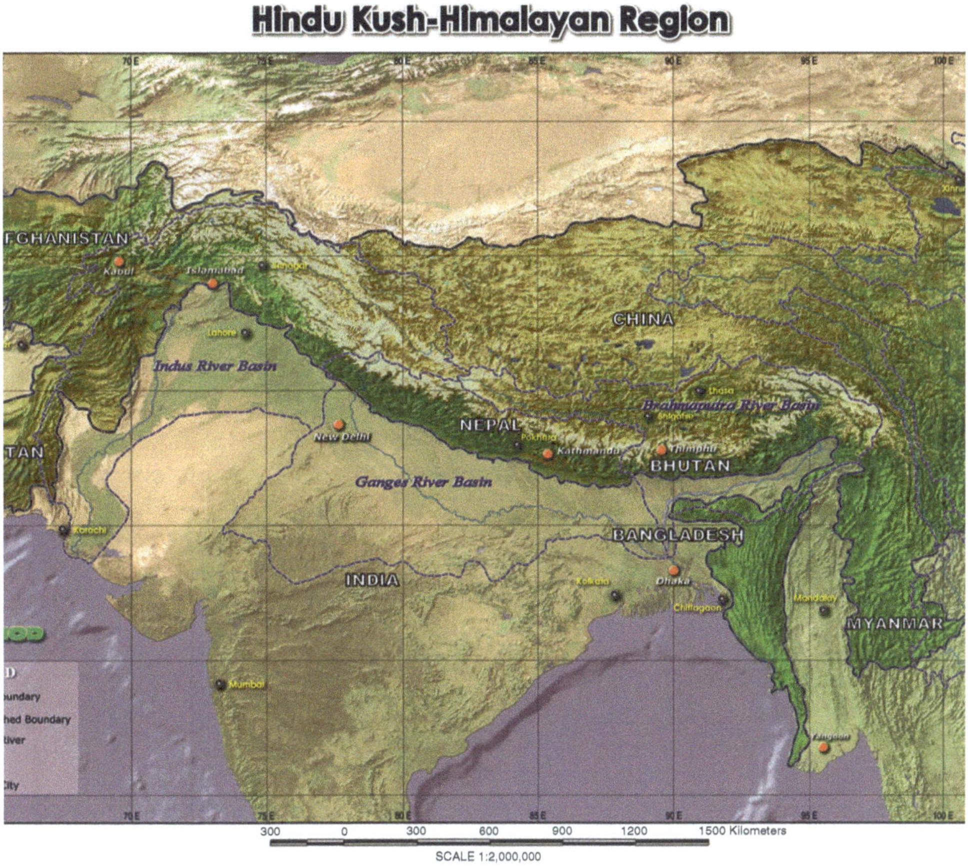

Figure 31.1. Hindu Kush–Karakoram–Himalayan region.

place in one locale, can effect changes in systems elsewhere. The underlying activity need not be widespread or global in scale, but its potential impact is global in that it influences the operation and functioning of the whole system as manifested through subsequent adjustments in the system. Emissions of CO_2 from limited activities that have impacts on the ecosphere biosphere system of the Earth and cause global warming are a prime example. Cumulative change refers to localized but widely replicated activities where changes in one place do not affect changes in other distant places. When accumulated, however, they may acquire sufficient scale and potential to influence the global situation in various ways. Widespread deforestation, extractive land-use practices, groundwater pollution and depletion, biodiversity losses and their potential impacts on the global environment are some of the examples. Both types of change are the product of nature–human interactions and are linked to each other in several ways. Mountain areas are subject to both types of change. The impact of systemic change is more readily visible (e.g., glacial melt due to warming; upward shift of certain plant species; and distortion of flowering seasons). Cumulative environmental changes result in resource degradation and depletion, and have adverse effects on biophysical flows and functions of the ecosystem. Hence, sustainable management of mountain resources is essential to minimize the outcomes of both these types of changes.

3. ICIMOD's initiative in sustainable mountain development

Sustainable development of the mountain people in the HKKH has been the main agenda of the International Center for Integrated Mountain Development (ICIMOD). Mountains are highly vulnerable to human and natural ecological imbalance. Mountains are the areas most sensitive to all climatic changes in the atmosphere. Specific information on ecology, natural-resource potential, and socio-economic activities is essential. There is, however, a lack of sufficient knowledge of mountain ecosystems for the reasons of understanding of mountain specificities; ICIMOD's effort has been to establish itself as a hub for mountain specific knowledge. ICIMOD has been working on developing methodologies for applying the technology in portraying the livelihoods of the people in relation to the resource base and infrastructure, thus helping for better planning and implementation of development programs. Described below are the studies conducted on climate change towards this initiative.

4. Areas of intervention

4.1. Glacier and climate change

Climate change due to the enhanced greenhouse effect has increased the rate of glacier melting and thereby the frequency of GLOFs in recent years. Since industrialization, human activities have resulted in steadily increasing concentrations of greenhouse gases in the atmosphere, leading to fears of enhanced greenhouse effect. The world's average surface temperature has increased between 0.3–0.6°C over the past hundred years. The Intergovernmental Panel on Climate Change (IPCC), in its third assessment report, revealed that the rate and duration of warming in the 20th century is larger than at any other time during the last one thousand years. The 1990s was likely to be the warmest decade of the millennium in the Northern Hemisphere, and the year 1998, the warmest year (IPCC, 2001). According to the World Meteorological Organization (WMO), years 2002 and 2003 have been the 2nd and the 3rd hottest years, respectively, ever since climate statistics have been monitored and documented.

Climate change is causing the net shrinkage and retreat of glaciers and the increase in size and number of glacial lakes, especially in the high mountains. Analysis of air temperature trends across 49 stations in Nepal from 1977 to 1994, for example, revealed a clearly rising trend, with the change much more pronounced in the higher altitude regions of the country (Shrestha et al., 1999).

Numerous studies carried out during 1999–2001 lend credence to the link between climate change and glacier melting. Most of the glaciers in the Himalaya have retreated by approximately a kilometer since the Little Ice Age (1550–1850) (Mool et al., 2001).

The Dokriani Bamak Glacier in the Indian Himalaya retreated by 20 m in 1998, compared to an average retreat of 16.5 m over the previous 5 years, (Matny, 2000). A long-term study entitled, 'The Chinese Glacier Inventory', by the Chinese Academy of Sciences has reported that during the last 24 years there has been a 5.5% shrinkage in volume of China's 46,928 glaciers, equivalent to the loss of more than 3000 km^2 of ice. The study predicts that if climate continues to change at the present rate, two-thirds of China's glaciers would disappear by 2050, and almost all would be gone by 2100 (China Daily, September 23, 2004). Evidence has been conclusive enough to make glacier melting and retreat an important indicator for climate change.

One of the noteworthy examples of glacier retreat, melting of glacier ice, and formation of a glacial lake is Tsho Rolpa in Nepal. The development trend of the Tsho Rolpa Glacial Lake from 1957 to 2000 is shown in Fig. 31.2, which is based on

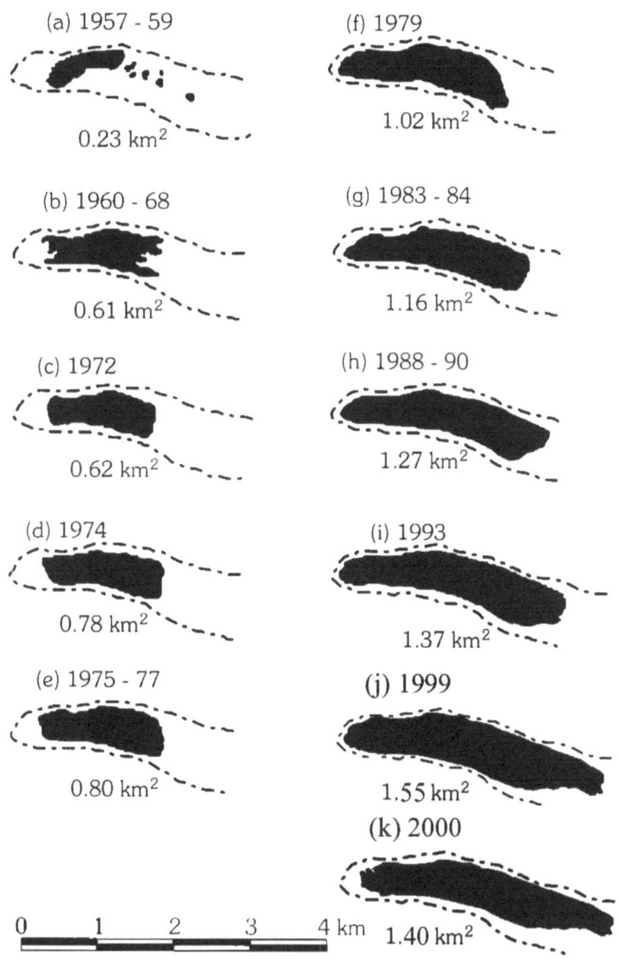

Figure 31.2. Successive development of the Tsho Rolpa glacial lake from 1957 to 2000; modified after WECS, 1993A (Mool et al., 2001).

the available maps, aerial photographs, satellite images, and field surveys. The lake was first mapped in the topographic map by the Survey of India with an area 0.23 km^2 in 1957 and it has grown successively. The average growth of the lake area from 1974 to 2000 was 0.024 $km^2 yr^{-1}$. The growth rate of the Tsho Rolpa Glacial Lake per year was more than the size of the defined major glacial lake (0.02 km^2) having the volume of more than 0.6 cubic m^3 of water (Bajracharya and Mool, 2005). The graph of Fig. 31.3 shows that the lake had increased abruptly from 0.23 km^2 to 0.61 km^2 in the year 1959 to 1960. According to the surveys of 1993 and 1994, the lake is about 3.2 km long and about 0.6 km wide with the volume of 76.6 million m^3. The maximum depth of the lake was 132 m, the stored water volume was 71 million m^3 in 1993, and the volume was 76.6 million m^3 in 1994. The lake volume has been increasing every year due to glacier retreat. Due to this retreat, deepening of the lake and narrowing of the damming moraines have also occurred.

The glaciers of the Himalayan Region are nature's valuable source of fresh water for the present and future needs of millions of people living in this region as well as downstream. These frozen reservoirs release large amounts of ice melt water to many of the major rivers of the region. The glaciers, some of which consist of huge amount of perpetual snow and ice, are found to have created many glacial lakes over the centuries. However, these glaciers are retreating in the face of accelerating global warming.

ICIMOD, along with the support from various sources, the United Nations Environment Program's regional resource center for Asia and Pacific (UNEP RRC.AP), the Asia–Pacific Network for Global Change (APN), the Global Change System for Analysis, Research and Training (START), and the Global Land Ice Measurements from Space (GLIMS) project and collaboration from national institutions have been carrying out comprehensive inventories of glaciers and glacial lakes using remote sensing (RS) and geographic information systems (GIS) in the PR China, Pakistan, India, Bhutan, and Nepal.

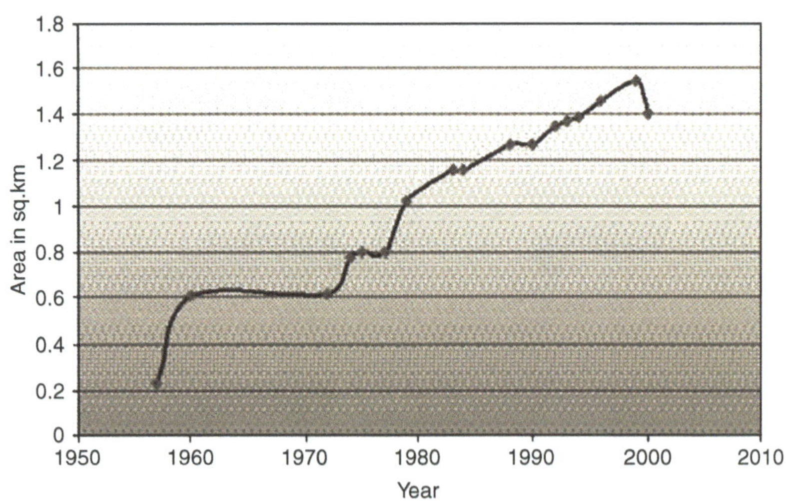

Figure 31.3. Development trend of Tsho Rolpa glacial lake area from 1957 to 2000.

The results and database generated from the present studies provide the baseline information on the region for the development of monitoring and early warning systems, and for planning and prioritizing disaster mitigation efforts. In addition, the study is anticipated to help lay down the planning guidelines for infrastructure and development, and water resources and land-use in the region in the individual countries of the HKKH. Along with information on climate change and future monitoring of glaciers, glacial lakes, and GLOFs, this database will provide the basis for estimating future available water resources and their planning and management.

Regular assessment monitoring of glaciers and glacial lakes, and adaptation measures including engineering structure and policy linkage for potential dangerous glacial lakes are required. Field validation of these glaciers and glacial lakes are also needed. Advanced scientific knowledge of potential GLOF hazards, building capacity among local institutions and agencies for undertaking or monitoring these studies, and linking this knowledge to policy and planning are, therefore, of immense importance to the lives of millions of mountain dwellers and their downstream neighbors. (Mool et al., 2001)

4.2. Trans-boundary air pollution

Mountains are often associated with clean crisp air; however, such topography can receive more pollution at high elevation than in adjacent valleys due to winds moving faster and less impeded aloft. Thus the mountains may look beautiful but may be cloaked in air-quality issues. The degraded air quality may directly affect the mountains' flora and fauna and indirectly affects our quality of life. Therefore, it is necessary to monitor the quality of air and its recipient direction of pollutants so that long-term mitigation measures can be planned locally or with regional cooperation. Already two programs are in operation, viz., Malè Declaration on Control and Prevention of Air Pollution and Its Likely Trans-boundary Effects for South Asia, and Project Atmospheric Brown Cloud.

The Malè Declaration is mainly focused on the effects of acidification on soil, water, and vegetation caused by trans-boundary air pollution. The program was endorsed in the seventh Governing Council of SACEP held in Malè, Maldives, in 1998, setting a prime example of tackling air pollution through regional cooperation. The program is divided into three phases: Phase 1 – Agreement and Awareness, Phase II – Capacity building, Phase III – Tackling Air Pollution Problems.

Each South Asian country has started monitoring both air quality and rainwater chemistry. The air-quality measurements include particulate matter up to 10 μm, sulfur dioxide and nitrogen oxide. The rainwater analysis includes pH, electrical conductivity, and concentration of anions and cations. Though data collection has been started, more data will be needed to make meaningful interpretation of possible trends in the trans-boundary movement of the pollution.

The Indian Ocean Experiment (INDOEX) carried out in February 1999, in the islands of the Maldives, revealed that a 3 km-thick toxic umbrella of Brown Cloud, stretches over Afghanistan, Pakistan, Bangladesh, Bhutan, India, Maldives, Nepal, and Sri Lanka, which are among the most densely populated places in the world.

The finding comes from observations gathered by more than 200 scientists supplemented by satellite readings and computer modeling. The Brown Cloud (haze) puts millions of people at risk not only for various respiratory diseases but also for severe natural disasters as weather patterns are radically altered and become more extreme and unpredictable. The Project ABC is built upon the INDOEX completed in 1999.

The long-range transport of the haze was an important finding. Releasing the intensity of the problem, with the initiation from UNEP and coordinated by ICIMOD, Nobel Laureate Paul Crutzen and Prof. V. Ramanathan visited Kathmandu on March 2001. Aerial survey was conducted to study the magnitude of the problem in Fig. 31.4, photos of which were taken approximately 30 km south of Mt. Everest, from a flight altitude of about 3 km. Both photographs were taken from the same location, one viewing north (top) and the other south (bottom). During the dry season from January to April, the brown sky seen over Nepal is typical of many areas in South Asia. The dry northeast monsoon winds carry this anthropogenic haze thousands of kilometers south and south-eastwards, and spread it over most of the tropical Indian Ocean.

The photographs revealed the need to conduct further studies on this regard. The initial finding shows that Kathmandu air is heavily loaded with anthropogenic aerosols (Fig. 31.5) (Ramana et al., 2004). Transport of pollution was also clearly noted (Fig. 31.6); however, much more information and study is required for

Figure 31.4. Haze in the Himalaya. View ~30 km south of Everest looking north (top) and looking south (bottom).

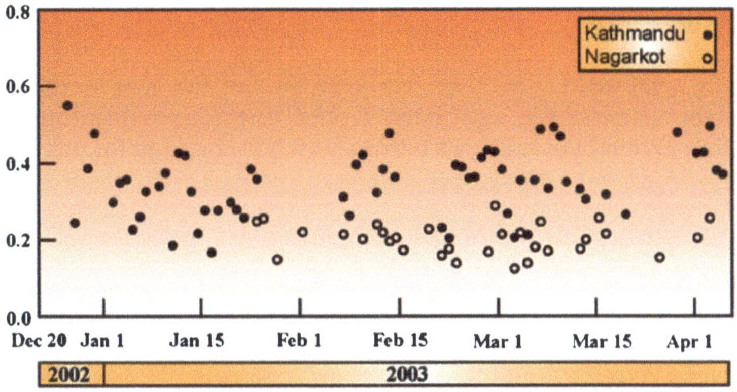

Figure 31.5. Aerosol optical depth measurement.

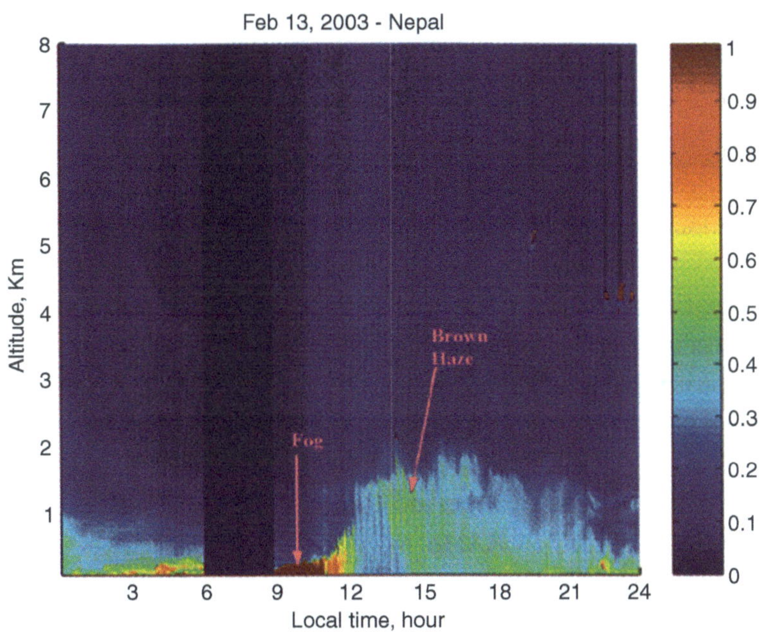

Figure 31.6. LIDAR observation.

unambiguous sources of origin. The findings will be helpful in designing policy reforms both in the country as well as in the region. They will answer some of the major environmental challenges facing the Indo–Asia–Pacific region in the coming decades, specifically the environmental consequences of rising air pollution levels due to rapid industrialization and population growth. A permanent monitoring station has been installed in Kathmandu, to study radiation, various components in aerosols, and rainwater chemistry.

Acknowledgement

We gratefully acknowledge Samjwol Bajracharya and Pradeep Mool for providing information on glaciers. We also thank UNEP, the Scripps Institute of Oceanography and Stockholm University, Department of Meteorology for providing support in this study.

References

Bajracharya, S.R. and Mool, P.K., 2005. Growth of hazardous glacial lakes in Nepal. Proceedings of the JICA Regional Seminar on Natural Disaster Mitigation, 131–148.

IPCC, 2001. IPCC Third Assessment Report – Climate Change 2001. Working Group I: Technical Summary. Geneva, WMO and UNEP.

Matny, L., 2000. Melting of Earth's Ice Cover Reaches New High. Worldwatch News Brief, 06 March 2000.

Mool, P.K., Bajracharya, S.R., and Joshi, S.P., 2001. Inventory of Glaciers, Glacial Lakes and Glacial Lake Outburst Floods, Monitoring and Early Warning Systems in the Hindu Kush-Himalayan Region: Nepal, ICIMOD & UNEP RRC-AP.

Ramana, M.V., Pradhan, B.B., Ramanathan, V., et al., 2004. Direct observations of large aerosol radiative forcing in the Himalayan region – Geophys. Research Letters 31, L05111. doi:10.1029/2003GL018824.

Shrestha, A.B., Wake, C.P., Mayewski, P.A., and Dibb, J.E., 1999. Maximum temperature trends in the Himalaya and its vicinity: an analysis based on temperature records from Nepal for the period 1971–94. Journal of Climate, 12: 2775–2767.

Mountains Witnesses of Global Changes
R. Baudo, G. Tartari, and E. Vuillermoz (Editors)

32

Climate research in the Nepal Himalaya

Saraju K. Baidya

Abstract

Nepal is a mountainous country with extreme topographical variation. This paper discusses some climatic research works in Nepal, mainly focusing on climate change and its impact, as well as the monsoon in Nepal. Studies have indicated increasing trends of maximum temperature in most parts of the country, with high warming trends in the Himalayan and middle mountain regions compared to lower altitudes and plains. This result is also supported by the retreating glacier trend in the Nepal Himalaya. Studies show that these glaciers are retreating at an alarming rate, with a potential negative impact on water resource management and threat of glacier lake outburst floods. General circulation models were used in 1994 and climate change scenarios were developed to study the impact of climate change on water resources and agriculture. The Department of Hydrology and Meteorology in Nepal has now started experiments on a high-resolution regional climate model, the RegCM3 to study the impact of climate change and to assess vulnerability. The role of the Himalaya on the Asian Summer Monsoon (ASM) has been well recognized in many modeling studies. In 1999, the Monsoon Himalayan Precipitation Experiment was carried out to study the interaction of the Himalaya and the ASM.

1. Introduction

Nepal is a small mountainous country situated on the southern slope of the Himalaya. Although it lies near the northern limit of the tropics, the climate varies from tropical in the southern plain area to polar and arctic in the high Himalaya due to intense north/south topographical variations (60 m a.s.l. in the south to 8848 m a.s.l. in the north) within a short horizontal distance of about 200 km.

The main source of precipitation in Nepal is the summer monsoon (June to September), which results in response to the large thermal gradient between the warm Asian continent to the north and the cooler Indian Ocean to the south (Slingo et al., 2002), and due to seasonal shifting of thermally induced planetary belts of pressure and winds under continental influences (Pant and Kumar, 1997). About 80% of the annual rainfall occurs during this period (Nayava, 1974; Shrestha, 2000). This monsoon current enters Nepal from the southeast direction. Once it enters Nepal, topography

ISSN: 0928-2025 DOI: 10.1016/S0928-2025(06)10032-2

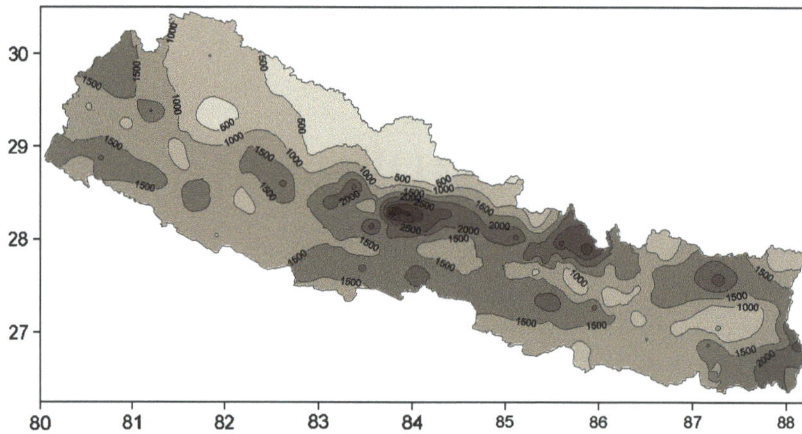

Figure 32.1. Spatial distribution of monsoon rainfall.

determines the spatial distribution of rainfall. The windward side of the mountain barrier receives a lot of rainfall while the leeward side receives comparatively less rainfall and is almost dry. But there are some high rainfall pockets, which are favored by the topography (windward side) and its orientation (Fig. 32.1).

The rugged mountain topography, the fragile geology of the young Himalayan Mountains, and high intensity of monsoon rainfall makes Nepal prone to natural disasters such as floods, landslides, droughts, and other problems. With global warming and climate change in the limelight for quite some time, Nepal is no exception. This paper discusses some climatic research work in Nepal, mainly focusing on climate change and its impact; as well as the monsoon in Nepal.

2. Climatic trends in Nepal

The subject of global warming and climate change has become an issue of great concern to the scientific as well as political communities. Studies of long-term global temperature change (Jones et al., 1986; Hansen and Lebedeff, 1987) show a meaningful warming trend in the past century. Similarly, Hingane et al. (1985) found a slight but definite warming trend in the mean annual Indian temperatures. The first study of such kind for Nepal was done by Shrestha et al. (1999) based on temperature data from a network of surface observations. Shrestha et al. (1999) found increasing trends of maximum temperature in most parts of the country after 1977 (Fig. 32.2), with high warming trends in the Himalayan and middle mountain regions (0.06°C to 0.12°C) and low warming (less than 0.03°C) over most of the Siwalik and Terai (southern plains) regions. The capital city, Kathmandu, shows the increase in mean annual temperature by 0.05°C/yr.

The monsoon rainfall of Nepal for the period 1971–2000 (Fig. 32.3) shows a large year-to-year variation. Although there is evidence of a slight increasing trend, the result is statistically insignificant. So, it can not be concluded with certainty that the rainfall over Nepal is increasing, and besides, a regional variation in rainfall

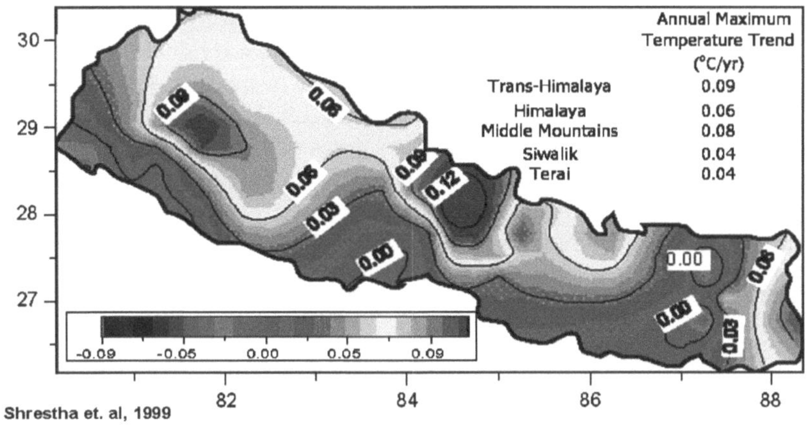

Figure 32.2. Spatial distribution of annual maximum temperature in Nepal.

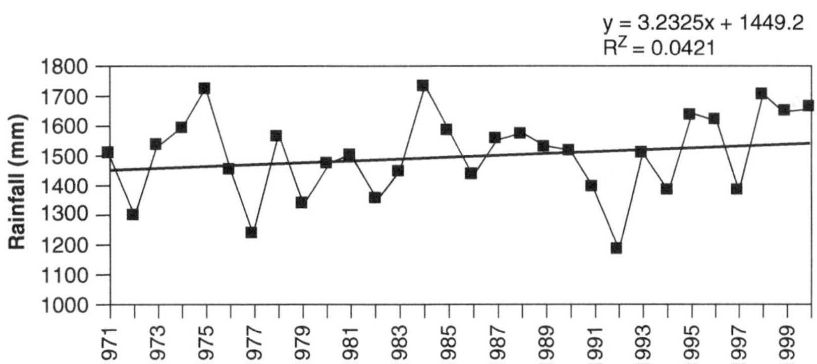

Figure 32.3. The monsoon rainfall trend in Nepal.

pattern exists. The rainfall usually decreases from south to north and east to west (Shrestha, 2000).

3. Climatic research

The glaciological and meteorological observations of the glaciers and climate in the Nepal Himalaya started in the early 1970s as the "Glaciological Expedition to Nepal (GEN)" Higuchi (1976) with the aim of obtaining data on the state of glaciers and their variations to understand the relation between glaciers, climate, and water resource development. Since then, a lot of other research has been carried out in the Nepal Himalaya. The country is rich in water resources, with over 6000 rivers, and major ones are fed by glaciers and glacial lakes located at the source of these rivers (Mool et al., 2001). The rivers finally drain into the Ganges. So any major changes in the Nepal glaciers will have regional scale impact on water resources (WWF, 2005). The Department of Hydrology and Meteorology of Nepal has installed six regular high altitude stations in the Nepal Himalaya for such study purpose. Glaciers are also important indicators of climate and climate change (Higuchi, 1976; Oerlemans, 1994). Though the Himalayan glaciers have been retreating for a long time, the studies show that there is compelling evidence of these glaciers retreating at an alarming rate and the rate of retreat is accelerating in conformity with the high rate of warming over the Himalaya. This leads to a heightened concern as to what effect such warming will have on the seasonal and long-term water resource management on glacier-fed rivers. Therefore, glaciological studies are important to study the impact of climate change and assess the water resources of this region.

The inventory compilation of glaciers in the Nepal Himalayas started in the early 1970s (Müller, 1970; Higuchi et al., 1976). The latest study on glaciers of the Nepal Himalayas revealed 3252 glaciers (Mool et al., 2001). Many of these glaciers are found to be in a general recession. Glacier AX010 is one of the most studied such glaciers in Nepal. Monitoring of the glacier terminus started from 1978. Photographs of the glacier terminus in different years (Fig. 32.4) clearly depict the retreat of the glacier. The glacier retreated by 30 m from 1978 to 1989, which is equivalent to a 12 m thinning of the glacier surface (WWF, 2005). The retreat of such glaciers results in the formation of the treacherous glacier lakes. Most of these lakes in the Nepal

Figure 32.4. Retreat of the AX010 glacier in the Nepal Himalaya.

Himalayas are moraine dammed. The natural moraine dams are structurally weak and unstable, and are susceptible to breaching as their banks release the enormous amount of water causing the catastrophic glacier lake outburst floods (GLOF) downstream. Such GLOF events cause extensive damage to constructions, infrastructures, environment, and even claim human lives. There are records of several such disastrous events in Nepal, the latest being the Dig Tsho GLOF on August 4, 1985, in the Langmoche Valley, Khumbum (Ives, 1986; Yamada, 1998). A survey by ICIMOD in Nepal revealed 2323 glacier lakes and identified 20 of them as potentially dangerous lakes. Imja is one of such potentially dangerous lakes. The lake started to form as small ponds in the late 1950s (Fig. 32.5). The size of the lake has grown to 0.86 km^2 by 2002, which was 14.7% more than in 1999. Many researches in the Nepal Himalaya highlight the impacts on water resource management and GLOF.

Nepal ratified the June 1992 Rio Earth Summit, committing to the objective of United Nations Framework Convention for Climate Change to take necessary steps and measures to reduce greenhouse gas emission in the atmosphere for mitigating the climatic change process and adopt national policies. The Nepal Himalaya being most vulnerable to the climate change and its impact, His Majesty's Government of Nepal instituted a Country Study program on climate change in 1994. For the first time in Nepal, results of general circulation models (GCMs) were used and climate change scenarios were developed. Four models were used (Shrestha, 1997), the Canadian Climate Centre Model (CCCM), Geophysical Fluid Dynamics Laboratory R-30 Model (GFD3), United Kingdom Meteorological Office Model (UK89), and Goddard Institute for Space Sciences (GISS). According to this study, out of the four, only the CCCM and GFD3 were able to simulate the climatology of the region satisfactorily and were therefore used to develop the climate change scenario for Nepal with a doubling of the carbon dioxide. The climate change scenario developed for temperature was more realistic than for precipitation. CCCM and GFD3 estimated the temperature to increase on an average by 2.9°C and 3.1°C, respectively. Both models showed more increase in temperature at higher elevations than at lower elevations. Precipitation-wise the estimation was 36% and 67% increase from the

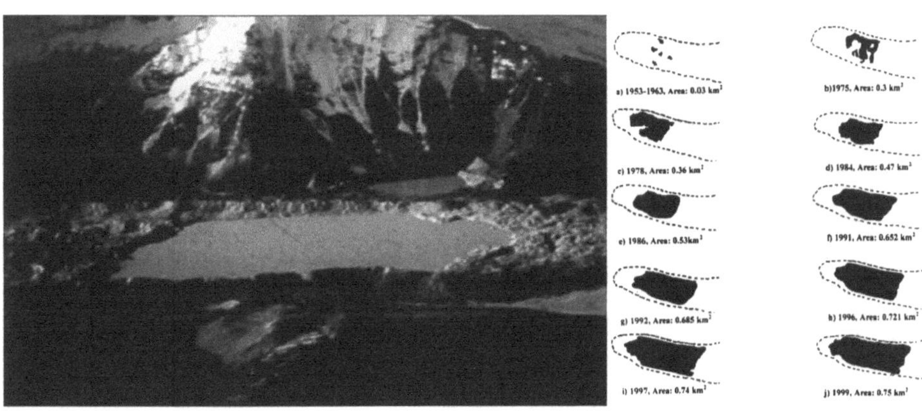

Figure 32.5. Growth of Imja glacier lake (Sources: WECS and DHM).

baseline (1971–1990) for CCCM and GFD3, respectively. The models results show a general tendency of more precipitation in summer months than in other months. This indicates that the wet seasons will be even wetter, i.e., extreme rainfall events will be more frequent in the future warm climate. This study provided the indication of the change in climate by doubling of carbon dioxide. But, because of the complex topography and low resolution of GCMs, the results of such models were not realistic for regional and local impact studies.

In order to study climate change on a regional scale, especially in a complex topographic region like Nepal, high-resolution models are required. Currently, Nepal is experimenting with a high-resolution Regional Climate Model (RCM), RegCM3 developed by NCAR. The main aim of this model is to study the climate change impact in the country. Preliminary results of RegCM3 show that the spatial precipitation pattern improved when the resolution of the model increased from 75 km to 30 km. But it is still not able to reproduce the observed pattern. Using 15 km nesting (Fig. 32.6), the model was able to simulate the observation precipitation amount and pattern fairly well. But the high rainfall pockets in the simulation are slightly shifted southward more than their actual observed location. This shows that the model still needs to be refined. The main problem in using the high-resolution model is the lack of high-speed computing facilities and computational time. Figure 32.7 shows the computational time required for different resolutions for a 1-month simulation with the current available facilities in the department.

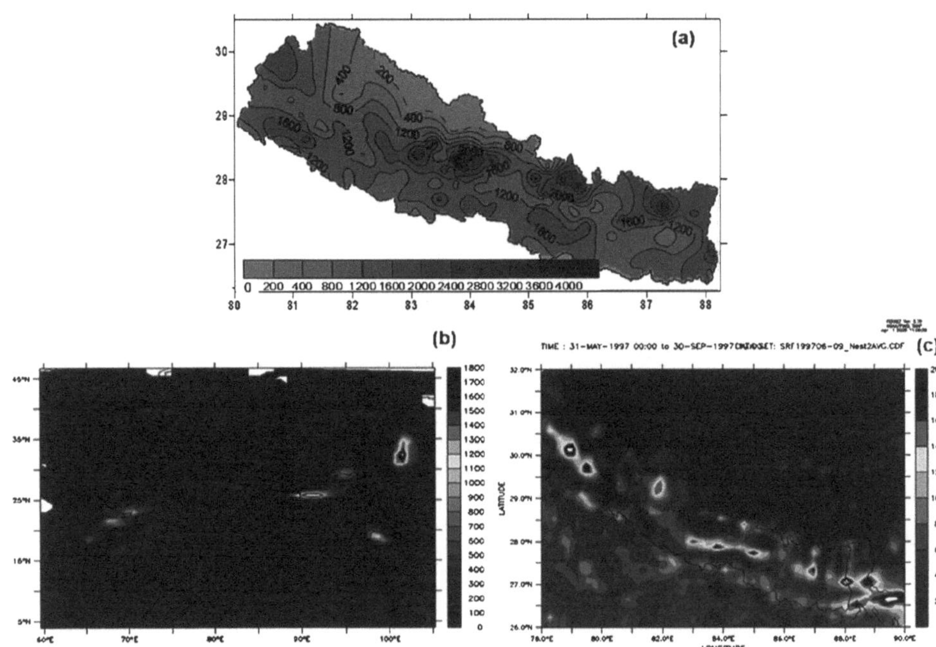

Source: Archana Shrestha, DHM

Figure 32.6. Comparison of 1977 monsoon precipitation with (a) observation and simulation at resolutions (b) 75 km and (c) nested 15 km.

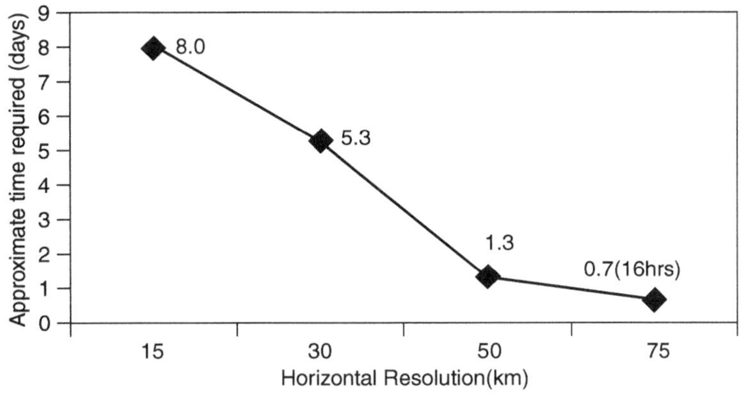

Figure 32.7. Consumption of computing time.

Hahn and Manabe (1975), using numerical simulation methods, demonstrated the role of the Himalaya and Tibetan plateaus in generating and maintaining the South Asian monsoon. In spite of the importance of the high mountains and the Himalaya in Asian summer monsoon, only a few studies have been done over these regions, especially in Nepal. In 1999, a network of rain-gauge stations at a variety of elevations from 528 to 4435 m a.s.l. was installed in the Marsyangdi river basin in central Nepal in order to study the orographic effects on precipitation over the south-facing slopes of the Himalaya. Using the first set of observation data from 1999 and the Tropical Rainfall Measuring Mission derived 3D precipitation radar (PR) rain rates, Barros et al. (2000) showed that there exists a strong interaction between mesoscale convective systems and steep terrain at elevations of 1 to 2 km, which agreed with the observational rainfall of more than 300 cm at these elevations ($\geq$2000 m). They also found a substantial variation in seasonal rainfall within less than a few km distance suggesting the importance of orographic effects on rainfall.

The most important study so far in this region probably is the monsoon Himalayan precipitation experiment (MOHPREX) that occurred during June 2001 along the southern slopes of the Himalaya in central Nepal. The main aim of the project was to better understand the spatial and temporal variation of precipitation along the south-facing slopes of the Himalaya (Barros and Lang, 2003). Analyzing the MOHPREX data, they showed that the total moisture column and convective instability gradually built up during the onset phase of monsoon. The middle (500 hPa) and upper (200 hPa) level westerly winds weakened considerably. They found that the 500 hPa westerly wind shifted to a more easterly track. A significant reduction in 200 hPa westerly wind from $>20\,ms^{-1}$ to a light and variable westerly is quite remarkable. The lower level winds were, however, modulated by the diurnal cycle of upslope (during day) and downslope (during night). In harmony with the diurnal cycle of moisture and instability, they found the post-midnight maximum in rainfall and postulated a mechanism to explain this phenomenon (Fig. 32.8). The southeasterly monsoon flow from the Bay of Bengal encounters the mountains that act as barriers to the flows and consequently low-level convergence occurs. But during the day time, upslope and upvalley flows reduce this convergence. The upslope flow, however, leads to high-level convection and a secondary

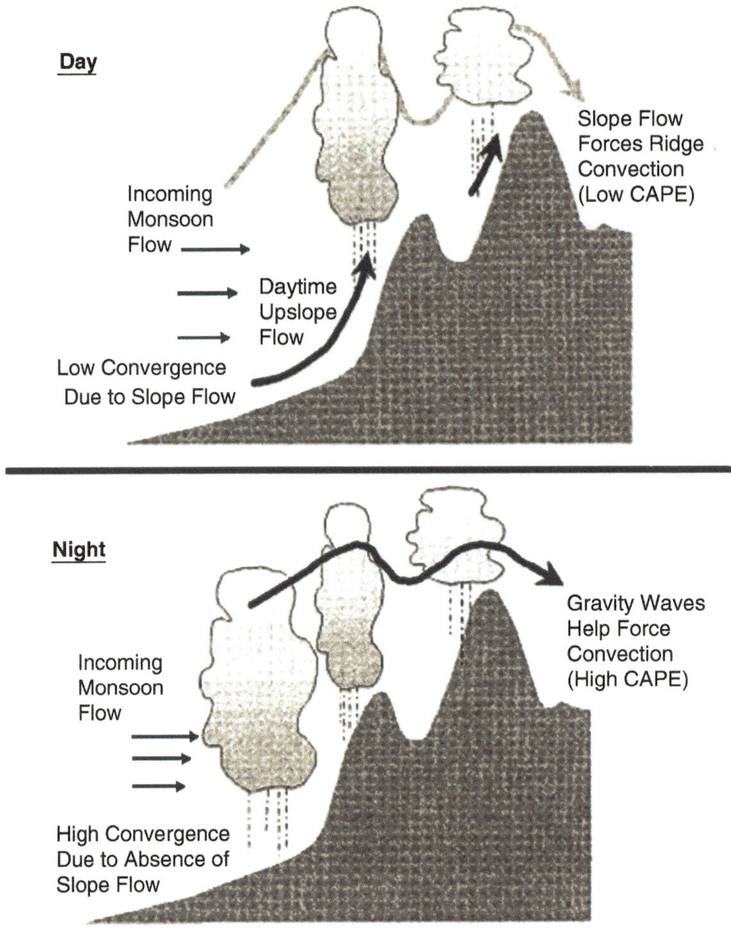

Figure 32.8. Schematic diagram of the proposed mechanism to explain nocturnal rainfall in the Himalaya (Barros and Lang, 2003).

peak in precipitation. In contrast, during the night, with the absence of upslope winds, and prevalent downslope winds, strong convergence of the moist advected monsoon flow occurs that acts to force convection. Atmospheric instability is also maximum during the night. All these factors lead to the nocturnal peak in rainfall in these areas. They also indicated that interaction between mountain-forced gravity waves and the thermodynamics of the Himalayan atmosphere could enhance nocturnal convergence, thus favoring nocturnal precipitation maxima.

4. Conclusion

The Himalaya of Nepal provide a unique opportunity for scientists and researchers to study climate change impacts and the effect and role they play in the regional monsoon circulation. Because of the complex topography of the region, climate change studies require the use of high-resolution models such as the RCM. While deglaciation is considered to be a worldwide problem, the retreat of Himalayan glaciers at an alarming rate and rapid growth of several Himalayan glacier lakes in the recent decades could create serious consequences for regional water-resource management and threat of catastrophic GLOF. Studies have shown that the local scale modification by the mountains and the Himalaya on the large-scale monsoon flow play an important role in spatial and temporal distribution of seasonal rainfall. Nepal therefore deserves special attention in climate-related research.

References

Barros, A.P., Joshi, M., Putkonen, J., and Burbank, D.W., 2000. A study of the 1999 monsoon rainfall in a mountainous region in central Nepal using TRMM products and raingauge observations. Geophysical Research Letter 27, 3683–3686.

Barros, A.P. and Lang, T.J., 2003. Monitoring the monsoon in the Himalayas: observations in Central Nepal, June 2001. Monthly Weather Review 131, 1408–1427.

Hahn, D.G. and Manabe, S., 1975. The role of mountain in the South Asian Monsoon Circulation. Journal of the Atmospheric Sciences 32, 1515–1541.

Hansen, J. and Lebedeff, S., 1987. Global trends of measured surface air temperature. Journal of Geophysical Research 92, 13345–13372.

Higuchi, K., 1976. Outline of the glaciological expedition to Nepal. Journal of the Japanese Society of Snow and Ice 38, 1–5.

Higuchi, K., Iozawa, T., and Higuchi, H., 1976. Flight observations for the inventory of glaciers in the Nepal Himalayas. Journal of the Japanese Society of Snow and Ice 38, 10–16.

Hingane, L.S., Kumar, K.R., and Murti, B.V.R., 1985. Long term trends of surface air temperature in India. Journal of Climatology 5, 521–528.

Ives, J.D., 1986. Glacial lake outburst floods and risk engineering in the Himalaya. Occasional Paper No. 5, ICIMOD.

Jones, P.D., Wigley, T.M.L., and Wright, P.B., 1986. Global temperature variations between 1861 and 1984. Nature 322, 4030–4434.

Mool, P.K., Bajracharya, S.R., and Joshi, S.P., 2001. Inventory of Glaciers, Glacial Lakes and Glacial Lake Outburst Floods. ICIMOD.

Müller, F., 1970. Inventory of glaciers in the Mount Everest region. "Perennial ice and snow masses", Technical papers in Hydrology 1, UNESCO/IASH: 47–59.

Nayava, J.L., 1974. Heavy monsoon rainfall in Nepal. Weather 29, 443–450.

Oerlemans, J., 1994. Quantifying global warming from the retreat of glaciers. Science 264, 243–245.

Pant, G.B. and Kumar, R., 1997. Climates of South Asia. Wiley, New York.

Shrestha, M.L., 1997. Development of Climate Change Scenarios with Reference to Nepal. Proceedings of the Workshop on Climate Change in Nepal, pp 16–32.

Shrestha, M.L., 2000. Interannual variation of summer monsoon rainfall over Nepal and its relation to the Southern Oscillation. Meteorology and Atmospheric Physics 75, 21–28.

Shrestha, A.B., Wake, C.P., Mayewski, P.A., and Dibb, J.E., 1999. Maximum temperature trends in the Himalaya and its vicinity: an analysis based on temperature records from Nepal for the period 1971–94. Journal of Climate 12, 2775–2787.

Slingo, J., Inness, M., and Sperber, K.R., 2002. Monsoon overview. Encyclopedia of Atmospheric Sciences, pp. 1365–1370.

WWF, 2005. An overview of glaciers, glacier retreat, and its subsequent impacts in the Nepal, India and China. WWF, pp. 68.

Yamada, T., 1998. Monitoring of glacier lake and its outburst floods in Nepal Himalaya. Journal of the Japanese Society of Snow and Ice. Monograph No. 1.

Mountains Witnesses of Global Changes
R. Baudo, G. Tartari, and E. Vuillermoz (Editors)

33

Development of a mesoscale convective system over the foothills of the Himalaya into a severe storm

Qamar-uz-Zaman Chaudhry and Ghulam Rasul

Abstract

The occurrence of weather-related extremes has been increased considerably over low elevation plains as well as over the high altitudes in recent years. One such extreme precipitation event was recorded in Pakistan that produced 620 mm rainfall in the capital during only 10 h. Diagnostic analysis and numerical simulations have been carried out using surface and NCEP reanalysis data along with radar and satellite images for the development of a mesoscale convective system (MCS) resulting in a severe storm. It has been revealed that the sudden evolution of the MCS was the direct result of strong surface convection in moist and unstable lower layers of the atmosphere. The subsequent rapid development was the combined effect of the presence of a mid-latitude trough in the westerlies in the north and a moisture supply through monsoonal flow along the Himalaya. The westward shifting of the sub-tropical high from the north of India, and the strong divergence zone on its east edge played a significant role in developing the upward motion. Movement of the system was controlled by the steering current in the middle troposphere. The model captured the location of heavy precipitation well at 15 and 30 km resolution but failed to predict the amount of rainfall. The scale analysis shows that the MCS was the combination of α, β, and γ mesoscale systems.

1. Introduction

Pakistan is located in South Asia and neighbors India on the east. It draws the western boundary to the South Asian summer monsoon region. Pakistan extends northeast to southwest from the Arabian Sea between latitudes 24° N to 37° N and longitudes 60° E to 75° E. It can also be mentioned that the peculiar orographic features, such as the Himalaya and Hindu–Kush ranges in the north and northwest, respectively, along with the Tibetan Plateau on the northeast, play an important role in modifying the weather systems. The high mountains act as a "blocking high" to the movement of low pressure systems from the south. When there is some moisture

ISSN: 0928-2025 DOI: 10.1016/S0928-2025(06)10033-4

feeding from the Bay of Bengal or the Arabian Sea in association with some active westerly systems, the heavy rainfall is produced sometimes in the northern parts of Pakistan. This is primarily the situation in which devastating floods occur in the rivers and cause heavy losses downstream to agriculture and infrastructure. The contribution of monsoon rainfall is more that 55% of Pakistan's total annual rainfall (Chaudhry, 1991).

Islamabad and Rawalpindi are twin cities, situated north–south in the sub-mountainous region of the province of Punjab. Islamabad is located in the area with an inverted "V" pattern terrain (trumpet shape terrain) along the Margalla Hills, which are foot hills of Himalaya. On July 23, 2001, the twin cities of Islamabad and Rawalpindi experienced in 24 h a record-breaking heavy rainfall for any locality in Pakistan during the past 100 years for that time period. At the central observatory of Islamabad, 620 mm of precipitation was recorded in only 10 h. A continuous down-pour in somewhat cloudburst proportions, lasted for about 10 h from 0100 to 1100 UTC and caused the worst-ever flash flood in the living memory of the people in the local stream called "Nullah Lai" and its tributaries, which swept away low-lying areas of the twin cities (Rasul et al., 2004). The banks of the Nullah could not be identified and its core speed flux was not less than 30 km/h in general. Bi-standing housings were swept leaving no signs of their past existence along the banks of the Nullah Lai. The Pakistan Meteorological Department issued a heavy rainfall forecast in its regular Weather Broadcast Bulletin a day earlier but never expected this much mass of water to occur during such a short interval of time. Although the flash flood warning was issued almost 5 h in advance, yet still the people could not respond in a timely fashion to the extent of devastation and severe damages still occurred.

The heavy downpour occurring between 0600 and 0900 UTC accumulated a huge amount of water together with the rain water from the tributaries draining into Nullah Lai, which passes through Rawalpindi city and resulted into severe flooding. Water levels rose to 11 m causing the Nullah Lai into overflow. The depth of the flood water in certain low-lying areas of Rawalpindi was as much as 6 m. In Islamabad, some areas were also badly affected and civic life was totally paralyzed. Loss to the business community, including valuable medicines, was estimated up to tens of millions of dollars. The dwellers of the twin cities were the main sufferers as their dwellings were destroyed and the supply of public utilities was totally stopped for a considerable period of time.

According to the surface weather map drawn at the local Met Office at 0300 UTC on July 22, 2001, a high pressure system existed over Tajikistan as well as China, and low pressure prevailed over Pakistan between the two highs. A trough of the westerly wave was also passing across north of Pakistan. Another strong low pressure center was located on the West Bay of Bengal and adjoining coastal parts of India. A mesoscale low was just located to the northwest of Islamabad whose horizontal extent was initially about 200 km. The mesoscale low which appeared on July 22 produced a heavy downpour on its way to moving to the twin cities. It was quite different from typical monsoon depressions of the Bay of Bengal and the lows in the Meiyu front (Zhao and Mills, 1991; Zhang and Zhao, 2002; Zhao, 1988; Zhao et al., 2002) because it developed abruptly as a result of intense local convection. The above-mentioned mesoscale low pressure system formed over the Hazara–Malakand

Division between 34°–37° N and 70°–75° E, and intensified quickly with a central pressure less than 996 hPa due to its favorable orographic location ahead of the westerly trough at 500 hPa, which existed between 40°–50° N and 60°–70° E. At the same time there were also strong monsoon incursions from the Arabian Sea in the south and the Bay of Bengal. The southwesterly monsoon current from the Arabian Sea showed strong convergence over Islamabad and the Hazara–Malakand Division located in the north of Pakistan. The strong southeasterly monsoon current from the Bay of Bengal reached right up to the northern parts of Pakistan after traveling along the foothills of the Himalaya. During ENSO events, rainfall becomes highly erratic over the Indian peninsula (Ailikun and Yasunari, 2001).

2. Mesoscale systems and water vapor supply

The mesoscale low was formed in the northern hilly terrain, including the foot hills of the Himalaya as a result of local convection, and was further developed under the favorable conditions of moisture supply and the influencing westerly wave passing across the north of Pakistan.

In Fig. 33.1(a) and (b), the vertical cross sections of relative humidity along 73° E longitude (the longitude of the twin cities) are shown on July 22 and 23, 2001, respectively. It can be noticed from Fig. 33.1(a) that the maxima ranging from 80 to 90% prevailed between 850 and 700 hPa from 33° to 35° N. The lower atmosphere was relatively drier. The minima of relative humidity bearing values even less than 20% up to mid-troposphere existed between 35° N and 38° N. On the following day as shown in Fig. 33.1(b), the maxima was further accentuated and extended down to the surface. The already existing minima between 35° N and 38° N experienced further expansion from the surface to 500 hPa. The core of the minima-bearing values of less than 10% humidity existed between 37° and 38° N from 850 to 700 hPa. The prevailing conditions of relative humidity resulted into the development of highly unstable atmosphere.

To know the reasons behind this movement of the system, the streamline analyses were carried out at the lower levels of the atmosphere. The difference of the streamlines at 700 and 850 hPa was compared as shown in Fig. 33.2(a) and (b). It can be seen that there were southerly and southwesterly wind in northern parts of Pakistan at 850 hPa whereas the northwesterly wind in the region mentioned above was at 700 hPa. Obviously, the movement of cloud cluster echoes was controlled by the steering current in the middle troposphere (700 hPa), rather than by that in the lower troposphere as at 850 hPa. This resulted in a southeastward movement of cloud clusters. It does not mean that the current in the lower troposphere was not of any importance. On the contrary, it played a very significant role in supplying of water vapor to the system at its development stage. From Fig. 33.2(a) and (b), it can be noticed that there were strong monsoon incursions from the Arabian Sea and the Bay of Bengal converging over northern parts of Pakistan. It must be noticed that when these highly unstable and moist currents struck the east–west extended mountain terrain at right angles, a strong orographic lifting took place in this already convectively unstable region north of Islamabad. The direction of the southwesterly

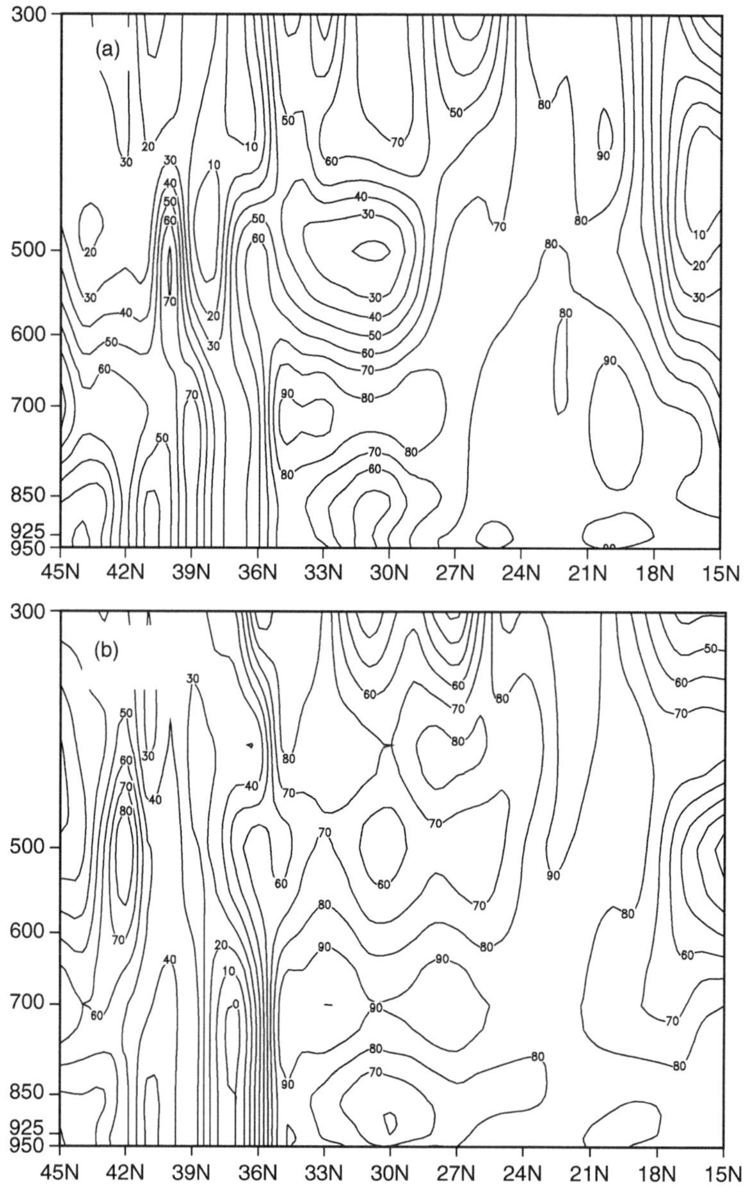

Figure 33.1. (a) Relative humidity along 73° E on July 22, 2001 at 0000 UTC (b) Relative humidity along 73° E on July 23, 2001 at 0000 UTC.

Arabian Sea current from the surface to 2000 m a.s.l. was oriented in such a way that its maximum impact was focused over northern parts of Pakistan comprising the area of development of mesoscale low.

The existence of cyclonic circulation can be seen in the streamline field at 850 hPa on July, 22 2001 shown in Fig. 33.3(a) in the northern parts of Pakistan. In the

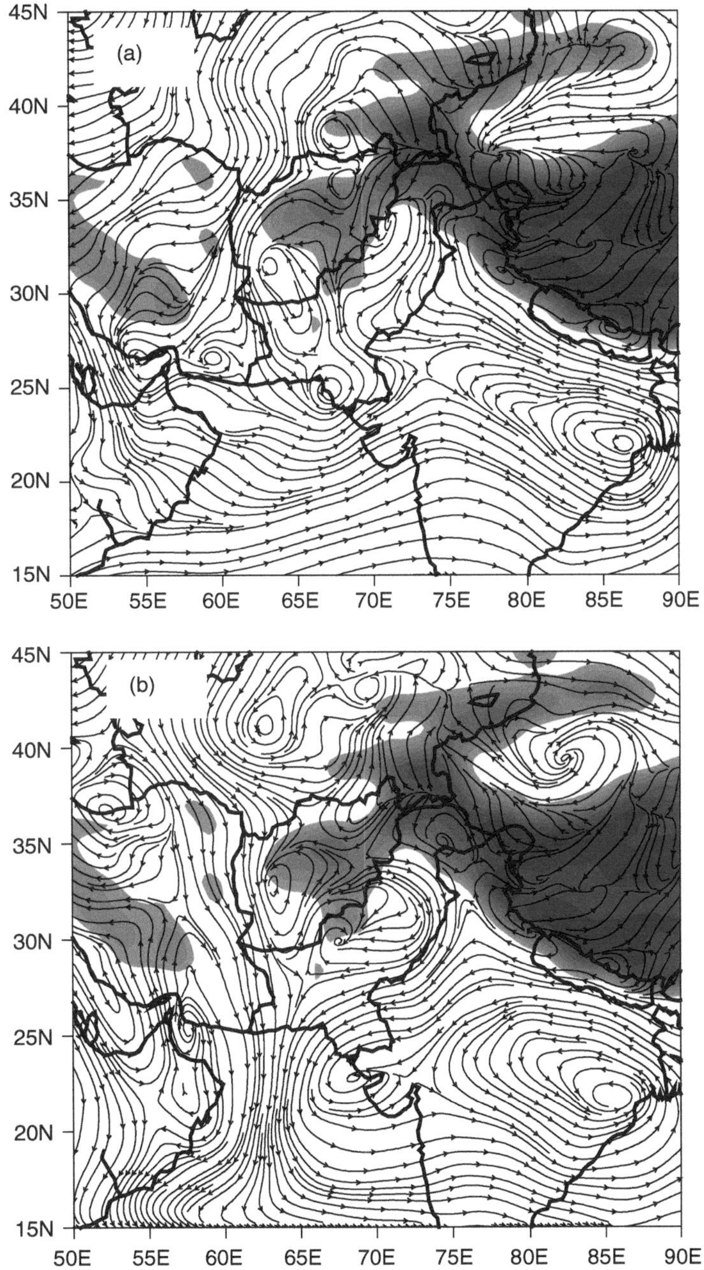

Figure 33.2. Streamlines at 0000 UTC (a) 850 hPa on July 22, 2001 and (b) 700 hPa on July 22, 2001. Moisture transport from the Arabian Sea and the Bay of Bengal is evident in the figure.

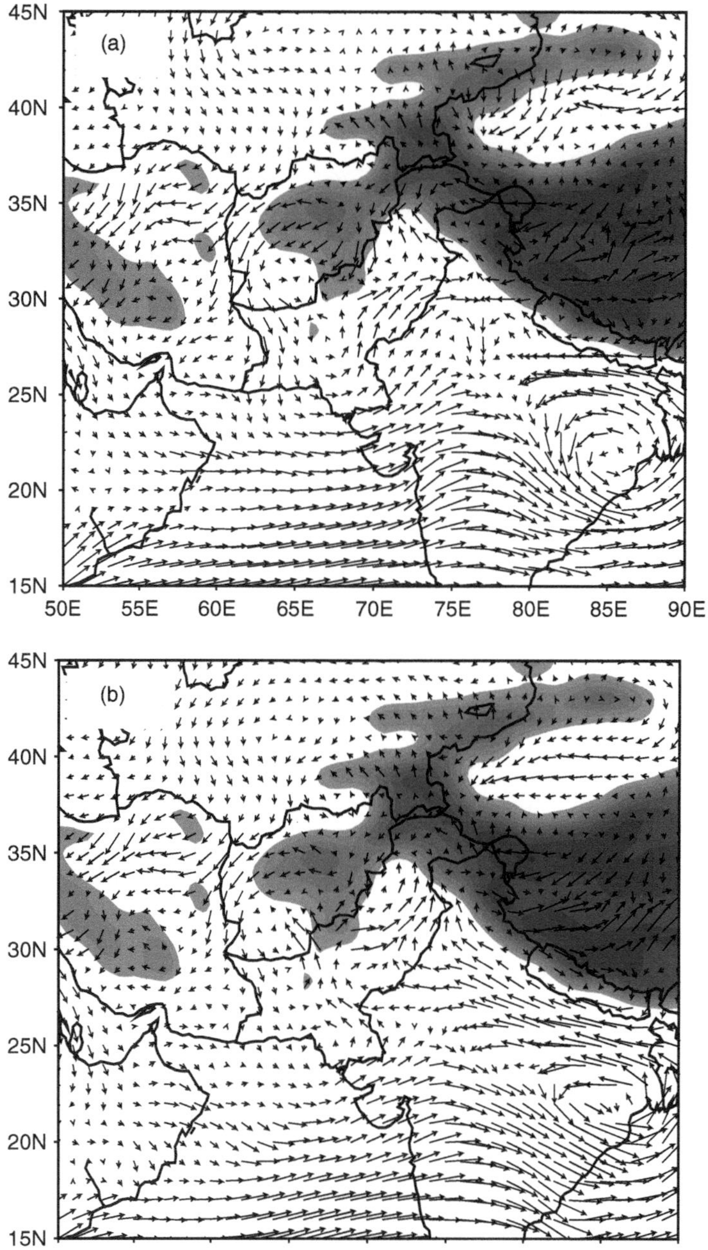

Figure 33.3. (a) 850 hPa divergence of moisture flux (10^{-7} g/hPa cm^{2}s) at 0000 UTC on July 22 (a) and 23 (b), 2001.

northeastern part of the cyclonic circulation there existed a strong convergence zone of moisture flux, i.e., the negative value region of divergence over Islamabad. From Fig. 33.3(a) and (b), it is clear that there existed a very strong convergence zone of moisture flux at 850 hPa, especially a day before occurrence of the record heavy rainfall in Islamabad and Rawalpindi on July, 23.

It can be noticed that minima of divergence of the moisture flux (DMF), i.e., the negative value region in Fig. 33.3(a), was located in the northern parts of Pakistan with values as low as -40×10^{-7} g/hPa cm^2 s. Strong convergence of moisture along with other above-mentioned parameters provided highly favorable large scale conditions for the formation and development of the heavy rainfall system. The next day, on July 23, the negative value region of DMF was still situated in the same region, although the intensity was slightly weaker than that on July 22. The continued convergence of moisture resulted in the production of highly unstable conditions and triggered the further development and intensification of the system.

3. The influence of cold air and terrain lifting

It is also worth mentioning that cold air from the middle latitudes also played an important role in setting a conducive stage in the occurrence of heavy rainfall in the presence of active monsoon currents. The analyses of geopotential height, as well as temperature on July 22 and 23, 2001 (Fig. 33.4) revealed that the western part of subtropical high extended up to western parts of Pakistan letting cold air to penetrate into northern parts of Pakistan. Monsoon air masses are warmer and convectively unstable in nature (Ohsawa et al., 2001), therefore its interaction with cold air of higher latitudes results in interesting developments of weather in the regions of their interaction. A trough of the westerly wave could be seen passing across Afghanistan in Fig. 33.4a and b. Such an interaction can produce heavy, to very heavy rain over the track of the system (Rao et al., 1970; Tao, 1980). It can be noticed that the –4°C isotherm was passing across the north of Afghanistan and the –2°C isotherm across north boundary of Pakistan at 500 hPa at 0000 UTC on July 22, 2001, as shown in Fig. 33.4(a). However, on July 23, cold air moved southward, the −4°C isotherm invaded northern Afghanistan and also the –2°C isotherm extended further south and prevailed over the northern parts of Pakistan (Fig. 33.4b).

In this situation, the stratification would be more unstable due to the cold air invasion at the middle-upper troposphere in the region, and therefore it added to triggering the formation of a severe weather system. Of course, it was not enough if only large-scale environmental conditions were favorable. The triggering mechanism also played a key role in producing heavy rainfall in association with other factors discussed in earlier sections. A strong convergence zone in northern Pakistan, especially near Rawalpindi and Islamabad existed as shown in Fig. 33.3a. Due to warm, as well as moist air below and cold air advection above, a strong convective instability was generated in the vertical column of the atmosphere. With the continued supply of moisture, an intense weather system originated abruptly under the prevailing conditions which was thereby stimulated throughout its development process. Rasul et al. (2005) studied a heavy downpour event in the arid climate of

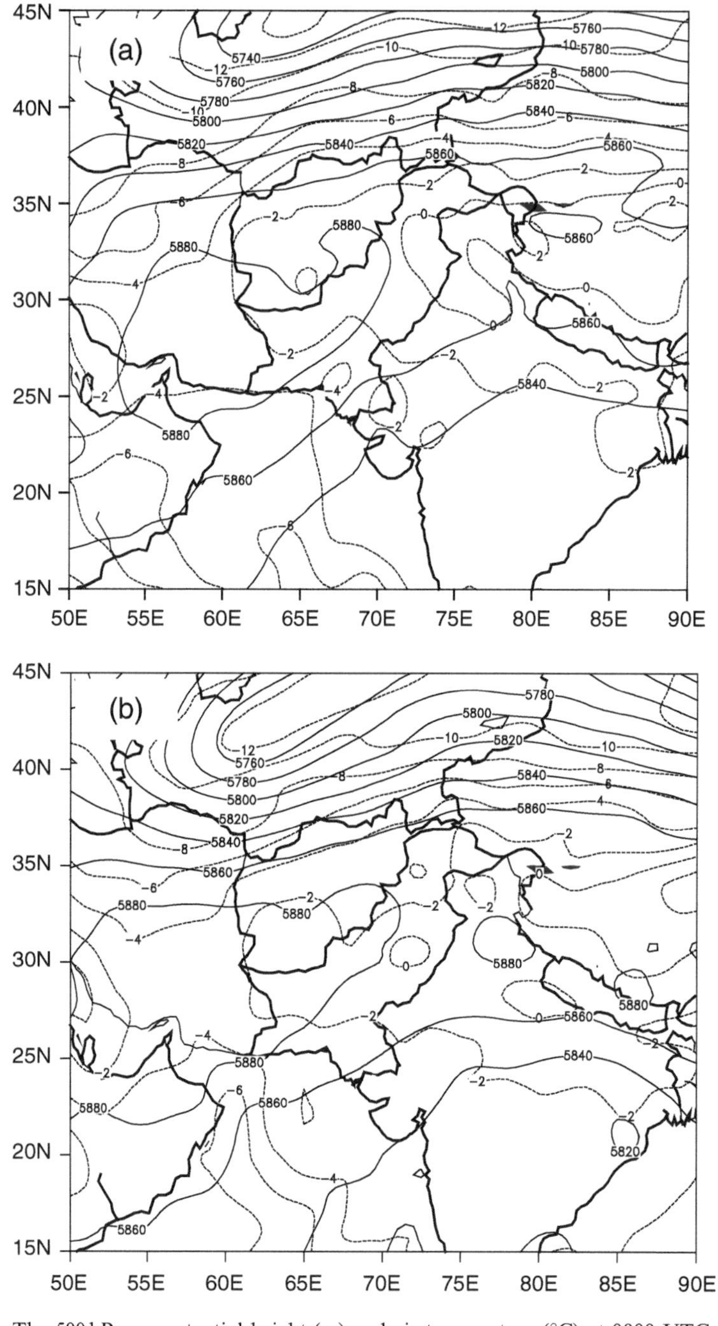

Figure 33.4. The 500 hPa geopotential height (m) and air temperature (°C) at 0000 UTC on July 22 (a) and July 23 (b) 2001. Bold lines are contours and dotted lines show isotherms.

India and Pakistan where a mesoscale low merged with a diffused tropical depression and reactivated into a strong system due to convective instability and strong moisture convergence.

In addition, Islamabad and Rawalpindi are just located in the inverse "V" shaped mountainous area composed of the foothills of the Himalayas opening to the south. The southerly wind perpendicular to the slope lifted due to peculiar orographic features and produced stronger vertical motion. The augmentation of rainfall on the windward side of the mountain ranges is an established fact that is primarily related to orographic lifting. Convergence of both the monsoon currents from the Bay of Bengal and the Arabian Sea and the terrain-lift forcing might have played an important role in triggering the appearance of the heavy rainfall. This could be confirmed from the increasing intensity of precipitation as the system moved from its origin towards the twin cities. The maxima of precipitation reached over the twin cities, and this might be the additional effect of terrain lifting forcing on converging monsoon currents.

This extreme precipitation of the mesoscale convective system (MCS) was simulated by the Mesoscale Meteorological Model (MM5) at different resolutions but did not produce encouraging results regarding location and amount of precipitation. It seems quite reasonable to simulate such events by using Regional Climate Model RegCM3 with Emanuel parameterization scheme (Kar et al., 2001). ERA40

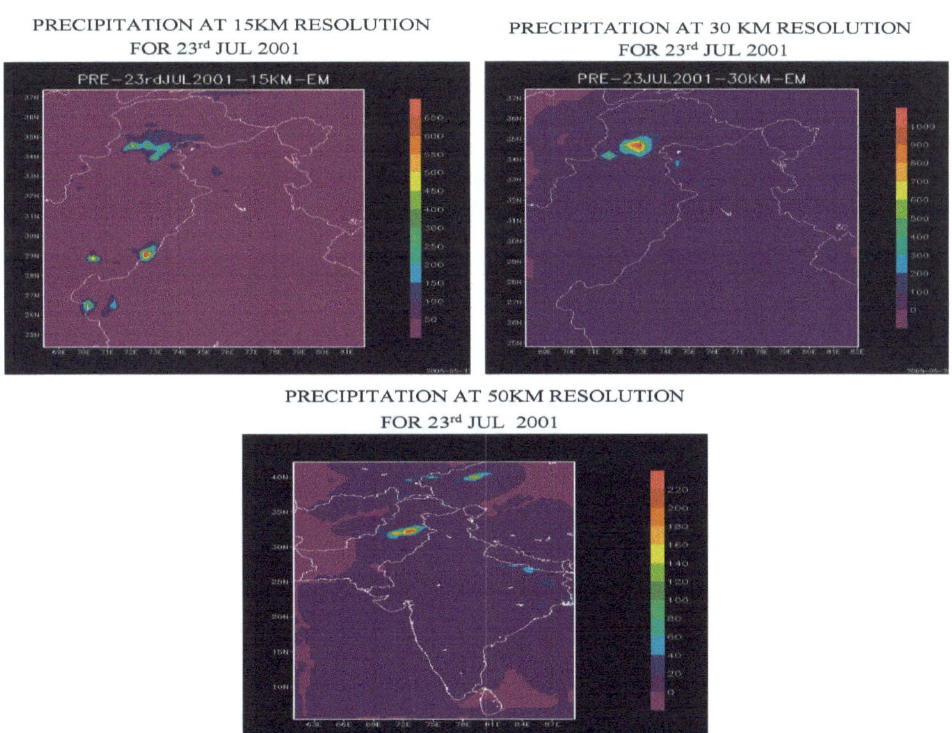

Figure 33.5. Output of the Reg CM3 model at different resolutions and domains.

(the 40-year European re-analysis data of the global atmosphere) has been used to get the results of the model. The model was run at different resolutions ranging from fine to coarse (15 km, 30 km, and 50 km) in varied domains. The simulation results are presented in Fig. 33.5. It can be seen that finer resolution (15 km) could only reveal the evolution site but gave no indication of the intensity of the system. On the other hand, 30 km resolution produced satisfactory results related to the location and intensity features whereas 50 km resolution presents unrealistic output.

4. Conclusions

In this paper the record-breaking 620 mm rainfall produced by a mesoscale convective system evolved along the foothills of the Himalayas has been studied. The main results are:

(1) The summer monsoon trough over South Asia intensified, the current of the northern branch of the monsoon trough, transported huge moisture to the north of Pakistan along the foothills of Himalaya. The easterly current played a very important role in the formation of very severe heavy rainfall. In addition, the southern current coming from the Arabian Sea also contributed significantly to the occurrence of very heavy rainfall. The above-mentioned two currents converged north of Islamabad and the maxima of convergence of moisture flux just appeared in the region. It could be mentioned here that the Bay of Bengal and the Arabian Sea were the two major sources for moisture supply to this weather system.

(2) In the middle troposphere, the colder air at 500 hPa invaded north Pakistan. The advection of cold air made the atmospheric stratification more unstable in the vertical column and produced a conducive environment for upward extension of convective cumulus clouds.

(3) The sudden heavy rainfall producing severe flooding is associated with the formation of a low pressure system in the lower troposphere that was just located in the inverse "V-shaped" mountain area north of Islamabad. The dynamic lifting of the inverse "V-shaped" terrain to the south and east currents, undoubtedly enhanced the upward motion in the heavy rain area which greatly increased the further development resulting in the heavy downpour.

(4) Some mesoscale β systems, even mesoscale γ systems evolved simultaneously. They further developed in the favorable environment of the lower tropospheric low pressure system. These mesoscale convective systems influenced directly the intensity and position of heavy precipitation. It was the unique phenomenon termed as a "cloud burst." The radar-echo data showed that the mesoscale system moved southeastward along the steering current in the middle troposphere along the foothills to the east of the inverse "V-shaped" terrain, rather than the surface current direction to shift northwestward.

(5) The position of the high pressure over South Asia at 200 hPa was situated over the Iran Plateau and Tibetan Plateau. Its eastern edge was just over the boundary between India and Pakistan, rather than over East Asia. The strong divergence

zone in the upper troposphere was located to the east of Islamabad and Rawalpindi, which provided favorable conditions for intensification of upward motion and hence for heavy rainfall.

(6) Although the numerical models could not fully simulate this historic extreme event of the MCS precipitation, nevertheless encouraging results were drawn regarding location and intensity of this weather system.

References

Ailikun, B. and Yasunari, T., 2001. ENSO and summer monsoon; persistence and transitivity in the seasonal march. Journal of Meteorological Society of Japan 79 (1), 145–159.

Chaudhry, Q.Z., 1991. Analysis and seasonal prediction of Pakistan summer monsoon. Ph. D. dissertation at University of Philippines.

Kar, S.C., Sugi, M., and Sato, N., 2001. Interannual variability of Indian summer monsoon and internal variability in the JMA global model simulations. Journal of Meteorological Society of Japan 79 (2), 607–623.

Ohsawa, T., Ueda, H., and Hayashi, T., 2001. Diurnal variations of convective activity and rainfall in tropical Asia. Journal of Meteorological Society of Japan 79 (1), 333–352.

Rao, Y.P., Srinivasan, V., and Raman, S., 1970. Effect of middle latitude Westerly systems on Indian monsoon. Symposium Tropical Meteorology Hawaii, pp. N, IV 1–4.

Rasul, G., Chaudhry, Q.Z., Zhao, S.X., and Zeng, Q.C., 2004. A diagnostic study of record heavy rain in twin cities Islamabad–Rawalpindi. Advances in Atmospheric Sciences 21 (6), 976–988.

Rasul, G., Chaudhry, Q.Z., Zhao, S.X., et al., 2005. Diagnostic analysis of a heavy rainfall event in South Asia. Journal of Advances in Atmospheric Sciences 22 (3), 375–391.

Tao, S.Y., 1980. Rainstorm in China, Beijing, Scientific Press, 1–225.

Zhao, S.X., 1988. Energetics of Cyclogenesis on Meiyu (Baiu) Front. Proceedings of Palmen Memorial Symposium on Extratropical Cyclones, Helsinki, Finland, 205–208.

Zhang, F. and Zhao, S.X., 2002. Study on one kind of cyclone on Meiyu (Baiu) front. Proceedings of International Conference on Mesoscale Convective Systems and heavy rainfall/snowfall in East Asia, Tokyo, Japan. 117–122.

Zhao, S.X., Bei, N.F., Sun, J.H., et al., 2002. A study of heavy rainfall systems in mid-lower latitude zone of Asian–Australian monsoon area. Climatic and Environmental Research 7 (4), 377–385.

Zhao, S.X. and Mills, G.A., 1991. A study of a monsoon depression bringing record heavy rainfall over Australia. Part II; Synoptic-Diagnostic description. Monthly Weather Review 119 (9), 2074–2094.

Mountains Witnesses of Global Changes
R. Baudo, G. Tartari, and E. Vuillermoz (Editors)

34

Study of land surface heat fluxes and water cycle over the Tibetan plateau

Yaoming Ma, Tandong Yao, Hirohiko Ishikawa and Toshio Koike

Abstract

The energy and water cycles over the Tibetan Plateau play an important role in the Asian monsoon system, which in turn is a major component of both the energy and water cycles of the global climate system. Using field observational data observed from the GAME/Tibet (Global Energy and Water cycle Experiment [GEWEX] Asian Monsoon Experiment on the Tibetan Plateau) and the CAMP/Tibet (Coordinated Enhanced Observing Period [CEOP] Asia–Australia Monsoon Project [CAMP] on the Tibetan Plateau), some results of the local surface energy partitioning (diurnal variation and inter-monthly variation etc.) are presented in this paper. The study on the regional surface energy partitioning is of paramount importance over the heterogeneous landscape of the Tibetan Plateau and is also one of the main scientific objectives of the GAME/Tibet and the CAMP/Tibet. Therefore, the regional distributions and their inter-monthly variations of surface heat fluxes (net radiation flux, soil heat flux, sensible heat flux and latent heat flux) are also derived by combining NOAA-14/AVHRR data and Landsat-7 ETM data with field observations. The derived results are validated with field observation, and by using the methods proposed in this study it shows that the derived regional distributions and their inter-monthly variations of land surface heat fluxes are reasonable. Further improvement of the method and its application field are also discussed. In order to bring up to scale the land surface heat fluxes to the whole Tibetan Plateau area, the Institute of Tibetan Plateau Research (ITP) of the Chinese Academy of Sciences (CAS) is establishing a Monitoring and Research Platform (MORP) for the study of land surface and atmospheric processes on the Tibetan Plateau. The establishing and monitoring plans of long-term scale (5–10 years) of the MORP are also introduced here.

1. Introduction

As the most prominent and complicated high terrain on the globe, the Tibetan Plateau (Ye and Gao, 1979; Ye, 1981; Yanai et al., 1992; Ye and Wu, 1998), with an

ISSN: 0928-2025 DOI: 10.1016/S0928-2025(06)10034-6

elevation of more than 4000 m on average above sea leave (a.s.l.) makes up approximately one-fourth of the land area of China. Long-term research on the Tibetan Plateau has shown that the giant prominence exerts thermal effects on the atmosphere, thus greatly influencing circulations over China, Asia and even the globe (Ye and Gao, 1979; Ye, 1981; Yanai et al., 1992; Ye and Wu, 1998; Ma et al., 2002a; Ma and Tsukamoto, 2002b). Due to its topographic character, the plateau surface absorbs a large amount of solar radiation energy (much of which is redistributed by cryospheric processes), and undergoes dramatic seasonal changes of surface heat and water fluxes (Ye and Gao, 1979; Yanai et al., 1992; Ye and Wu, 1998). The lack of quantitative understanding of interactions between the land surface and atmosphere makes it difficult to understand the complete energy and water cycles over the Tibetan Plateau and their effects on the Asian monsoon system with numerical models. Therefore, it has increased the number of land surface processes studies over the Tibetan Plateau in the past 30 years. But the previous experiments were only carried out in a short period, the observational items were limited, and the previous investigations were only in the summer period and on a few points or at the local level (Ye and Wu, 1998; Zhang et al., 1988; Ma et al., 2002a; Ma and Tsukamoto, 2002b).

The intensive observation period (IOP) and long-term observation of the GAME/Tibet (Global Energy and Water cycle Experiment [GEWEX] Asian Monsoon Experiment on the Tibetan Plateau, 1996–2000) and the CAMP/Tibet (Coordinated Enhanced Observing Period [CEOP] Asia–Australia Monsoon Project [CAMP] on the Tibetan Plateau, 2001–2005) have been done successfully in the past 8 years. Large amounts of data have been collected, which is the best dataset so far for the study of energy and water cycle over the Tibetan Plateau. It gives us a chance to investigate the energy and water cycle over the Tibetan Plateau in detail.

The purpose of this paper is to analyse the characteristics of local and regional surface energy partitioning by using field observational data, NOAA-14/AVHRR data, and Landsat-7 ETM data collected during the GAME/Tibet and the CAMP/Tibet.

2. Experiment

2.1. Experiment of the GAME/Tibet

The overall goal of the GAME/Tibet is to clarify the interaction between the land surface and the atmosphere over the Tibetan Plateau in the context of the Asian monsoon system. To achieve this goal, the scientific objectives of the GAME/Tibet are to improve the quantitative understanding of land-atmosphere interactions over the Tibetan Plateau, to develop process models and methods for applying them over large spatial scales, and to develop and validate satellite-based retrieval methods. The GAME/Tibet is an inter-disciplinary, coordinated effort by field scientists, modellers and remote-sensing scientists in meteorology and hydrology to address these objectives.

Figure 34.1a is the layout of the sites during the IOP of the GAME/Tibet. A meso-scale observational network (150 × 250 km, 91°–92.5°E, 30.5°–33°N) was

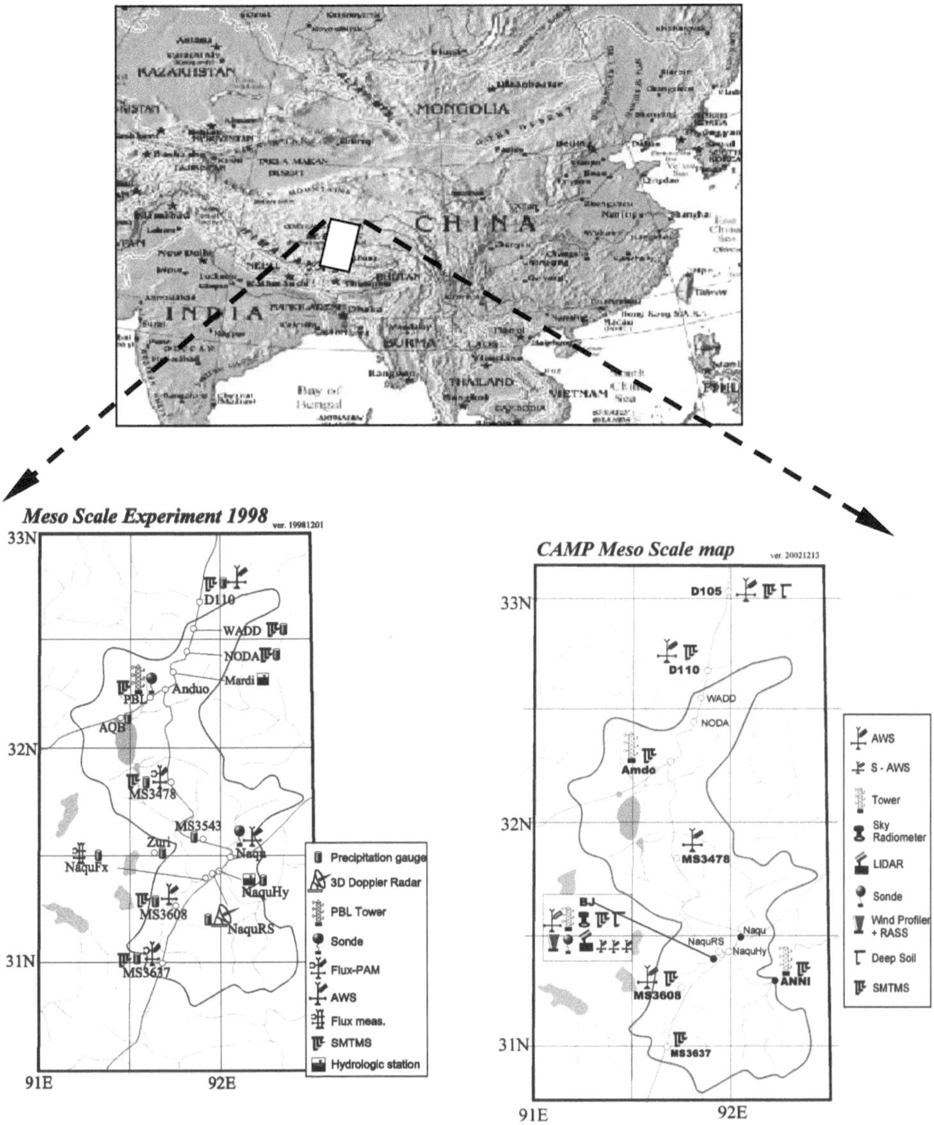

Figure 34.1. The geographic map and the sites layout during the GAME/Tibet (a) and the CAMP/Tibet (b).

implemented in the central plateau: (1) Anduo PBL station (91°37′30″E, 32°14′28″N, elevation: 4700 m a.s.l.). A PBL tower (17 m a.s.l., wind speed, wind direction, air temperature and humidity at three-levels), radiation observational system, the turbulent flux measurements (sonic anemo-thermometer), soil temperature and moisture measurement, radiosonde observation system etc. have been set up in this station; (2) AWS networks. Five AWS (Automatic Weather Station, D66, TTH, Naqu, D110, MS3608) and two sets of flux-PAM (Portable Automated Meso-net,

MS3478-NPAM and MS3637-SPAM) stations have been deployed along the Tibetan (Qinghai-Xizang) highway; (3) A basic PBL observational station (NaquFx). A PBL tower (3.5 m a.s.l., four-levels), radiation observational system, the turbulent flux measurements (sonic anemo-thermometer), soil temperature, moisture measurement and soil water content have been set up in this station in 1998; (4) Barometer network (nine sites: Anduo, AQB, Naqu, Noda, North-mt, South-mt, Sexi, Ziri, Wadd), the ground truth observation sites for validation of satellite data, the sampling network for isotope study on water cycle; (5) Soil temperature and soil moisture network (SMTMS). Nine SMTMS sites (D66, TTH, D110, WADD, NODA, Anduo, MS3478, MS3608 and MS3637) have been deployed along the Tibetan highway; (6) Three-dimensional Doppler radar and precipitation gauge network. The Doppler radar system has been deployed about 10 km south of Naqu in 1998 (91°56′20″E, 31°22′59″N). Seven rain gauges were set up around the radar station; (7) Precipitation gauge net (D110, WADD, NODA, AQB, MS3478, MS3543, Zuri, NaquFx, MS3608, Naquhy, NaquRS and MS3637) has been deployed along the Tibetan high way; and (8) two radiosonde observational systems were set up in Anduo PBL station and Naqu Meteorological Bureau. The IOP of GAME/Tibet was done successfully during May through September 1998.

2.2. *Experiment of the CAMP/Tibet*

The objectives of CAMP/Tibet are: (1) Quantitative understanding of an entire seasonal hydro-meteorological cycle including winter processes by solving surface energy "imbalance" problems in the Tibetan Plateau; (2) Observation of local circulation and evaluation of its impact on plateau-scale water and energy cycle; and (3) Establishment of quantitative observational methods for the entire water and energy cycle between land surface and atmosphere by using satellites. To achieve the scientific objectives of CAMP/Tibet, a meso-scale observational network (150 × 250 km, 91°–92.5°E, 30.7°–33.3°N) was implemented in the central plateau (Fig. 34.1b): (1) A basic observational station (BJ). A flux measuring tower (20 m a.s.l., two levels), a Sky Radiometer, a LIDAR system, a Wind Profiler and RASS, a radiosonde system, and four AWSs have been set up at this station; (2) AWS networks. Six AWS (D105, D110, MS3478, BJ, MS3608 and ANNI) stations have been deployed in this area; (3) Soil moisture and soil temperature measurement systems (SMTMS) networks. Seven SMTMS sites (D105, D110, Anduo, BJ, MS3608, ANNI and MS3637) have been deployed in this area; (4) Two deep soil temperature measurements were put in D105 and NaquDS (nearby the BJ basic station); (5) Anduo PBL station (91°37′30″E, 32°14′28″N). A PBL tower (17 m a.s.l., wind speed, wind direction, air temperature and humidity at three levels) and radiation observational system have already been continued for 6 years from 1997 and were continued for another 2 years until the end of 2005. The enhanced automated observing period (EAOP) and IOP of the CAMP/Tibet were carried out from October 1, 2002 to September 30, 2004.

3. Data analysis and results

The field data observed at three AWS stations (D105, BJ and MS3478) of the CAMP/Tibet under conditions of a clear day will be used in this section. The reason is that each component of surface radiation budget (downward short- and long-wave radiation and upward short- and long-wave radiation) and land surface heat flux budget (net radiation flux, soil heat flux, sensible heat flux and latent heat flux) will be in "truth-value" under the condition of a clear day, and fortunately, there is one clear day each month at least during the measurement period. The geographical coordinates of three AWS stations are, D105: (91°54′22″E, 33°4′2″N), BJ: (91°48′59″E, 31°18′42″N) and MS3478-NPAM: (91°42′10″E, 31°54′15″N). The elevations of the three AWS stations are, D105: 5020 m a.s.l., BJ: 4580 m a.s.l. and MS3478-NPAM: 5063 m a.s.l. The selected data period is from September 2000 to August 2001.

3.1. Local land surface energy budget

Using data observed at the three AWS stations (D105, BJ and MS3478) of the CAMP/Tibet, the turbulent fluxes will be determined as

$$\text{Sensible heat flux } H = \rho C_p C_{HN}(u_z - u_s)(T_{sfc} - T_z), \tag{1}$$

$$\text{Latent heat flux } LE = \rho L C_{EN}(u_z - u_s)(q_{sfc} - q_z) = HB^{-1}, \tag{2}$$

$$\text{Bowen ratio B} = \frac{H}{LE} = \frac{C_p(T_{z_1} - T_{z_2})}{L(q_{z_1} - q_{z_2})}, \tag{3}$$

where ρ is air density, C_p is air specific heat at constant pressure, L is latent heat of vaporization, E is evaporation flux, u_z, T_z and q_z are wind speed, air temperature and specific humidity at the height z; u_s, T_{sfc} and q_{sfc} are wind speed, air temperature and specific humidity on the land surface; C_{HN} and C_{EN} are the bulk transfer coefficients in the neutral state. Normally, C_{HN} and C_{EN} are taken to be identical in the neutral state, and

$$C_{HN} = \frac{k^2}{\left[\text{In}(z/z_{0m})\right]^2}, \tag{4}$$

where k is the Von Karman constant, and z_{0m} is aerodynamic roughness length. Z_{0m} over the Tibetan Plateau area can be determined by using the turbulence data observed with a sonic anemometer-thermometer, PBL tower data and AWS data (Ma et al., 2002a; Ma and Tsukamoto, 2002b).

Figure 34.2 shows the diurnal variations and inter-monthly variations of surface heat fluxes at the stations (D105, BJ, and MS3478) of the CAMP/Tibet, including net radiation flux (R_n), sensible heat flux (H), latent heat flux (LE) and soil heat flux (G_0). One clear day dataset was selected in each month, and the point of the diurnal cycle for 1 month stops and the diurnal cycle for the next month starts just at the centre between the 2 days curves in Fig. 34.2. The results show that: (1) The diurnal

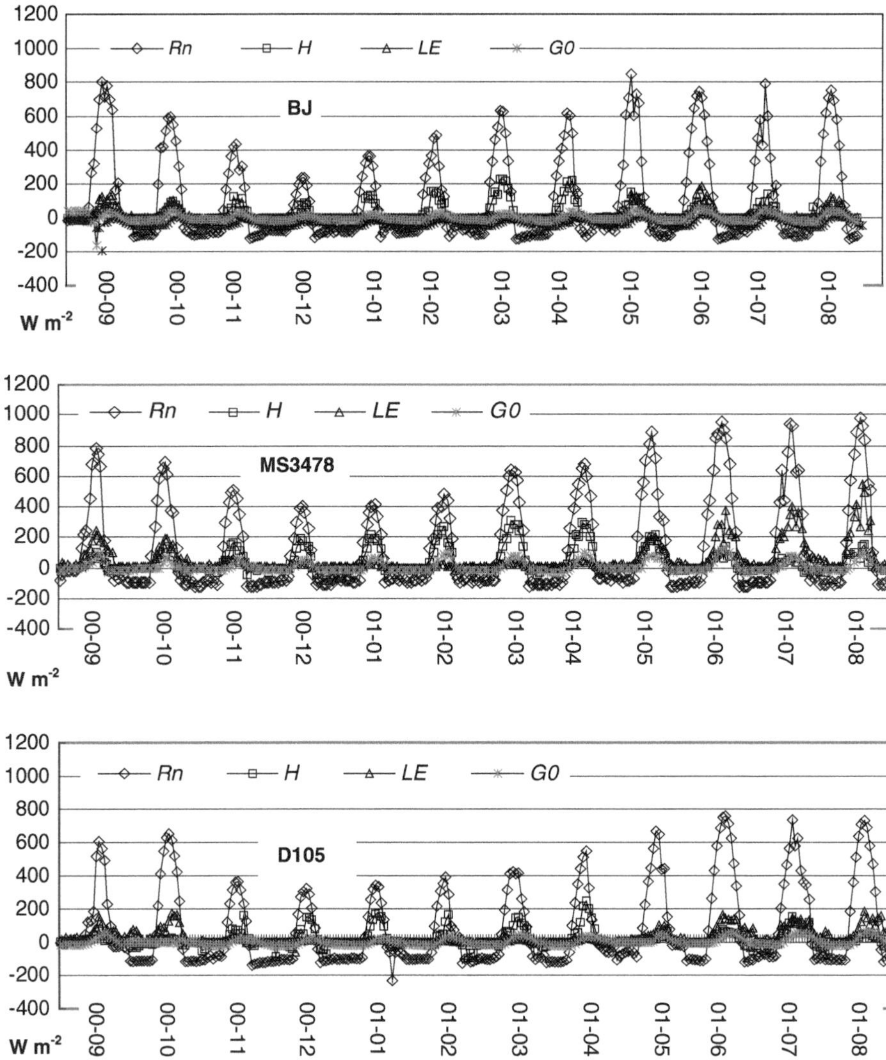

Figure 34.2. The diurnal variation and inter-monthly variations of land surface heat fluxes at the CAMP/Tibet stations.

variations for surface heat fluxes in the Tibetan Plateau area are very clear. The net radiation flux in each month at MS3478 station is larger than that at D105 station and BJ station due to the good vegetation coverage at this station (lower upward long-wave radiation); (2) Sensible heat and latent heat fluxes play different roles in the partition of the net radiation flux in different month in the Tibetan Plateau. In other words, the sensible heat flux plays the main role in winter and the latent heat flux plays the main role in summer and autumn; (3) The surface energy budget was, however, not well closed from the observed data. In the $CR = (H+LE)/(R_n-G_0)$ equation, the present results show $CR = 0.7$ and sometimes as low as 0.67. These kind of "imbalances" were also found not only from PBL data analysis, but also from sonic anemometer data analysis (Ma et al., 2000; Tanaka et al., 2001). And the "imbalance" is larger in summer than in winter. There are various possibilities for this "imbalance": (1) measurement problems (Ma et al., 2000); (2) advections around the experimental stations (Ma et al., 2000); (3) there exists a discussion that a very weak systematic vertical flow can cause such an imbalance (Lee, 1998).

3.2. *The distribution and inter-monthly variation of regional land surface heat fluxes*

The study on the regional surface energy partitioning and its inter-monthly variation is of paramount importance over the heterogeneous landscape of the Tibetan Plateau and it is also one of the main scientific objectives of the GAME/Tibet and the CAMP/Tibet. Therefore, we will try to derive here the regional land surface heat fluxes by using the NOAA-14/AVHRR data, Landsat-7 ETM data and the field observational data. Here, "Regional" land surface heat fluxes are not "aggregated" fluxes (Batchvarova et al., 2001), but surface fluxes fields or surface heat fluxes on each pixel of NOAA-14/AVHRR data and Landsat-7 ETM data.

The general concept of the methodology is as follows (Ma et al., 2002c; Ma et al., 2003): the surface reflectance for short-wave radiation (r_0), land surface temperature (T_{sfc}), is retrieved from NOAA-14/AVHRR data and Landsat-7 ETM data with the atmospheric correction by using the radiative transfer model MODTRAN (Berk et al., 1989) and aerological observational data. The radiative transfer model also computes the downward short- and long-wave radiation at the surface. With these results the surface net radiation flux (R_n) is determined. The soil heat flux (G_0) is estimated from R_n, T_{sfc}, r_0 and Modified Soil Adjusted Vegetation Index (MSAVI) of Qi et al., 1994), which is also derived from NOAA-14/AVHRR data and Landsat-7 ETM data. Sensible heat flux (H) is estimated from T_{sfc}, surface and aerological data with the aid of so-called "blending height" approach (Mason, 1988), and latent heat flux is the residual of the energy budget theorem for the land surface.

3.2.1. *Net radiation*

The regional net radiation flux is expressed as

$$R_n(x,y) = (1 - r_0(x,y))K_{\downarrow}(x,y) + L_{\downarrow}(x,y) - \varepsilon_0(x,y)\sigma T^4_{sfc}(x,y) \qquad (5)$$

where σ is the Stefan Boltzmann constant and surface reflectance $r_0(x,y)$ can be derived from NOAA-14/AVHRR data and Landsat-7 ETM data (Ma et al., 2002c, Ma et al., 2003). The surface emissivity $\varepsilon_0(x, y)$ is a function of the vegetation coverage, and is also derived from Valor and Caselles's model (1996). The incoming short-wave radiation flux $K_{\downarrow}(x, y)$ and the incoming long-wave radiation flux $L_{\downarrow}(x, y)$ in Eq. (6) can be derived from radiative transfer model MODTRAN (Kenizys et al., 1996) directly. The land surface temperature $T_{sfc}(x, y)$ is retrieved from the brightness temperature of channels 4 and 5 of NOAA-14/AVHRR and channel 6 of Landsat-7 ETM data (Ma et al., 2002c, Ma et al., 2003).

3.2.2. Soil heat flux

The regional soil heat flux $G_0(x, y)$ is usually determined by (Chodhury and Monteith, 1988)

$$G_0(x,y) = \frac{\rho_s C_s[(T_{sfc}(x,y) - T_s(x,y)]}{r_{sh}(x,y)}, \tag{6}$$

where ρ_s is soil dry bulk density, C_s is soil specific heat, $T_s(x, y)$ stands for soil temperature of a determined depth, $r_{sh}(x, y)$ represents soil heat transportation resistance. However, the regional soil heat flux G_0 (x, y) cannot directly be mapped from satellite observations through Eq. (6) for the difficulty to determine the soil heat transportation resistance $r_{sh}(x, y)$ and the soil temperature at a reference depth $T_s(x, y)$. Many investigations have shown that the mid-day G_0/R_n fraction is reasonably predicted from special vegetation indices (Daughtry et al., 1990). Some researchers have shown that $G_0/R_n = \Gamma(\text{NDVI})$ (Clothier et al., 1986; Chodhury et al., 1987; Kustas and Daughtry, 1990). An improved fraction of $G_0/R_n = \Gamma(r_0,\ T_{sfc},$ NDVI) was proposed (Menenti et al., 1991; Bastiaanssen, 1995). However, problems exist in the NDVI definition equation because of the effects of external factors, such as soil background variations (Huete et al., 1985; Huete, 1989). In order to reduce the soil background effect in NDVI, a parameterisation based on MSAVI is proposed over the Tibetan area in this study as

$$G_0(x,y) = R_n(x,y)\left(\frac{T_{sfc}(x,y)}{r_0(x,y)}\right)(a + b\bar{r}_0 + c\bar{r}_0^2)[1 + dMSAVI(x,y)^e] \tag{7}$$

where the constants a, b, c, d and e are determined by using the field data observed at six observation stations (AWS110, Anduo, NPAM-MS3478, Naqu, AWS3608 and SPAM-MS3637) during the IOP of GAME/Tibet; $\bar{r}_0$ is a daily mean reflectance value obtained from field observations. MSAVI (x, y) is derived from the band reflectance of NOAA-14/AVHRR data and Landsat-7 ETM data (Qi et al., 1994)

3.2.3. Sensible and latent heat fluxes

The sensible heat flux $H(x, y)$ can be derived from

$$H(x,y) = \rho C_p k^2 u(x,y) \frac{[T_{sfc}(x,y) - T_a(x,y)]}{[\text{In}(z - d_0(x,y)/Z_{0m}(x,y)) + kB^{-1}(x,y) - \psi_h(x,y)][\text{In}(z - d_0(x,y)/Z_{0m}(x,y)) - \psi_m(x,y)]}. \tag{8}$$

The straightforward way to model sensible heat flux in a large area is to sum up the contribution from different surface elements. If the local scale advection is comparatively small, it is desired that the development of convective boundary layer may smooth the local heterogeneity of surface disorganized variety at the so-called "blending height", where atmospheric characteristics become approximately independent of horizontal locations. The corresponding "effective" surface variables can be determined accordingly (Mason, 1988). This approach has been proved to be successful to calculate regional averaged surface fluxes recently (Lhomme et al., 1994; Bastiaanssen, 1995; Wang et al., 1995; Ma et al., 1999a; Batchvarova et al., 2001; Ma et al., 2002c). Based on this approach, the regional sensible heat flux $H(x, y)$ is expressed as

$$H(x,y) = \rho C_p k^2 u_B \frac{[T_{sfc}(x,y) - T_{air-B}]}{[\mathrm{In}(z_B - d_0(x,y)/Z_{0m}(x,y)) + kB^{-1}(x,y) - \psi_h(x,y)]\,[\mathrm{In}(z_B - d_0(x,y)/Z_{0m}(x,y)) - \psi_m(x,y)]} \quad (9)$$

where Z_B is blending height, u_B and T_{air-B} are wind speed and air temperature at the blending height, respectively. Z_B, u_B and T_{air-B} are determined by using field measurements or numerical models. In this study, these variables will be determined with the aid of radiosonde measurements. $Z_{0m}(x, y)$ is effective aerodynamic roughness length including the effect of topography and low vegetation (e.g. grass), and is determined by the Taylor's model (Taylor et al., 1989). In other words, local aerodynamic roughness length can be determined by using the observed turbulent data (Chen et al., 1993; Ma et al., 2002a), and Table 1 is local aerodynamic roughness length at two stations (Anduo and NPAM-MS3478) derived from turbulent data (in order to compare each other, local aerodynamic roughness length on other land surface are also given here). Then, the effective aerodynamic roughness length $Z_{0m}(x, y)$ can be derived by considering the effect of grass and mountains (Taylor et al., 1989). The excess resistance to heat transfer, kB^{-1}, is shown as a function of surface temperature over the Tibetan Plateau area (Ma et al., 2002a). d_0 is zero-plane displacement length, which can be calculated from Raupach's model (Raupach, 1994) over this area. $\psi_h(x, y)$ and $\psi_m(x, y)$ are the integrated stability functions in Eq. (9). They can be derived by using the models of Paulson (1970) and Webb (1970).

The regional latent heat flux $LE(x, y)$ is the residual of the energy budget theorem for land surface, i.e.

$$LE(x,y) = R_n(x,y) - H(x,y) - G_0(x,y). \quad (10)$$

3.2.4. *Data and results*

It is better to select the satellite data of clear days to study the distribution and inter-monthly variation of energy budget components. Unfortunately, it is difficult to select this kind of satellite data over the Tibetan Plateau area because of the strong convective clouds when NOAA-14/AVHRR and Landsat-7 ETM observations took place. Only three scenes of the NOAA-14/AVHRR (June 12, 1998, July 16, 1998 and August 21, 1998) and five scenes of the Landsat-7 ETM (June 9, 2002, July 24, 2001, August 28, 2002, December 2, 2002 and March 24, 2003) could be selected during the whole IOP of the GAME/Tibet and the whole enhanced observation period (EOP) of the CAMP/Tibet.

The regional distribution of surface reflectance, surface temperature, NDVI, MSAVI, vegetation coverage, LAI, net radiation flux, soil heat flux, sensible heat flux and latent heat flux around the GAME/Tibet and CAMP/Tibet area were derived by using the methods proposed by Ma et al. (2002c, 2003) (figures are omitted). The derived land surface heat fluxes were validated by field measurements. Since it is difficult to determine where the exact locations of the experimental sites are, the values of a 5×5 pixel rectangle, surrounding the determined Universal Transfer

Macerator (UTM) coordinate, are compared with the field measurements and they are shown in Fig. 34.3. The mean absolute percent difference (MAPD) was computed as a quantitative measure of the difference between the derived results on no. *i* point ($H_{\text{derived}(i)}$) and measured value on no. *i* point ($H_{\text{measured}(i)}$) of one scene as,

$$\text{MAPD} = \frac{100}{\text{n}} \sum_{i=1}^{n} \left(\frac{\left| H_{\text{derived}(i)} - H_{\text{measured}(i)} \right|}{H_{\text{measured}(i)}} \right). \tag{11}$$

The results over the GAME/Tibet area show that: (1) the derived land surface heat fluxes were in good accordance with the land surface status. These parameters show a wide range of variations due to the strong contrast of surface features in the study area; (2) not only on June 12, but also on July 16 and August 21, the derived net radiation flux, soil heat flux and sensible heat flux were close to the field measurements. The difference between the derived results and the field observation MAPD was less than 10% (Fig. 34.3); (3) during the experimental periods, the derived net radiation flux was larger than that in the HEIFE area (Ma et al., 2002c) due to the high altitude (the higher value of downward short-wave radiation) and land surface coverage of grassy marshland (the lower value of the upward long-wave radiation) in this area. For example, the regional average value of net radiation flux was 470 W/m^2 over the HEIFE area in July 9, 1991 and that was 750 W/m^2 over the GAME/Tibet area in July 16, 1998; (4) the values of soil heat flux and sensible heat flux in June over this area were larger than these values in July and August. Net radiation flux and latent heat flux in June were lower than their values in July and August. The reason is that although land surface in this area is covered by the same grassy marshland in these days, June 12 was the day before the Asian monsoon coming, and the land surface was dry on that day, July 16 and August 21 were within and after Asia monsoon, the land surface was wet and the grass was high and in growing; (5) The derived regional soil heat fluxes based on MSAVI were reasonable in different months in this area with MAPD less than 10% (Fig. 34.3); and (6) all elements of heat balance equation at NPAM site on June 12 corresponded well to the satellite data. On the other hand, all but latent heat flux corresponded to the satellite data in other seasons and other stations. The conclusions derived from above facts were: (a) net radiation, sensible heat flux and soil heat flux could be derived from satellite; (b) in the case of NPAM on June 12, because the surface energy balanced in the surface observation so that the latent heat flux estimated by surface observation corresponded well to that estimated by the residual of the satellite data analysis. A one dimensional energy budget did not balance due to large error of the latent heat flux through the surface observation and advection in this area. The large error of the measurement of latent heat flux may depend on the accuracy of the turbulence measurement sensors (Ma et al., 1999b; Ishikawa et al., 1999; Ma et al., 1999c; Ma et al., 2000; Tanaka et al., 2001). This point and conclusion (a) clearly show the disagreement of latent heat flux between surface observations and satellite ones.

The results over the CAMP/Tibet area have shown that: (1) the derived surface variables (land surface reflectance and surface temperature) and surface heat fluxes (net radiation flux R_n, soil heat flux G_0, sensible heat flux H and latent heat flux LE) in five different months over the study area are in good accordance with the land

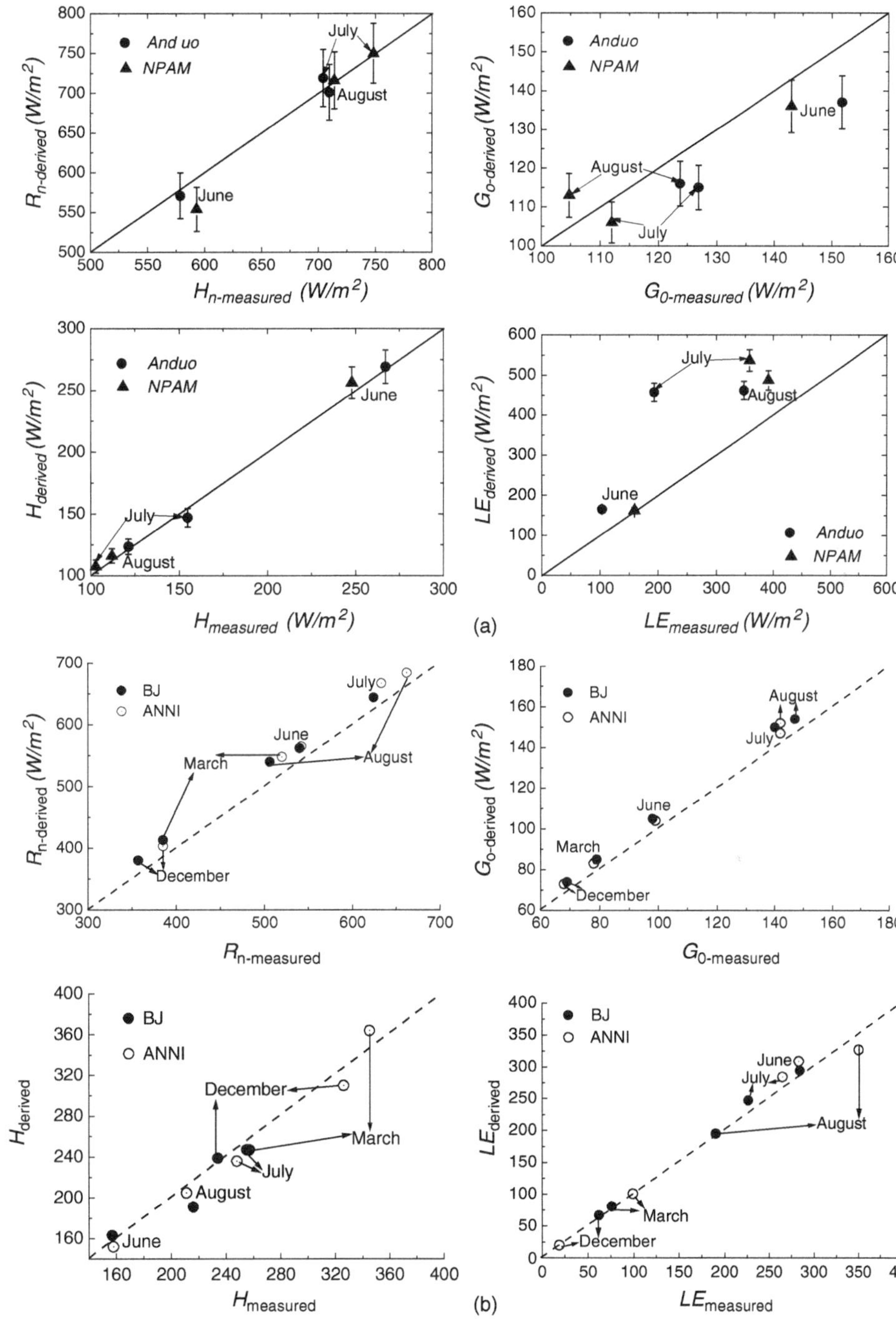

Figure 34.3. Comparison of the derived results with the field measurements for the land surface heat fluxes over (a) the GAME/Tibet area and (b) the CAMP/Tibet area, together with 1:1 line.

surface status. These parameters show a wide range due to the strong contrast of surface features. Surface reflectance, surface temperature and sensible heat flux around the lake in the distribution maps are much higher in summer (June, July and August), and at the same time, net radiation flux, soil heat flux and latent heat flux are lower in the area. The reason is that most of the vegetation around lake areas is desertification grass land; (2) the derived pixel value and average value of surface temperature, net radiation flux, soil heat flux and latent heat flux in June, July and August are higher than that in December and March. This means that there is much more evaporation in summer than it is in winter (December) and spring (March) in the central Tibetan Plateau area. And it is also pointed out that the heating density ($H+LE=R_n-G_0$) in summer is much higher than it is in winter and spring in the central Tibetan Plateau area. But the sensible heat flux is the main role in the distribution of the net radiation flux in December and March; (3) not only in summer (June 9, July 24 and August 28), but also in winter (December) and spring (March), the value of the vegetation coverage in this area was almost the same due to the fact that the land surface in this area is covered by the same grassy marshland on those days. The only difference was that the grass was dry on December, March and June (with small MSAVI). It became green and wet on July 24 and August 28 (with higher MSAVI) for this area; (4) during the experimental periods, the derived net radiation flux was larger than that in the HEIFE area (Ma et al., 2002c) due to the high altitude (the higher value of downward short-wave radiation) and land surface coverage of grassy marshland (the lower value of the upward long-wave radiation) in this area. For example, the regional average value of net radiation flux was 470 W/m^2 over the HEIFE area in July 9, 1991 and that was 725 W/m^2 over the CAMP/Tibet area in July 24, 2001; (5) Even the resolution of Landsat-7 ETM is different with the NOAA/AVHRR, the derived results here are also comparable to those derived from NOAA/AVHRR data around the relative homogeneous sites of the CAMP/Tibet area (Ma et al., 2003). For example, the derived results of net radiation flux, soil heat flux, sensible heat flux and latent heat flux from Landsat-7 ETM data in the BJ site (Naqufx or Naqu) site during the GAME/Tibet period) in June are 562, 105, 163 and 294 W/m^2, respectively, and at the same time, the derived values from NOAA/AVHRR data are 578, 108, 178 and 292 W/m^2, respectively. But it should be validated over the heterogeneous sites (e.g. ANNI site). Unfortunately, we have no measurements at ANNI site during the GAME/Tibet period; (6) the derived surface reflectance and surface temperature in this research are in good accordance with the field measurements with MAPD less than 9.0% (Fig. 34.3); (7) the derived net radiation flux over the study area is very close to the field measurement with MAPD less than 8.0% (Fig. 34.3); (8) the regional soil heat flux derived from the relationship between soil heat flux and net radiation flux is suitable for heterogeneous land surface of the CAMP/Tibet area, and the MAPD is less than 8.0% at validation sites due to the relationship itself was derived from the same area (Ma et al., 2003) (Fig. 34.3); and (9) the derived regional sensible heat flux and latent heat flux with MAPD less than 9.5% at validation site in the CAMP/Tibet area is in good agreement with field measurements (Fig. 34.3). It is pointed out that the proposed parameterisation methodology for sensible heat flux and latent heat flux is reasonable, and it can be used over the central Tibetan Plateau area.

4. Concluding remarks

In this paper, the field experiments of the GAME/Tibet and the CAMP/Tibet are introduced and some results on the local surface energy partitioning (diurnal variation and inter-monthly variation) are presented by using the field observational data. The surface energy budget was not well closed from the observed data. The reason for this "imbalance" is pointed out, but further systematic research is necessary to figure out the cause of surface flux imbalances.

The regional distributions and inter-monthly variations of land surface heat fluxes (net radiation flux, soil heat flux and sensible heat flux) over the heterogeneous area of the GAME/Tibet and the CAMP/Tibet area are derived by using NOAA-14/AVHRR data, Landsat-7 ETM data and field observations. The results are in good agreement with field observations. The approach of deriving regional latent heat flux as the residual of the energy budget may not be a good method due to the measured "imbalance" of energy and the strong advection over the study area. Future improvements need to be made to derive more accurate regional latent heat flux over such areas. It is also worth trying SEBS (Su, 2002) to derive regional land surface heat fluxes, especially for latent heat flux.

All the results in this paper were obtained only from the meso-scale area. In order to upscale the land surface heat fluxes to the whole Tibetan Plateau area, the Institute of Tibetan Plateau Research (ITP) of the Chinese Academy of Sciences (CAS) is establishing a Monitoring and Research Platform (MORP) for the study of the land surface and atmospheric processes on the Tibetan Plateau. Eight comprehensive observation and study stations and 10 observational sites will be established and set

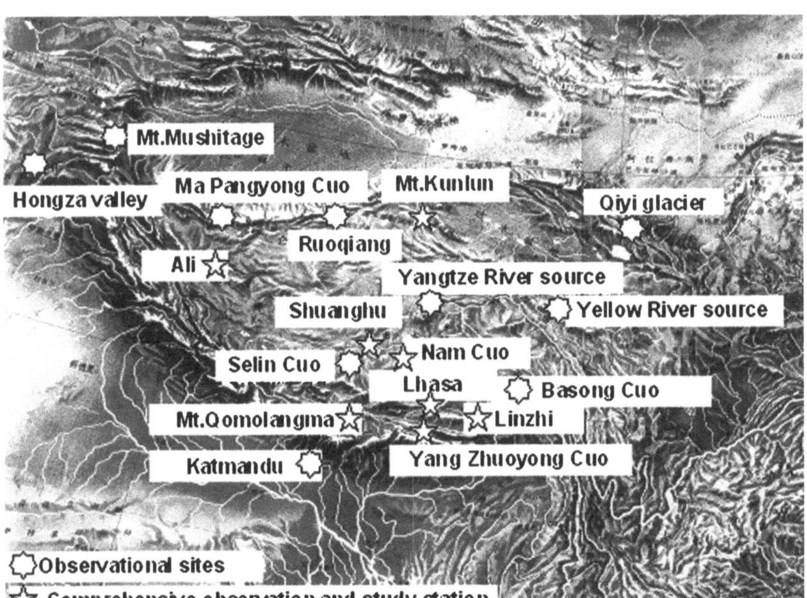

Figure 34.4. Monitoring and research platform (MORP) over the Tibetan Plateau.

up over the Tibetan Plateau area (see Fig. 34.4). The atmospheric boundary layer (ABL) tower, radiation system, SMTMS, GPS radiosonde system, wind profiler and RASS system, sonic turbulent measurement systems and CO_2/H_2O flux measurement systems will be laid out in these stations and sites. Three stations (Mt. Qomolangma [Everest], Nam Cuo and Linzhi) have already been established in the end of August, 2005, and they are working well now. Large amounts of data have been collected during the past 4 months.

The land-surface heat fluxes and water-cycle processes can be analyzed with the aid of the stations and sites data. Then, using the atmospheric models and remote-sensing parameterisation method, the point or local results can be scaled up to the Plateau scale. The large-scale results can also be validated by the stations and sites data. The recycle will be done again and again, and we believe that we can contribute much to the study of land surface heat fluxes and water cycle over the Tibetan Plateau.

Acknowledgements

This research is under the auspices of the Chinese National Key Programme for Developing Basic Sciences (2005CB422003), the National Natural Science Foundation of China (40520140126), and the Innovation projects of Chinese Academy of Science (KZCX3-SW-339 and KZCX3-SW-231). This study was done as a cooperative research work in the International Institute for Geo-Information Science and Earth Observation, the Netherlands. The first author would like also to acknowledge Prof. Z. Su and Prof. M. Menenti for their help in the procedure of the paper. The authors thank all the participants from China and Japan in the field observations of the GAME/Tibet and the CAMP/Tibet.

References

Bastiaanssen, W.G.M., 1995. Regionalization of surface fluxes and moisture indicators in composite terrain. Ph.D. Thesis. Wageningen Agricultural University, 273 pp.

Batchvarova, E., Gryning, S.E., and Hasager, C.B., 2001. Regional fluxes of momentum and heat over a sub-artic landscape during late winter. Boundary-Layer Meteorology 99, 489–507.

Berk, A., Bernstein, L.S., and Robertson, D.C., 1989. MODTRAN, A moderate resolution model for LOTRAN 7, GL-TR-89-0122.

Chen, J.Y., Wang, J.M., and Mitsuta, Y., 1993. An independent method to determine the surface roughness length. Chinese Journal Atmospheric Sciences 17 (1), 21–26.

Chodhury, B.J., Idso, S.B., and Reginato, R.J., 1987. Analysis of an empirical model for soil heat flux under a growing wheat crop for estimating evaporation by infrared-temperature based energy balance equation. Agricultural and Forest Meteorology 39, 283–297.

Chodhury, B.J. and Monteith, J.L., 1988. A four-layer model for the heat budget of homogeneous land surfaces. Quarterly Journal of the Royal Meteorological Society 114, 373–398.

Clothier, B.E., Clawson, K.L., Pinter, P.J., et al., 1986. Estimating of soil heat flux from net radiation during the growth of alfalfa. Agricultural and Forest Meteorology 37, 319–329.

Daughtry, C.S.T., Kustas, W.P., Moran, M.S., et al., 1990. Spectral estimates of net radiation and soil heat flux. Remote Sensing of the Environment 32, 111–124.

Huete, A.R., 1989. Soil influences in remotely-sensed vegetation-canopy spectra. In: Asrar, G. (Ed.) Theory and Applications of Optical Remote Sensing. Wiley, New York, pp. 107–141.

Huete, A.R., Jackson, R.D., and Post, D.F., 1985. Spectral response of a plant canopy with different soil backgrounds. Remote Sensing of the Environment 17, 37–53.

Ishikawa, H., Hayashi, T., Tanaka, K., et al., 1999. Summary of the boundary layer observation and the preliminary analysis. Third International Scientific Conference on the Global Energy and Water Cycle, Beijing, China, 16–19 June 1999, 45–46.

Kenizys, F.X., Abreu, L.W., Anderson, G.P., et al., 1996. The MODTRAN3/2 report and LODTRAN 7 Model. (Abreu L.W. and Anderson, G.P., Eds), prepared by Ontar Corp., North Andover, MA, for Phillips Laboratory, Geophysical Directorate, Hanscom AFB, MA., Contract No.F19628-91-C-0132.

Kustas, W.P. and Daughtry, C.S.T., 1990. Estimation of the soil heat flux/net radiation ratio from spectral data. Agricultural and Forest Meteorology 39, 205–223.

Lee, X.H., 1998. On micrometeorological observations of surface-air exchange over tall vegetation. Agriculture and Forest Meteorology 91, 39–49.

Lhomme, J.P., Chehbouni, A., and Monteny, B., 1994. Effective parameters of surface energy balance in heterogeneous landscape. Boundary-Layer Meteorology 71 (3), 297–310.

Ma, Y.M. and Tsukamoto, O., 2002b. Combining satellite remote sensing with field observations for land surface heat fluxes over inhomogeneous landscape. China Meteorological Press, Beijing, China.

Ma, Y.M., Ishikawa, H., Tsukamoto, O., et al., 2003. Regionalization of surface fluxes over heterogeneous landscape of the Tibetan Plateau by using satellite remote sensing. Journal of the Meteorological Society of Japan 81, 277–293.

Ma, Y.M., Tsukamoto, O., Ishikawa, H., et al., 2002c. Determination of regional land surface heat flux densities over heterogeneous landscape of HEIFE integrating satellite remote sensing with field observations. Journal of the Meteorological Society of Japan 80 (3), 485–501.

Ma, Y.M., Tsukamoto, O., Wang, J.M., et al., 1999c. Transfer and micrometeorological characteristics in the surface layer of the atmosphere above Tibetan Plateau area. Third International Scientific Conference on the Global Energy and Water Cycle, Beijing, China, 16–19 June 1999, 55–56.

Ma, Y.M., Tsukamoto, O., Wang, J.M., et al., 1999b. The characteristics of micrometeorology in the northern Tibetan Plateau area. Proceedings of the 1st International Workshop on GAME-Tibet, Xi'an, China, 11–13 January 1999, 99–102.

Ma, Y.M., Tsukamoto, O., Wang, J.M., et al., 2002a. Analysis of aerodynamic and thermodynamic parameters over the grassy marshland surface of Tibetan Plateau. Progress in Natural Science 12 (1), 36–40.

Ma, Y.M., Tsukamoto, O., Wu, X.M., et al., 2000. Characteristics of energy transfer and micrometeorology in the surface layer of the atmosphere above marshland of the Tibetan Plateau area. Chinese Journal Atmospheric Sciences 24 (5), 715–722 (in Chinese).

Ma, Y.M., Wang, J.M., Menenti, M., and Bastiaanssen, W.G.M., 1999a. Estimation of fluxes over the heterogeneous land surface with the aid of satellite remote sensing and field observation. ACTA Meteorological Sinica 57, 180–189 (in Chinese).

Mason, P., 1988. The formation of areally averaged roughness lengths. Quarterly Journal of the Royal Meteorological Society 114, 399–420.

Menenti, M., Bastiaanssen, W.G.M., Hefny, K., et al., 1991. Mapping of ground water losses by evaporation in the Western Desert of Egypt. DLO Winand Staring Centre, Report no. 43. Wageningen, The Netherlands, 116 pp.

Paulson, C.A., 1970. The mathematic representation of wind speed and temperature profiles in the unstable atmospheric surface layer. Journal of the Meteorological Society of Japan 9, 856–861.

Qi, J.G., Chehbouni, A., Huete, A.R., et al., 1994. A modified soil adjusted vegetation index. Remote Sensing of the Environment 48, 119–126.

Raupach, M.R., 1994. Simplified expressions for vegetation roughness length and zero-plane displacements as functions of canopy height and area index. Boundary-Layer Meteorology 71, 211–216.

Su, Z.B., 2002. The surface energy balance system (SEBS) for estimation of turbulent heat fluxes. Hydrology and Earth System Sciences 6 (1), 85–99.

Tanaka, K., Ishikawa, H., Tamagawa, I., and Ma, Y.M., 2001. Surface energy budget at Amdo on Tibetan Plateau using GAME/Tibet IOP'98 data. Journal of the Meteorological Society of Japan 79 (1B), 505–517.

Taylor, P.A., Sykes, R.I., and Mason, P.J., 1989. On the parameterization of drag over small-scale topography in neutrally stratified Boundary flow. Boundary-Layer Meteorology 48, 409–422.

Valor, E. and Caselles, V., 1996. Mapping land surface emissivity from NDVI: application to European, African, and South American areas. Remote Sensing of the Environment 57, 167–184.

Wang, J.M., Ma, Y.M., Menenti, M., and Bastiaanssen, W.G.M., 1995. The scaling-up of processes in the heterogeneous landscape of HEIFE with the aid of satellite remote sensing. Journal of the Meteorological Society of Japan 73 (6), 1235–1244.

Webb, E.K., 1970. Profile relationships: the log-liner range and extension to strong stability. Quarterly Journal of the Royal Meteorological Society 96, 67–90.

Yanai, M., Li, C.Y., and Song, Z., 1992. Seasonal heating of the Tibetan Plateau and its effects on the evolution of the Asian summer monsoon. Journal of the Meteorological Society of Japan 70, 319–351.

Ye, D.Z., 1981. Some characteristics of the summer circulation over the Qinghai-Xizang (Tibet) Plateau and its neighborhood. Bulletin of the American Meteorological Society 62, 14–19.

Ye, D.Z. and Gao, Y.X., 1979, The Meteorology of the Qinghai-Xizang (Tibet) Plateau. Science Press, Beijing, 278 pp. (in Chinese).

Ye, D.Z. and Wu, G.X., 1998. The role of the heat source of the Tibetan Plateau in the General circulation. Meteorology and Atmospheric Physics 67, 181–198.

Zhang, Q. J., Zhu, B. Z., Zhu, F. K., et al., 1988, Advances in the Qinghai-Xizang Plateau Meteorology. The Qinghai-Xizang Plateau Meteorological Experiment (1979) and Research. Science Press, Beijing, 268 pp. (in Chinese).

Mountains Witnesses of Global Changes
R. Baudo, G. Tartari, and E. Vuillermoz (Editors)

35

Research on global changes in Pakistan

Rakhshan Roohi

Abstract

The research on climate change in Pakistan is in its initial stages and has started with the commencement of the new century. Realizing the importance and potential impact of climate change on natural resources and population and their sustainable development, several projects have been initiated in collaboration with international institutions such as APN, ICIMOD, UNEP, START, and others. The focus of these initiatives is on the impact of climate change on various components of the complex system such as water resources, ecology, agriculture, socio-economics, and others. So far these activities have been centered on the resource inventories/vulnerability assessment and identification of potential threats based on which coping mechanisms could be identified and policy recommendations could be formulated. Using historical climatic data it has been concluded that since 1931 in general the extreme temperatures in the arid environments have increased while in sub-humid and humid environments these have been decreased. Several studies have been carried out to use the simulation models for climate change impact on water resources, forest ecosystems, and socioeconomic conditions. Recently an inventory of glaciers and glacial lakes in the Hindu Kush – Karakoram – Himalayan (HKKH) region of Pakistan has been completed. A Global Change Impact Study Center has also been established.

1. Introduction

Pakistan lies between 24° and 37° N latitude and 61° and 75° longitude covering an area of 88.2 million hectares including the northern areas. Inherently the country has highly variable topography, climate, and culture. The three major mountain ranges, namely Hindu Kush, Karakoram, and Himalaya (HKKH) border in the north followed by plateaus, plains, and coastal areas. The climatic variability is expressed by humid zones in the northeast to hyper-arid in the southwest and west. The major part of the country could be classified as arid to hyper-arid.

The country's high mountains comprise the western end of the 2400 km long Himalayan range and some parts in the Hindu Kush and Karakoram ranges. The northern areas spread over 72,496 km^2 amidst towering snow-clad peaks with heights varying from 1000 m a.s.l. to over 8000 m a.s.l. Of the 14 peaks >8000 m a.s.l. on

ISSN: 0928-2025 DOI: 10.1016/S0928-2025(06)10035-8

Earth, 4 occupy an amphitheater at the head of Baltoro glacier in the Karakoram Range. In addition to these, there are 68 peaks >7000 m a.s.l. and hundreds that are >6000 m a.s.l.

Some of the lower mountain ranges in the northeast receive high monsoon rainfall in summer and snow precipitation during winter. The forest covers are dense in this mountain region. The high northern and northwestern areas are generally out of the reach of monsoon so the climate is dry and precipitation occurs only due to depressions moving in from the west during the spring and summer. The western dry mountains are lower and more arid with the highest peak of 3374 m a.s.l.

2. Global change research

The global change research in the country has just started and is mainly focused to understand the past trends and current variabilities. Few capacity building and awareness workshops have been organized at the national level. In general the focus of climate change research in the country could be categorized as follows:

2.1. Physical indicators

2.1.1. Glaciers and glacial lakes

Glaciers and glacial lakes are the barometers of climate change. The rising global temperatures (greenhouse effect) that causes glacial retreat in many mountain regions are responsible for the formation of glacial lakes. In a comprehensive study using high quality toposheets, remote-sensing data of the Landsat-7 Enhanced Thematic Mapper Plus (ETM+) and digital elevation models the inventory of glaciers and glacial lakes was completed and the potentially dangerous glacial lakes were identified (Roohi et al., 2005).

2.1.1.1. Glaciers. Over a vast area of Indus River Basin (128,730.8 km^2) 5218 glaciers are identified (Fig. 35.1). These glaciers are mainly distributed in the northern part of HKH region covering the higher Karakoram and Hindu Kush ranges. These glaciers are also present at the higher elevations of Himalaya.

The northern sub-basins have relatively large size glaciers while in southern sub-basins the glaciers are small in size. These glaciers contribute to a total glaciated area of about 15,040.8 km^2 which is 11.68% of the total area. These glaciers contribute a total of 2738.5 km^3 of ice reserves. The total length of glaciers in the glaciated area is more than 9718 km (Table 35.1). The glaciers in the northern basins, especially in the Karakoram Range, have maximum length.

2.1.1.2. Glacial lakes. In summary, a total of 2420 glacial lakes have been identified in the HKKH region of Pakistan. Figure 35.2 shows the distribution of glacial lakes.

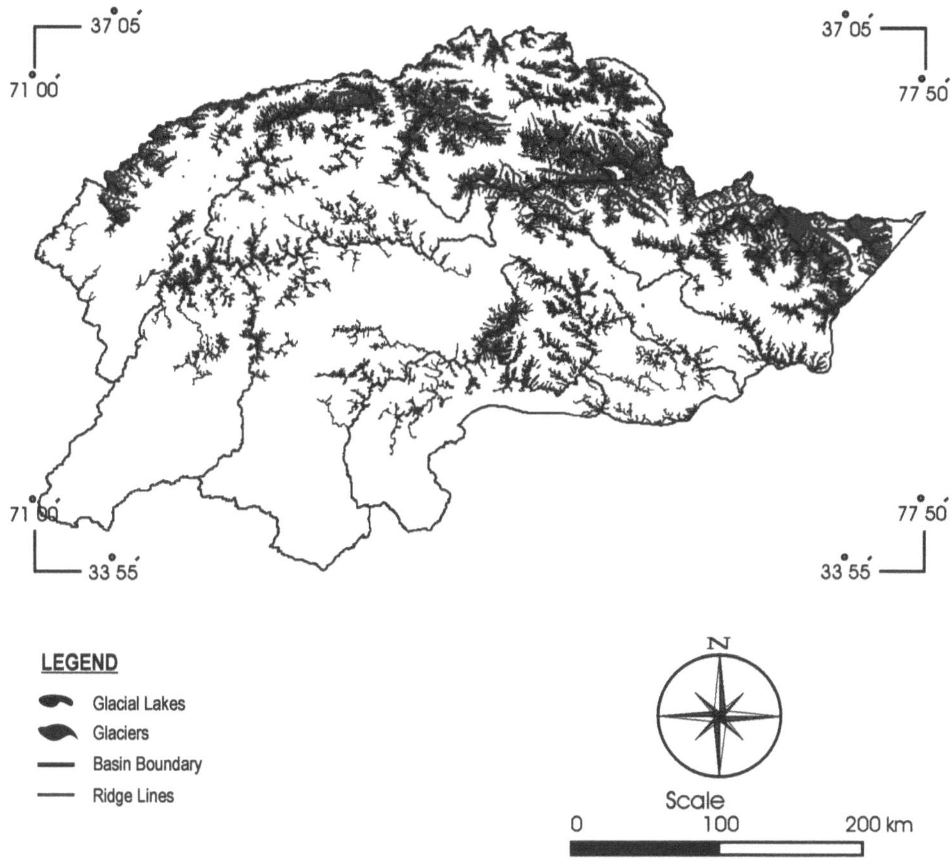

Figure 35.1. Glacier distribution in HKKH region of Pakistan.

Table 35.1. Summary of glacier inventory.

Total basin area (km^2)	128,730.80
Number of glaciers	5,218
Glaciated area (km^2)	15,040.70
Total length (km)	9,718.21
Ice reserves (km^3)	2,738.51

Generally the lakes are distributed all over the northern basins but mostly concentrated in different pockets. These lakes contribute about 126.35 km^2 of lake area. Out of a total of 2420 glacial lakes, 1328 lakes are characterized as major lakes. Among the major lakes 52 are characterized as potentially dangerous lakes. Generally the lakes identified as dangerous lakes belong to cirque, end moraine and valley-type lakes. Dangerous lakes are generally distributed in the southern sub-river basins.

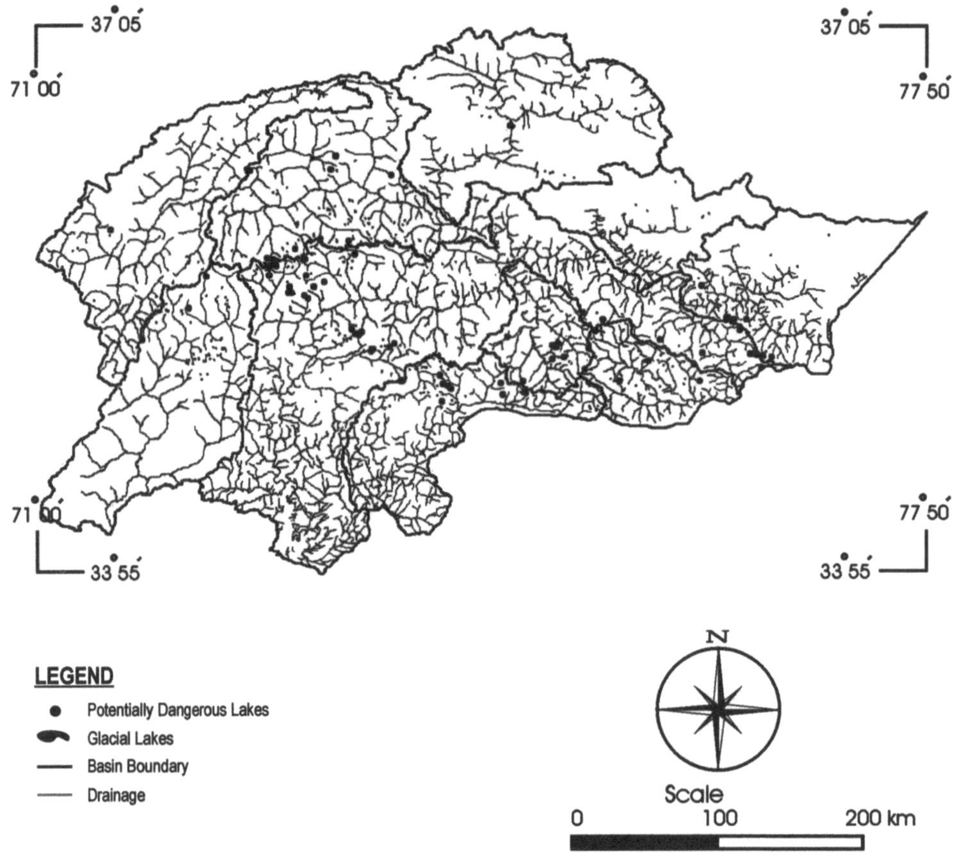

Figure 35.2. Distribution of glacial lakes in HKH region of Pakistan.

2.1.2. Climate and its role in glaciated areas

Historical records of glacier fluctuations in the Himalaya and Karakoram indicate that in the late 19th and early 20th centuries the glaciers were generally advancing, followed by predominant retreat during 1910–1960 (Mason, 1930, 1935; Goudie et al., 1984). High summer radiation and steep barren slopes control the glacier ablation patterns. The maximum radiation balance measured at Batura Glacier was over 27.9 MWm^{-2} (Zhang and Zhongyan, 1980). It is estimated that melting accounts for 80% of the heat loss whereas only 20% is due to evaporation and convection (Goudie et al., 1984).

In a study, Hewitt (1998) observed in detail the behavior of glacier surges of Chiring Glacier of the Karakoram. Between 1994 and 1996 catastrophic movement of the 16-km-long Chiring Glacier transferred 1–1.5 km^3 of ice from its upper two thirds to its lower third, and into the main Panmah glacier of which it is a tributary. By October 1996, a lobe of Chiring ice some 3.2 km^2 in area had entered and compressed the main glacier, which was severely disturbed for 3 km above and 5 km below the junction of the glaciers. Ice streams and medial moraines were pushed into a series of looped or "tear-drop" forms, well-known in surging glaciers. Despite an observational record back to 1856, it was not previously recognized that this glacier could surge. In the last 100 years, 26 sudden, rapid advances have been reported involving 17 glaciers (Hewitt, 1969).

2.1.3. Surges and climate change

The Karakoram is of unusual interest and perhaps sensitive to climate change, since its glaciers lie within the variable influence of three major weather systems: the sub-Mediterranean regime of mainly winter, westerly storms; the summer monsoon; and the Tibetan anticyclone. Winter storms dominate glacier nourishment at present. However, nearly one-third of the high-elevation snow accumulation that has been measured occurs in summer (Hewitt, 1990). It has been argued that general patterns of advance and retreat in the region relate to the changing vigor of the summer monsoon (Mayewski et al., 1980). The possibility of shifts in the atmospheric sources, regime, and seasonal occurrence of glacier nourishment, do not seem to be a factor in other regions with surging glaciers. This seems to be a further exaggerated by the climate change scenarios in the region. It is therefore, necessary to give more attention to surging glaciers in a relatively neglected region.

2.2. Climatic parameters

2.2.1. Establishment of Global Change Impact Studies Centre

Realizing the importance of climate change research in the country, under Ministry of Science and Technology, Government of Pakistan a Global Change Impact Studies Center was established in 2002. The center has a mandate to do research related to the impact of climate change on water resources, agriculture, natural resources, socio-economics, and other factors. Initially, to understand the current situation and past trends, the analysis of historical climatic data has been initiated and these data have been correlated with agriculture and water resources, especially in the mountain environment. Furthermore, a strong component of the center is capacity building and enhancement. So far a number of training courses, workshops and seminars have been organized by the center.

2.2.2. Climate change in Pakistan

Analysis of past climatic data depicts that our climate is changing (Farooq and Khan, 2004). There is temporal as well as spatial variability to the rate of change and the nature of resulting impacts affecting all aspects of life. In the context of climate change there is a need to develop and implement incremental adaptation strategies and policies to face the realities associated with climate change. Macro- as well as micro-level strategies and a continued monitoring and analysis have been suggested. Climatic data of a period from 1951 to 2000 were used to assess the regional changes in climate (Sheikh, 2005). Pakistan was broadly subdivided into six regions. It was observed that the maximum and minimum temperatures have dropped in the north-eastern mountains while the temperatures increased in western parts of the country. Generally summer (April–May) maximum temperatures have increased in all the regions except in the coastal belt while minimum temperatures show a mixed trend with increasing trend in the Balochistan plateau, coastal belt and central and southern Punjab. Precipitation has increased in all the regions during the monsoon period except in the western Balochistan plateau and coastal belt where it has decreased.

Akhtar et al. (2005) studied the spatial and altitudinal variations in air temperature in the upper Indus basin of Pakistan using correlation and regression analyses. The results show a strong negative correlation between elevation and mean daily temperature in the northern half of the region. A significant increase of maximum temperature and decrease of minimum temperature most likely led to strong increasing trends in the diurnal temperature range (DTR) at most of the meteorological stations of the study area, which is in line with findings reported by Chaudhry and Sheikh (2003). There is a positive correlation between DTR and mean maximum temperature, while there is a negative correlation between DTR and mean minimum temperature. The results also show a strong negative correlation between elevation and normal mean temperature.

As mentioned earlier there is a high variability in the mountain region of Pakistan as well ranging from low hills to towering mountains of Karakoram Range. To study the climate change within this complex mountain system the entire area has been divided into two regions (Sheikh and Manzoor, 2004). First, the region above 35° N (above 1200 m a.s.l.) is where mostly winter rains dominate due to western disturbances passing from December to March. The second region is below 35° N (600 to 1200 m a.s.l.) which is mainly fed by monsoon depressions developed in Arabian Sea and Bay of Bengal from July to September. For trend analysis, the time series climate data of 1951–2000 and 30 years climate normal (1961–90), with parameters like monthly temperature (mean, maximum, and minimum) and precipitation (total monthly) were used.

It was observed that in the region above 35° N there is a positive trend in mean and maximum temperature during the winter season while there is a negative trend during monsoon season. The temperature increase is quite pronounced during the snow melt months (April–May) both for mean and maximum temperature, which favors enhanced melting of glaciers. In the region below 35° N decreasing trends in maximum and minimum temperature during the monsoon season are predominant. However, in both the regions greater diurnal variations were observed.

Precipitation trend analysis indicates that above 35° N eight out of nine stations show increasing trends in monsoon rains while the winter rains have been reduced but not significantly. Similar trends were there for region below 35° N where both monsoon and winter rains on an average have increased slightly. In addition, the months of October and November, already the driest months of the year, appear to have become drier.

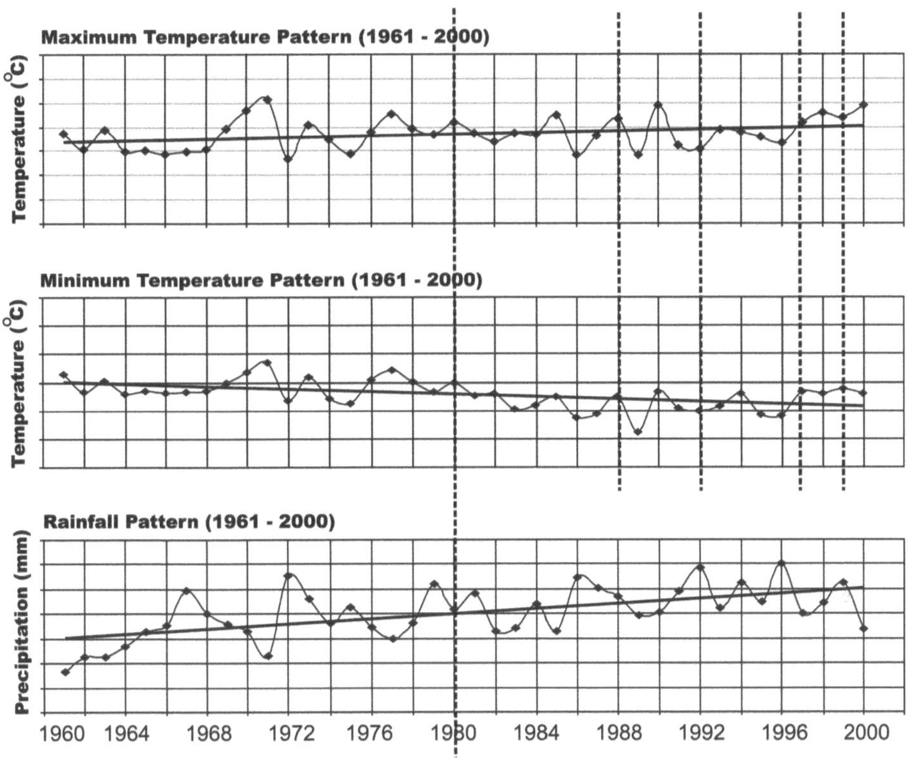

Figure 35.3. Historical maximum, minimum temperature and rainfall in northern areas.

In another study (Ahmad et al., 2003) the historical trends of maximum and minimum temperatures in the northern areas indicate that there is generally an increase in maximum temperature and a decline in minimum temperature (Fig. 35.3). The rainfall pattern shows a general increase in the area. At various points when the minimum and maximum temperatures were at higher ranges, this had an impact on rainfall, which then shows a decline.

Further, an analysis of spatial data of annual mean temperatures of 1961–1990 and 1931–1960 was done to study the impact of climate change on water resources and its vulnerabilities, as well as to identify coping mechanisms. The study revealed that, in general, there was a rise in temperature of 0.5°C to 1°C in the northern arid mountains, western dry mountains, and coastal areas. The northern and western dry mountains are outside the monsoonal zone, as they experience winter rainfall similar to the temperate zone. In the monsoonal zone, generally a decrease or no change in temperature was observed.

The same datasets were used to analyze the climate change in terms of extreme temperatures, rainfall and highest fall in 24 h. Generally, there was an increase in extreme temperatures in the arid environments comprising broadly the plains, dry mountains, and coastal areas. However, decrease in extreme temperatures was observed generally in the sub-humid and humid environments. Increase in extreme annual rainfall was observed in the hot humid, sub-humid and semi-arid environments, whereas decrease was observed generally in cool humid/sub-humid regions, dry mountains and coastal areas.

2.2.3. Climate change and water resources of Pakistan

The study of impact of past climate change and extreme events on water resources revealed that there is a decreasing trend in river flows of the Indus and Kabul, whereas a mixed trend was observed for the rivers Jhelum and Chenab (Ahmad et al., 2003). For the western rivers, the analysis of both Rabi and Kharif seasons indicated a reduction in river flows during 1968–1997 from that of 1937–1967. Extremes of water availability were also reduced during the same period. Due to climate change the snow avalanches and GLOF events were observed at high elevations (>3500 m a.s.l.), landslides, debris flows, and flash floods at medium elevations (500–3500 m a.s.l.), and floods in lower valleys and plains were common. Furthermore, due to climate change the sea level will raise and the freshwater–seawater interface would be shifted further inland and this shifting would lead to a reduction in the volume of fresh-groundwater resources (Murtaza and Iqbal, 2005). Based on these detailed analysis the key issues were highlighted and the policy recommendations were made. Furthermore, it is identified that because of rapid glacial melting, the base flow of Indus River will be affected. A macro-scale hydrological model for river flow simulations suggested that the runoff of the Indus will decrease by 27% by the year 2050 (Ahmad, 2004).

A distributed hydrology-soil-vegetation model (DHSVM) has been used to evaluate the hydrologic effects of land-cover change in the Siran River Basin (Saeed and Jhengir, 2005). Different land cover scenarios have been formulated. The analysis showed that the change in land cover due to human interventions could significantly affect the water resources. Such a model can ably predict the effect of that change on the stream flow, flood peaks, sediment control, and other factors.

2.3. Biological indicators

Vulnerability to climate change has been characterized as being a function of both exposure to climatic conditions and the adaptive capacity of the population at risk (Smit and Pilifosova, 2001). The impacts of climate change on some vulnerable regions can be expected to cause conditions of scarcity, land-use change, habitat destruction, resource degradation, displacement of large numbers of people, and even lead to mass movements of "environmental refugees." If climate model projections are accurate, climatic changes in Pakistan would likely exacerbate present environmental conditions that give rise to land degradation, shortfalls in food production, rural poverty, and urban unrest. Circular migration patterns following extreme weather events could be expected. Such changes would likely affect not only internal migration patterns, but also migration movements to other countries.

2.3.1. Natural ecosystems

The impact of climate change on natural ecosystems was studied using BIOME3 model (Siddiqui et al., 1999). Simulation of nine dominant plant types, or biomes was done taking the assumptions as 0.3°C rise in temperature, precipitation change of 0, +1, −1% per decade, 1990 as the base year, and atmospheric CO_2 concentration of 350 ppmv assumed to increase to:

425 ppmv in 2020,
500 ppmv in 2050, and
575 ppmv in 2080

The outcome of the model indicates that of the nine biomes, three biomes (alpine tundra, grassland/arid woodlands, and deserts) showed a reduction in their area while five biomes (cold conifer/mixed woodland, cold conifer/mixed forests, temperate conifer/mixed forests, warm conifer/mixed forests, and steppe/arid shrub lands) showed an increase in their area that may be as a result of climate change. Net primary productivity exhibited an increase in all biomes and scenarios. There is a possibility of forest dieback occurring and of time lag before the dominant plant types could adjust to changed climate and migrate to new sites. There are possibilities that during the intervening period the biomes may be vulnerable to environmental and socio-economic disturbances. It is concluded from this study that overall impact of climate change on the forest ecosystems of Pakistan could be negative and therefore, several adaptation strategies are proposed to cope with climate change impacts on forest ecosystems.

2.3.2. Agriculture

It is generally considered that climate change might have a positive impact on mountain agriculture (increase in yields of food crops) because of higher CO_2 concentrations, increase in growing season length (GSL) and thus wheat cultivation for grains would be possible above 1500 m a.s.l. Other crops and fruits might also be benefited. Based on this understanding the impact of climate change on agriculture especially in mountain areas was studied using the GSL of wheat crops (Hussain, 2004). The two locations selected were Chitral (high altitude) and Swat (low altitude). The regression analysis was also carried out assuming linear relationships. It was observed that during the past decade the GSL of wheat has increased in the mountainous areas, which is more prominent at higher elevations. Furthermore, future forecasts under various climate change scenarios suggest that both wheat yield and area will increase, except under a scenario of "no change in temperature and 10% decrease in rainfall." Under the climate change scenarios, Swat district will apparently be benefited more through temperature and rainfall increase while the Chitral district will benefit by area expansion. It is expected that high yielding varieties of plain areas would become more and more suitable for mountain areas.

Due to population increase the country is spending around US$ 1.0 billion annually for import of food items. Climate change would further aggravate these problems as the irrigated agriculture in semi-arid and arid environments is already operating under extremely high temperatures (Ahmad, 2005). It is stipulated that the increase in temperature of 0.9°C–1.8°C would further shorten the GSL of cereals (wheat, rice, and maize) leading towards shift in potential area of these crops. The increase in crop evapo-transpiration would require additional water for growth and cooling in an environment where water is already in short supply. The increase in utilization of marginal quality groundwater would further add to secondary salinization of soils.

In selected districts of Punjab, econometric models have been used for assessing the impact of changes in temperature and the corresponding changes in wheat yield in selected cropping systems over time (Malik et al., 2005). The results reveal that wheat yield was inversely related to rise in temperature. It is estimated that wheat yield could decline by 9 to 30% due to a temperature increase of 1 to 4°C. However, positive relationships of yield with irrigated area indicated the tendency of mitigating the impact of temperature to some extent by bringing more area under irrigation. These findings are in line with the study carried out by Hussain et al. (2005).

In marginal communities, particularly in the mountain environments as well as in arid areas, with increasing temperature and low rainfall, crops, livestock, and on-farm agro-forestry productivity is declining at even alarming levels (Amir, 2005). Droughts reduced the value associated with livestock and livestock products, which result in out-migration. In these communities farmers are unwilling to sell water during droughts, indicating the high value placed on this resource.

3. Recommendations

As climate change research is in its infancy in Pakistan and its known impacts on human societies, as well as on natural systems, are alarming, the following are some of the aspects that need immediate attention and can be considered as guidelines for future climate change research:

- Since the impact of climate change is so complex, a holistic approach is required to account in full for Earth and human activities and their interactions.
- Neither climate change nor its impacts follow international political boundaries so joint regional initiatives are required.
- Weather forecasting systems in the region must be improved and a regional early warning system must be established. At the same time continued monitoring and analysis of climatic variability and trends is required.
- Coordination of activities on climate change adaptation among countries in the region may be enhanced.
- The regional information/data sharing should be encouraged.
- Education on climate change may be introduced in educational curricula at various levels.
- Food supplies through agriculture are prone to a threat of changing climate. New varieties, especially in terms of wheat that directly relates to our food security,

need to be introduced. Breeding programs need to design wheat varieties under global environmental change for GSL, high temperature tolerance, drought resistance, and vigor against pests and diseases. Furthermore, agronomic taming of the prevalent problems includes re-appropriation of cropping patterns, adjustments in the planting dates, and changes in cultural practices viz. planting sequence; fertilizer use efficiency, and so forth is required.

- Irrigation management strategies for enhancing water-use efficiency are needed to be tailored under increasing crop-water demands.
- New tools and techniques such as geographic information systems, remote sensing, and simulation modeling need to be introduced for system analysis which is not only cost effective but efficient as well.
- Natural disasters can further amplify climatic change impacts. There is a need for preparedness and management of such events and development of mitigation/ management plans.

References

Ahmad, I., 2004. Global Climatic Change and Pakistan's Water Resources. Chief Executive Secretariat No. 2, Islamabad, Pakistan.

Ahmad, S., 2005. Assessment of climate change impacts on crops and water balance and adaptation strategies in Pakistan. In: Muhammad, A. and Hussain, S.S. (Eds), Proceedings of National Workshop on Global Change Perspective in Pakistan—Challenges, Impacts, Opportunities and Prospects, April 28–30, 2005, Islamabad, Pakistan.

Ahmad, S., Bari, A., and Muhammad, A., 2003. Climate change and water resources of Pakistan: Impact, vulnerabilities, coping mechanisms, Proceedings of the Year End Workshop on Climate Change and Water Resources in South Asia, January 7–9, 2003, Kathmandu, Nepal.

Akhtar, M., Ahmad, N., and Hussain, S.P., 2005. Climate change in Upper Indus basin of Pakistan: A case study. In: Muhammad, A. and Hussain, S.S. (Eds), Proceedings of National Workshop on Global Change Perspective in Pakistan—Challenges, Impacts, Opportunities and Prospects, April 28–30, 2005, Islamabad, Pakistan.

Amir, P., 2005. Socioeconomic aspects of drought in Bahawalpur and Mirpur Khas hydrological units of Pakistan. Science and Culture 71 (8), 273–283.

Chaudhry, Q. and Sheikh, M.M., 2003. Climate change and its impact on the water resources of mountain regions of Pakistan. Journal of Pakistan Meteterology 1, 28–34.

Farooq, A.B. and Khan, A.H., 2004. Climate change perspective in Pakistan, Proceedings of the Workshop on Capacity Building on Global Change Research, June 8–10, 2004, Islamabad, Pakistan.

Goudie, A.S., Brunsden, D., Collin, D.N., et al., 1984. The Geomorphology of Hunza Valley, Karakoram Mountains, Pakistan. In: Miller, K. (Ed.). Proceedings of the International Karakoram Project, London, Royal Geographical Society, pp. 359–410.

Hewitt, K., 1969. Glacier surges in the Karakoram Himalaya (Central Asia). Canadian Journal of Earth Sciences 6, 1009–1018.

Hewitt, K., 1990. Overall Report: Snow and Ice Hydrology Project, Upper Indus Basin SIHP, Cold Regions Research Centre, Wilfrid Laurier University, Canada, 179 pp.

Hewitt, K., 1998. Recent Glacier Surges in the Karakoram Himalaya, South Central Asia, http://www.agu.org/eos_elec/97016e.html, © 1998 American Geophysical Union.

Hussain, S.S., 2004. Impact of climate change on agriculture in mountain areas of Pakistan. Paper presented in the Workshop on Adaptation to Climate Change in Mountain Ecosysytems: Bridging Research and Policy, 3–5 March 2004, Kathmandu, Nepal.

Hussain, S.S., Goheer, A., Sultana, H., et al., 2005. Sensitivity of wheat yield to climate change in Punjab using DSSAT-based CERES wheat simulation model. In: Muhammad, A. and Hussain, S.S. (Eds),

Proceedings of National Workshop on Global Change Perspective in Pakistan—Challenges, Impacts, Opportunities and Prospects, April 28–30, 2005, Islamabad, Pakistan.
Malik, W., Hussain, M.S., Abid, S., et al., 2005. Impact of climate change on wheat productivity in selected cropping systems in Punjab, Pakistan. In: Muhammad, A. and Hussain, S.S. (Eds), Proceedings of National Workshop on Global Change Perspective in Pakistan—Challenges, Impacts, Opportunities and Prospects, April 28–30, 2005, Islamabad, Pakistan.
Mason, K., 1930. The glaciers of the Karakoram and neighborhood. Geological Survey of India, Records 63, 214–278.
Mason, K., 1935. The study of threatening Glaciers. The Geographical Journal 85, 24–41.
Mayewski, P., Pergeant, G.P., Jeschke, P.A., and Ahemad, N., 1980. Himalayan and Transhimalayan Glacier Fluctuations and the S-Asian Monsoon Record, Arctic and Alpine Research, 12/1, 171B82.
Murtaza, U. and Iqbal, S., 2005. Impact of Global Warming on Water Resources and Adaptation Measures for Management. In: Muhammad, A. and Hussain, S. S. (Eds), Proceedings of national workshop on global change perspective in Pakistan—Challenges, Impacts, Opportunities and Prospects, April 28–30, 2005, Islamabad, Pakistan.
Roohi, R., Mool, P., Ashraf, A., et al., 2005. Indus basin, Pakistan Hindu Kush–Karakoram–Himalaya. Inventory of glaciers and glacial lakes and the identification of potential glacial lake outburst floods (GLOFs) affected by global warming in the mountains of HKH Region. Final report developed in the form of a CD.
Saeed, F. and Jhengir, S., 2005. Study on the impacts of land cover changes in upper indus basin by using a distributed hydrological model. In: Muhammad, A. and Hussain, S.S. (Eds), Proceedings of National Workshop on Global Change Perspective in Pakistan—Challenges, Impacts, Opportunities and Prospects, April 28–30, 2005, Islamabad, Pakistan.
Sheikh, M.M., 2005. Region-wise Climate Change in Pakistan (1951–2000) In: Muhammad, A. and Hussain, S.S. (Eds), Proceedings of National Workshop on Global Change Perspective in Pakistan—Challenges, Impacts, Opportunities and Prospects, April 28–30, 2005, Islamabad, Pakistan.
Sheikh, M.M. and Manzoor, N., 2004. Climate Change in Mountain Regions of Pakistan, GCISC. Paper presented at National Workshop on Impact of Climate Change on Mountain Environment, July 8th, 2004 held at FAST, Islamabad.
Siddiqui, K.M., Mohammad, I., and Ayaz, M., 1999. Forest ecosystem climate change impact assessment and adaptation strategies for Pakistan. Climate Research 12, 195–203.
Smit, B. and Pilifosova, O., 2001. Adaptation to climate change in the context of sustainable development and equity. In: McCarthy, J.J., Canzianni, O.F., Leary, N.A., Dokken, D.J., and White, K.S. (Eds), Climate change 2001: Impacts, Adaptation and Vulnerability—Contribution of Working Group II to the Third Assessment Report of the Intergovernmental Panel on Climate Change. Cambridge University Press, Cambridge.
Zhang, J. and Zhongyan, B., 1980. The surface ablation and its variation on the Batura Glacier. In: Shi, Y. (Ed.), Papers on the Batura Glacier, Karakoram Mountains. Science Press, Beijing, pp. 83–98.

Subject index

FSC
www.fsc.org
MIX
Papier | Fördert
gute Waldnutzung
FSC® C083411

Kolibri 360 GmbH
Ferdinand-Jühlke-Straße 7
99095 Erfurt, Deutschland
produktsicherheit@kolibri360.de